Fitting Equations
to Data

FITTING EQUATIONS TO DATA

Computer Analysis of Multifactor Data

Second Edition

CUTHBERT DANIEL
Consultant, Rhinebeck, New York
FRED S. WOOD
Standard Oil Company (Indiana), Chicago, Illinois

with the assistance of
JOHN W. GORMAN
Amoco Oil Company, Naperville, Illinois

JOHN WILEY & SONS

New York · Chichester · Brisbane · Toronto

Library of Congress Cataloging in Publication Data:

Daniel, Cuthbert.
  Fitting equations to data.

  (Wiley series in probability and mathematical statistics)
  Bibliography: p.
  Includes index.
    1. Curve fitting—Data processing. 2. Least squares—
Data processing. 3. Multivariate analysis—Data processing.
I. Wood, Fred S., joint author. II. Gorman, John Wayne.
III. Title.

QA297.D35 1979   519.4   79-11110
ISBN 0-471-05370-8

Printed in the United States of America

10 9 8 7 6 5 4 3 2 1

To Edith and Janet

Preface to the First Edition

The best way to summarize a mass of multifactor data is by a simple equation or set of equations. The data, however, must be studied critically, and here the standard texts give little guidance beyond stern warnings to be cautious.

Routine use of standard computer programs to fit equations to data does not usually succeed. A large proportion of the failures is due, not to the programs, computers, or data, but to the analyst's approach. It is not caution that is missing. Boldness in conjecture and persistence in follow-up are much more important. When computer work is a substitute for hard thinking, it is of little use. Its value increases insofar as it leads the analyst to new thoughts about the data and hence about the underlying system.

This book is a summary, mainly by examples, of our efforts to study multifactor data critically. It reflects what we have learned in the course of a decade of work with research engineers and scientists. We know that we have much more to learn, and we look forward to suggestions from our readers.

Nearly all engineering and research data are assembled in unbalanced sets, with little or no attention paid to the standard requirements of statistical design of experiments. This book is entirely devoted to methods of studying such sets. But data analysis is itself a form of experimentation. As the reader will see, we often study the data by the methods of statistical design of experiments.

Our purposes are to help the analyst, scientist, or engineer (1) to recognize the strengths and limitations of his data; (2) to test the assumptions implicit in the least-squares methods used to fit the data; (3) to select appropriate forms of the variables; (4) to judge which combinations of variables are most informative; and (5) to state the conditions under which the equations chosen appear to be applicable.

In trying to cover frequent contingencies not mentioned in any text, we have had to develop many devices, all of unknown power, sensitivity, and robustness. These include ways to detect and handle nested data, ways to spot

vii

bad or critical values, and ways to examine and select from large collections of alternative equations, those which are most appropriate. We have made extensive use of indicator or dummy variables to handle several sets of data at once, and also of computer-made residual plots to learn more about the fit of the resulting equations. All of these proposals, not now in the statistical canon, have been usefully discussed in seminars at Florida State, Harvard, New York, North Carolina State, Princeton, Texas A & M, Wisconsin, and Yale Universities, as well as at the Bell Telephone Laboratories and the National Cancer Institute. We have adopted all that has survived, and emerged from, these discussions to avoid the widespread practice of assuming without check that the form of the fitting equation is known, that the observations are independently normal with constant variance, and that there are no bad or critical values.

A major innovation, due to our colleague Colin Mallows (Bell Telephone Laboratories), is the statistic C_p, which we use to judge the total mean square error (bias squared plus random error squared) for each of the whole set of 2^K equations generated by various choices from K candidate independent variables. This approach is particularly useful when the data do not suffice to narrow the choice to a single equation. In such cases it is important to see the whole set of equations that give equally good fits.

To achieve simplicity in presentation, we have relied heavily on references to standard texts, especially those by Brownlee, Davies (Ed.), Bennett and Franklin, Hald, and Anderson and Bancroft, for most statistical derivations.

The two computer programs that we use are based on the usual Gaussian (or Laplacian) least-squares method with the classical assumptions. However, many innovations have been incorporated, as will be apparent. All of the computer work can be done quite economically on an IBM 360-65 or on a CDC 6400. In order that the programs may be easily obtained, they have been placed in the SHARE and VIM libraries, together with full program specifications in FORTRAN IV.

To facilitate translation between the several texts, the computer printout language, and engineering usage, we have included a relatively comprehensive Glossary.

We owe a special debt to K. A. Brownlee. His *Industrial Experimentation* came as a revelation to the senior author when it appeared in 1946. The reader will note that his *Statistical Theory and Methodology in Science and Engineering* is our most frequent reference.

It is with pleasure and gratitude that we acknowledge the detailed criticism of an earlier draft by Harry Smith (Rensselaer Polytechnic Institute). Jerry Warren has used the programs and text in courses at North Carolina State University and has sent us valuable comments by all members of his class. Gus Haggstrom (University of California, Berkeley), David Jowett (Iowa

State), and Michael Godfrey (Massachusetts Institute of Technology) have given us useful criticisms. George Box and William Hunter (University of Wisconsin) have made many helpful comments on the techniques used in the nonlinear example. Harold Steinour (Riverside Cement) and Waldemar Hansen (Cement Consultant) have provided valuable insight into the conditions and results of the Portland cement experiment as well as knowledge of current cement technology. Neil Timm (University of Pittsburgh) has efficiently adapted both the linear and nonlinear computer programs for the CDC 6400. M. W. Hemphill (Health, Education, and Welfare) has performed valuable services in reproducing and circulating early drafts to government agencies. We are thankful for all these favors.

We are also indebted to Hertsell Conway, Elizabeth Grimm, Robert Lyon, and George Schustek (American Oil) for their review of the manuscript, to John Boyle and Edward Sands for the initial drafts of a number of the figures, and to Edith, Kathy, and Chris Wood for their help in typing and assembling the many drafts of the manuscript.

Last, but by no means least, we acknowledge the helpfulness of our colleagues John Gorman and Robert Toman who assembled the reference material used in the first courses given at American Oil on fitting equations to data. Their 1966 paper Selection of Variables bears familial resemblance to our Chapter 6. Gorman has continued to be consistently helpful in reviewing, discussing, and contributing ideas during the development of this manuscript.

CUTHBERT DANIEL
FRED WOOD

Rhinebeck, New York
Valparaiso, Indiana
January 1971

Preface to the Second Edition

This edition differs from the first principally in its extended use of "component and component-plus-residual plots." These plots, the invention of F. S. W., are an addition to our earlier list of *interior analyses*. By this we mean the study of the data point by point. We have found repeatedly that sets of multifactor data produce global statistics (fitted coefficients, error estimates from residual mean squares, overall F-values, etc.) that are sensitive to small subsets of exceptional observations—sometimes to a single point. Once they are spotted, these unrepresentative points can often be seen to correspond to identifiable unusual conditions. They should not be used in fitting equations meant to represent normal conditions. As a striking example (Chapter 7), we find a single point in one set of data—unnoticed by 17 authors who have used this set—that determines one fitted coefficient.

Other important revisions of the programs and User's Manuals allow the reader to: (1) read data from designated files; (2) delete different designated observations from each fit; (3) assign indicator variables to selected observations; (4) reduce the amount of printout; (5) calculate the C_p statistic based on the residual mean square (RMS) of the full equation, on the subset with the minimum RMS, or on his own estimate of variance; (6) increase the number of C_p searches from 4096 (12 variables) to 262,144 (18 variables); (7) list functions related to the variance of the fitted value; (8) calculate the required precision of independent variables; and (9) provide cross verification of the coefficients with a second sample of data.

As a result of the above revisions, 75 figures have been brought up to date and 108 figures of new examples have been added to reflect the use of these procedures in our present campaigns.

We have been pleased that for the past six years the LINWOOD and NONLINWOOD computer programs have been the most requested in both the SHARE Library (double precision for IBM computers) and the VIM Library (single precision for CDC computers). Improvements in handling memory storage now permit the programs to also run on much smaller computers.

We wish to acknowledge the assistance of Robert Brehmer, Ralph Henschel, Michael Kelly, and Eric Ziegel [Standard Oil (Indiana)] in updating the programs. We are also grateful to a number of others who have adapted the revised programs to Control Data, UNIVAC, DECsystem, Honeywell, and Burroughs computers and placed them in their appropriate libraries. They include Eli Cohen and Bruce Foster (Northwestern University), James Keith (Johnson Space Center), Robert Kohm (ALCOA), David Zarnow (Naval Avionics), James Murat and Tom Zeisler (University of Wisconsin), and David Rumsey (Burroughs).

Users in a wide variety of fields have told us of new applications. The LINWOOD program has been used in studies in communications, metallurgy, nuclear power plants, utility loads, and conservation of energy. It has also been used in econometrics, marketing sales studies and surveys, and in studies by government agencies of variables for pollution control. In the biological sciences it has been used in cancer research (searching for influential variables), in studies of ecology, fish and wildlife migration, photosynthesis, and pharmaceutical drugs. In agricultural sciences there have been studies of fertility and breeding of edible grains and grasses. In management sciences there have been studies of airline schedules, hospital costs, and factors affecting growth and profitability of companies and products. In the social sciences, the program has been used to study language learning, absenteeism, evaluation of personnel proficiency tests, and performance before, during, and after college. As the result of a survey of the precision of a large number of programs, it was chosen for studies in astronomy. The NONLINWOOD program has been particularly useful in studying processes in which the response is a function of time: the kinetic reactions of paints and plastics, orthodontics, drug response, and in botany and biological research. A recent far-out application has been the analysis of data while they were being collected in the "Men in Space" program.

Feedback on these applications has been informative. We shall appreciate hearing of other successes, problems encountered, and suggestions on ways to improve the text or the programs.

The senior author insists on putting into the record the fact that most of the work and most of the improvements in this Edition are due to the junior author (F. S. W.).

CUTHBERT DANIEL
FRED WOOD

Rhinebeck, New York
Valparaiso, Indiana
January 1980

Contents

Fitting Equations
to Data

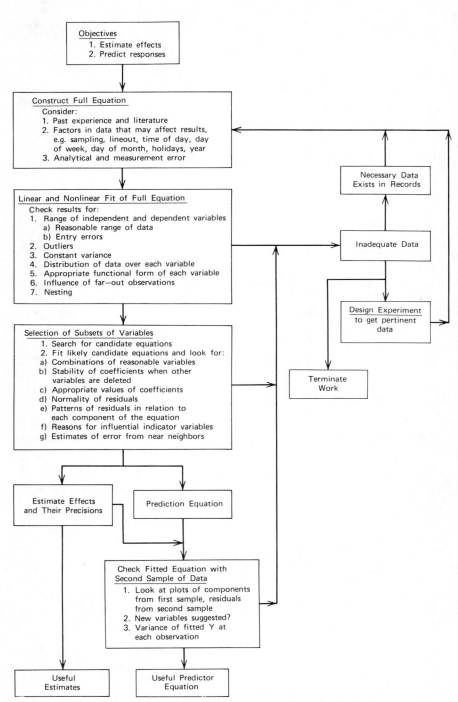

Figure 1.1 Flow diagram for fitting equations to data.

CHAPTER 1

Introduction

1.1 Flow Diagram of Procedures Used

Fitting equations to data is a step-by-step procedure. Figure 1.1 is a flow diagram of the procedure which we have found helpful in our campaigns; it indicates the work to be done and the decisions to be made at each stage. Usually there are a number of recycles to the computer as defects in the data and in equation form are recognized. The computer programs which we use are designed to help us recognize peculiarities in the data and deficiencies in the fitted equation. Each step is discussed in detail in later chapters when the associated problem first appears, be it with one, two, or more independent variables. As the number of variables increases, the problems become more complex.

1.2 Role of Computer

The computer Linear Least-Squares Curve-Fitting Program, LINWOOD, which is used throughout the book to provide examples, has been put into the libraries listed in Table 1.1 so that others* can obtain keypunched cards or tapes of the program and test problems. The User's Manual provides directions on the use of the program. The terms used in the computer print-out have been included in the Glossary for ease of reference.

A multiplicity of synonymous words and symbols has frequently evolved in various fields of work. We have tried to cross-reference these variations both in the text when the term is first defined and in the Index. The conventions in nomenclature and the symbols that we use are given in the first two sections of the Glossary.

* Not restricted to library membership. See page 396 on precision of program.

1

TABLE 1.1
LIBRARIES FOR COMPUTER PROGRAMS

Libraries	Computers	LINWOOD Program Numbers	NONLINWOOD Program Numbers
CUBE[a]	Burroughs	WIS/LINWOOD	WIS/NONLINWOOD
DECUS[b]	DECsystem	LINWOOD-10-257	NONLIN-10-258
		LINWOOD-11-419	NONLIN-11-420
HLSUA[c]	Honeywell	GES-1206	GES-1207
SHARE[d]	IBM	360D-13.6.008	360D-13.6.007
VIM[e]	CDC	G2-CAL-LINWOOD	G2-CAL-NLWOOD
—[f]	UNIVAC	LINWOOD	NLWOOD

[a] Computer Users of Burroughs Equipment Library, Department of Computer Science, U.S. Air Force Academy, Colorado 80840.

[b] Digital Equipment Computer User's Society, One Iron Way, Marlboro, Massachusetts 01752.

[c] Software Library, Mail Station K16, Honeywell Information Systems, P.O. Box 6000, Phoenix, Arizona 85005.

[d] Triangle Universities Computation Center, P.O. Box 12076, Research Triangle Park, North Carolina 27709.

[e] Software Distribution Department, ARH230, Control Data Corporation, 4201 North Lexington Avenue, St. Paul, Minnesota 55112. Also Vogelbeck Computer Center, Northwestern University, 2129 Sheridan Road, Evanston, Illinois 60201.

[f] Johnson Space Center, Mail Code TN74, Houston, Texas 77058.

1.3 Sequence of Subjects Discussed

The important but sometimes forgotten assumptions of the least-squares method are discussed in Chapter 2. Difficulties associated with fitting data with *one* independent variable are discussed in Chapter 3. Additional problems associated with *two or more* independent variables are attacked in Chapter 4. Each of these chapters builds on the information of the previous chapters and deals with variables *known* to be influential. In Chapter 5 we give an example of fitting an equation with three independent variables assumed to be known. New ways of studying the data produce conclusions quite different from those obtained by standard analysis.

When there are a number of variables that *may* be influential, a computer search is required to find, with the data at hand, which ones are the most influential. The success of the search depends on whether or not the influential variables have been observed. If they have, and if their movements have occurred over a range adequate to display their influence, then the suggestions made in Chapter 6 will be useful in making a selection. These proposals *replace* stepwise regression and other one-at-a-time selection methods that use only F- and/or t-tests as the criteria for selection.

Chapter 7 uses a data set with ten independent variables to introduce a number of devices for studying *internal* structure. A measure of distance is given that aids in spotting remote points and in finding near neighbors. The former are studied to see whether they indicate curvature. The latter are used to develop an estimate of random error that is often less biased than the usual residual mean square. A table of various functions of the variance of fitted values is used to spot and distinguish remote points that are influential as well as those that are not influential. A table showing the effect of each factor at each point facilitates study of the detailed operation of the fitting equation. Provision is made for judging the reality of single outliers. Finally a set of plots is developed that visualizes the component effect of each factor at each point, along with its corresponding residual. These plots (called "component and component-plus-residual plots") are used to learn in detail just how well each term in the fitting equation represents the actual data. This may suggest new terms for the fitting equation or it may suggest a discrepant observation. In some sets of data such plots support the best possible conclusion: every observation fits every term in the fitting equation as well as it possibly could.

The widespread presence of *nested* multifactor data is discussed in Chapter 8. So far as we know, this topic has not been treated in the statistical literature.

In Chapter 9 an example of nonlinear least-squares estimation is given. The Nonlinear Least-Squares Curve-Fitting Program, NONLINWOOD, which is used, is a modification of the University of Wisconsin's GAUSHAUS program. The changes introduced make it fairly easy to study a 43-parameter nonlinear equation, which is later simplified to a 17-coefficient equation with no loss in information. This program and test problems are also available in the libraries listed in Table 1.1. The User's Manual provides directions on the use of the program.

For ease of reading, mathematical derivations and examples have been placed in appendices at the end of each chapter. Although each chapter will be self-sufficient for some readers, we believe that even the experienced analyst will find it worthwhile to read the entire text at least once.

CHAPTER 2

Assumptions and Methods
of Fitting Equations

2.1 Assumptions

The purposes of fitting equations are:

a. To summarize a mass of data in order to obtain interpolation formulas or calibration curves.
b. To confirm or refute a theoretical relation; to compare several sets of data in terms of the constants in their representing equations; to aid in the choice of a theoretical model.

We assume that a mass of data is available. A number (N) of runs have been made in a laboratory, pilot plant, or commercial plant. For each run the experimental conditions (the values of each of the independent variables $x_1, x_2, \ldots, x_i, \ldots, x_K$) and the important properties of the system or products produced (the values of the dependent variables $y_1, y_2, \ldots, y_m, \ldots, y_Q$) have been recorded. All the conditions that are thought to be relevant must be included. Some may be *names* of broad classifications or groupings, not measurements. The conditions may have been varied deliberately, or unavoidably to adjust for changes in either the raw material or the environment.

A single equation or set of equivalent equations is desired for each dependent variable that relates its values to the corresponding conditions or *levels* of the independent variables. Statisticians usually call dependent variables *responses* and independent variables *factors*.

A good method of fitting should:

a. Use all the relevant data in estimating each constant.
b. Have reasonable economy in the number of constants required.
c. Provide some estimate of the uncontrolled error in y.

5

d. Provide some indication of the random error in each constant estimated.

e. Make it possible to find regions of systematic deviations ("function bias") from the equation if any such exist.

f. Show whether the conclusions are unduly sensitive to the results of a small number of runs, perhaps even of one run.

g. Help to spot sets of data that really are not from separate runs, but actually are from parts of one longer run. This defect is called plot-splitting, replication-degeneracy, duplicity, hierarchal arrangement, and nesting by various writers.

h. Give some idea of how well the final equation can be expected to predict the response—both in the overall sense and at important sets of conditions inside the region covered by the data.

In short, we want to obtain all the information possible from the data—both about the equation constants estimated and about the limitations on future use of the equation with these constants. It will be noticed that ease of computation has not been mentioned as a desideratum.

2.2   Methods of Fitting Equations

If the number of runs exceeds the number of constants to be estimated, the equation cannot be expected to fit all the data points exactly. There is then some necessity to choose values for the constants that are somehow "best."

One method, in use for over a century, is to choose values for the constants that make the data appear *most likely* when taken as a whole. Note that the data, which may be expected to contain some random variability, are held constant, whereas each of the "constants" which are being estimated is viewed as capable of taking a range of values. This range depends on the distribution of random fluctuation in the dependent variable. If we know both the form of equation to be fitted *and* the form of the distribution of random error in the observations, each independent piece of data has a "probability" that varies with each choice of a set of values for the constants. The *product* of the probabilities of all the data points, for a given set of values for all the constants, is called the *likelihood* of that set. There is usually a particular set that maximizes this product; it is termed the *maximum likelihood set*.

When the distribution of random disturbances is a so-called normal or Gaussian distribution, the maximum likelihood estimates can be found by a simpler procedure called the method of least squares.

2.3   Least Squares

The least-squares (LS) method says: "Find the values of the constants in the chosen equation that minimize the sum of the squared deviations of the observed values from those predicted by the equation."

The justification for this statement is given by the Gauss-Markov theorem (see Hald, Section 18.10), which states that the estimates obtained are the *most precise unbiased estimates* that are *linear* functions of the observations. It is assumed, of course, that the correct form of the equation has been chosen. By "most precise" we mean that the estimates in (imagined) successive sets of data taken under identical circumstances scatter around the true values with *minimum variance* or mean-squared deviation. By "unbiased" we mean that the LS estimates average out in the long run to the true values. By "linear" minimum variance estimates we imply that, although there may be other estimates that have even smaller variances, they will not be linear functions of the observations. This property of linearity is advantageous in its simplicity, but there may be nonlinear estimates with much smaller variance. However, when the random errors are *normally* distributed, the LS estimates are maximum likelihood estimates and are of minimum possible variance.

The first two of the following three key assumptions of the LS method are also crucial assumptions of maximum likelihood.

The first assumption is that we know, or have chosen, *the correct form of the equation*. If we have not used the correct form, some values given by the equation will be biased in the sense that, even if we could take new sets of data, the average of the estimated constants would not converge on the true values. The data will not necessarily give us a clear indication in this regard, but much more can be done than is shown in elementary textbooks. We will see later how we can sometimes use the data to detect systematic bias or lack of fit (also called function bias) in the equations used.

The second assumption is that the data are *typical*, that is, that they are a representative sample from the whole range of situations about which the data analyst wishes to generalize. It should be obvious that an equation which represents the typical behavior of a system cannot be derived from nontypical data. Yet this second assumption is probably the one most often violated. Each reader will have his favorite example. A large collection of data taken on a fluidized-bed catalytic reaction at only one set of temperatures may have no bearing at all on the operation of the system at some widely different set of temperatures. Likewise, data on recipe changes for angel food cake may tell us little about the effects of similar changes in a recipe for chocolate cake.

The third key assumption of the LS method is that the y-observations are *statistically uncorrelated*. Each y is taken to be made up of a "true" value, usually designated by η (eta), plus a random component, e. If we denote any two of these random components by e_j and $e_{j'}$, then this assumption may be stated: The population average value, or expectation, of $(e_j e_{j'})$ is zero. Thus $E(e_j e_{j'}) = 0$.

A more general requirement, which includes the one just mentioned, is *statistical independence.* If the probability of $e_{j'}$ appearing in some range of values is the same, regardless of the magnitude of e_j appearing in some range, then $e_{j'}$ is statistically independent of e_j. Statistical independence implies zero correlation, but the converse is true only when the distribution of the e_j is "normal" (Gaussian). It is pleasant to discover again and again that the random error in many sets of engineering data appears nearly normal. Gauss showed long ago that the presence of a *large number of small independent additive* causes of random disturbance would produce a normal distribution of errors. It is of course logically inadmissible to deduce from the appearance of near normality that we have indeed seen the operation of the four conditions stipulated by the words in italic, but in fact we do often suppose that such is the case.

Three less important assumptions of the LS method have been given extensive attention in the statistical literature. The first assumption is that all observations on y have the *same* (though unknown) *variance*. (When the variance of the dependent variable changes in a known way, then the more general *weighted* least squares method—weighting each point inversely by its variance—can be used. Such cases are discussed later and the computer program contains an option for weighting.) The second is that all the *conditions*, that is, the levels of all of the independent variables $(x_1, x_2, x_3, \ldots, x_i)$, are known without error. The third is that the distribution of the uncontrolled error is a *normal* one. The conclusions are not likely to be greatly weakened if the distribution is *nearly* normal or there is *nearly* constant variance. Substantial uncertainty in many of the x_i, however, will bias all the constants toward zero.

An unwritten assumption in all curve fitting is that the data used are "good" data. But most large collections of data and occasionally even small collections contain a few "wild points," sometimes called *mavericks* or *outliers*. What happened to make them nontypical cannot usually be reconstructed. They must be spotted, however, since to retain them may invalidate the judgments we make. Methods of detecting outliers will be discussed in later chapters.

When theory and practical background do not suffice to indicate the general form of the functional relation, the equations must be at least partly empirical. In other cases the data analyst will be able to develop definite forms of the equation to be fitted, basing these forms on theory and on experience.

Polynomial forms are among the easiest empirical equations to fit. The equation chosen is linear in some or all of its K independent variables. The calculated values of the response, Y, may be of the form:

$$Y = b_0 + b_1 x_1 + b_2 x_2 + \cdots + b_i x_i + \cdots + b_K x_K,$$

instead, the form may be second order:

$$Y = b_0 + b_1 x_1 + b_{11} x_1{}^2 + b_2 x_2 + b_{22} x_2{}^2 + b_{12} x_1 x_2 + \cdots,*$$

or it may even contain higher powers of the independent variables. The "coefficients," the b's, are calculated from the data. The independent variables may be the original recorded conditions, x_i, or they may be any *coefficient-free* functions of the x_i. For example, x_i might be $(x_1/x_2)^{1/4}$ or $x_1{}^2$, but not $1/(x_1 - c)$ unless c is known.

The ease of fitting comes from the fact that all the above equations are *linear in the coefficients* $b_0, b_1, b_{11}, b_{12}, \ldots$, hence the word linear in the term linear least squares.

2.4 Linear Least-Squares Estimation

By definition the added requirement for *linear* LS estimation is that the equations chosen be *linear in the coefficients*.

The mathematical assumptions and elementary theory of linear LS estimates of equations with one independent variable are given in Appendix 2A. The more general case and a description of the methods used in the computer Linear Least-Squares Curve-Fitting Program throughout this book are presented in Appendix 2B.

2.5 Nonlinear Least-Squares Estimation

In addition to the requirements already stated for LS estimation, nonlinear LS estimation requires that:

1. The equation by definition be *nonlinear in some of its coefficients*.
2. Preliminary estimates of all the parameters be available. If these estimates are not close "enough," the later machine-computed estimates will not converge on the best values.

For most nonlinear estimation problems, the services of a competent statistician should be available for a period as short as a day, or as long as several months, depending on the complexity of the equations under study. Able workers in nonlinear LS estimation always emphasize that careful preliminary analysis saves time in the long run. (See Box, 1960, page 803, for an informed discussion of this matter.) Here is one of those cases, actually very common, where hand work and taking thought before going on the machine are major prerequisites. As Augustus is said to have said *Festina lente* ("Make haste slowly").

Three types of nonlinear computer programs are now available: one,

---

* Here the double subscripts refer to the independent variables in each term. Thus b_{11} is read "b sub one one" and not "b sub eleven."

written by Booth and Peterson and discussed by Hartley, uses a "Taylor-series or Gauss-Newton" method to converge on a solution; another, written by Marquardt (1959), uses a "gradient or steepest-descent" method; the third, developed and discussed by Marquardt (1962, 1963), uses a "maximum neighborhood" method (an interpolation between the other two methods so that both the size and the direction of steps can be determined simultaneously). A mathematical summary of these methods is presented in matrix form by Draper and Smith. Examples of the first two methods have been published by Marquardt (1959) and by Behnken. Wood has revised a program written by Meeter of the University of Wisconsin which utilizes the "maximum neighborhood" method. An analysis of a nonlinear estimation problem using this program is discussed in Chapter 9.

APPENDIX 2A

LINEAR LEAST-SQUARES ESTIMATES— ONE INDEPENDENT VARIABLE

2A.1   Assumptions

We make four assumptions about the relationship between the value x_{1j} of the independent variable in the jth run and the observed value y_j of the dependent variable.

A. There is a linear relationship between the "true" value of a response, η, and the value of the independent variable:

$$\eta = \beta_0 + \beta_1 x_1,$$

where β_0 is the intercept and β_1 the slope.

B. $y_j = \eta_j + e_j = \beta_0 + \beta_1 x_{1j} + e_j$, where e_j is random error.

C. The e_j have the following properties:

    1. The expected value of e_j is zero; our observed y_j is an unbiased estimate of η_j.

    2. The variance of e_j is $\sigma^2(y)$, which remains constant for all values of x_1.

    3. The e_j are statistically uncorrelated, i.e. the expected (population) value of $e_j e_{j'}$ for any pair of points j and j' is zero.

D. The observed values of the independent variable are measured without error. All the error is in the y_j, and none is in the x_{1j}'s.

2A.2   Basic Idea and Derivation

We have the model $\eta = \beta_0 + \beta_1 x_1$ and must somehow estimate the unknown parameters, β_0 and β_1, by statistics (i.e., functions of the data) b_0

and b_1. We will obtain a fitted equation $Y = b_0 + b_1 x_1$. The least-squares estimates are those values of b_0 and b_1 which minimize the function

$$Q = \sum_{j=1}^{N}(y_j - Y_j)^2 = \sum_{j=1}^{N*}(y_j - b_0 - b_1 x_{1j})^2,{}^*$$

where Y_j is the fitted value at x_{1j}.

Upon setting

$$\frac{\partial Q}{\partial b_0} = 2\Sigma(y_j - b_0 - b_1 x_{1j})(-1) = 0$$

and

$$\frac{\partial Q}{\partial b_1} = 2\Sigma(y_j - b_0 - b_1 x_{1j})(-x_{1j}) = 0,$$

we obtain

$$b_0 = \bar{y} - b_1 \bar{x}_1$$

Notice that the fitted line must go through the point $(\bar{x}_1, \bar{y})$ and

$$b_1 = \frac{\Sigma(x_{1j} - \bar{x}_1)(y_j - \bar{y})}{\Sigma(x_{1j} - \bar{x}_1)^2} \equiv \frac{[1y]}{[11]}.$$

(For definitions of symbols see the Glossary.)

To obtain the variance of any fitted value Y, Var (Y), we look at our equation $Y = b_0 + b_1 x_1$ in a slightly different form, namely,

$$Y = \bar{y} + b_1(x_1 - \bar{x}_1),$$

so that $\bar{y}$ and b_1 are uncorrelated. Since variances under these conditions are additive, we now have

$$\text{Var }(Y_j) = \text{Var }(\bar{y}) + (x_{1j} - \bar{x}_1)^2 \text{ Var }(b_1).$$

First we need an estimate for the variance, σ^2. It is

$$s^2(y) = \frac{\Sigma(y_j - Y_j)^2}{N - 2}.$$

Notice that we divide by $(N - 2)$. The degrees of freedom are $(N - 2)$ because we have already used the data to obtain two estimates, namely, b_0 and b_1, from which we calculate the fitted values Y_j. Furthermore,

$$s^2(\bar{y}) = \frac{s^2(y)}{N}$$

$$s^2(b_1) = \frac{s^2(y)}{\Sigma(x_{1j} - \bar{x}_1)^2} \equiv \frac{s^2(y)}{[11]}.$$

* Henceforth, since most of the summations are over the total number of observations, from $j = 1$ to N, we will use the summation sign without indices, Σ, to indicate $\sum_{j=1}^{N}$.

(For derivation, see Brownlee, Section 11.2.) Putting the last two expressions into the equation given above for the estimated variance of Y, Est. Var (Y), we obtain some idea of how much uncertainty there is in our fitted values:

$$\text{Est. Var. } (Y) = \frac{s^2(y)}{N} + (x_1 - \bar{x}_1)^2 \frac{s^2(y)}{\Sigma(x_1 - \bar{x}_1)^2}$$

$$= s^2(y)\left[\frac{1}{N} + \frac{(x_1 - \bar{x}_1)^2}{[11]}\right], \quad \text{for any value of } x_1.$$

2A.3   Confidence Regions

To obtain a confidence region for our *line* we must have an appropriate multiplying factor to apply to the standard error of any point on the line. The multiplier is $[2F(0.90, 2, N - 2)]^{1/2}$ for a 90% confidence region (Brownlee, Section 10.3, and Scheffé, Section 3.5 and Problem 2.12). The use of t, rather than $(2F)^{1/2}$, as a multiplier is not appropriate because, when we make a confidence interval statement about the line, *two* parameters rather than one have been estimated, and we want to make a *joint* confidence statement about the two parameters. The first number after the F gives the "level of confidence desired, the second the number of parameters estimated, and the third the number of degrees of freedom for error. The expression

$$Y \pm [2F(0.90, 2, N - 2)]^{1/2} s(y)\left[\frac{1}{N} + \frac{(x_1 - \bar{x}_1)^2}{[11]}\right]^{1/2}$$

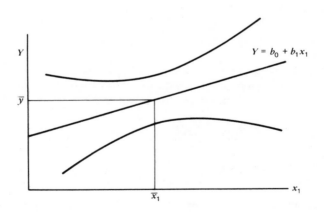

Figure 2A.1   The hyperbolic "90% confidence region" which covers the entire true line.

defines a 90% confidence region about the fitted line, inside which the true line will lie in 90% of the cases. Note that Y and x_1 are continuous variables, whereas [11] depends on the x-values at which the data were taken.

APPENDIX 2B

LINEAR LEAST-SQUARES ESTIMATES—GENERAL CASE

2B.1 Estimates of Coefficients

The coefficients of the equation are obtained by minimizing the sum of squares of the residuals, the differences between each observed value, y_j, and its corresponding fitted value, Y_j. The least-square solution will be illustrated for *three* independent variables. The formulas given are not the most convenient for computational purposes but serve well to illustrate the LS technique.

As in the example with one independent variable, we make four assumptions about the relationship between the jth observed values of the three independent variables, x_{1j}, x_{2j}, and x_{3j}, and the jth observed value of the dependent variable, y_j.

A. There is a linear relationship between the "true" value of a response, η, and the values of the independent variables, that is,

$$\eta = \beta_0 + \beta_1 x_1 + \beta_2 x_2 + \beta_3 x_3.$$

B. $y_j = \eta_j + e_j = \beta_0 + \beta_1 x_{1j} + \beta_2 x_{2j} + \beta_3 x_{3j} + e_j$, where e_j is random error.

C. As in Appendix 2A, the e_j's have the following properties:
  1. The expected value of e_j is zero; our observed y_j is an unbiased estimate of η.
  2. The variance of e_j is $\sigma^2(y)$, which remains constant for all values of x_i and η ($i = 1$ to K independent variables).
  3. The e_j are statistically uncorrelated, i.e. the expected (population) value of $e_j e_{j'}$ for any pair of points j and j' is zero.

D. Observed values of the independent variables are measured without error. All the error is in the y_j's, and none is in the x_{ij}'s.

The estimated equation is

$$Y = b_0 + b_1 x_1 + b_2 x_2 + b_3 x_3,$$

which can be rewritten as:

$$Y - \bar{y} = b_1(x_1 - \bar{x}_1) + b_2(x_2 - \bar{x}_2) + b_3(x_3 - \bar{x}_3),$$

where $\bar{y}$, $\bar{x}_1$, $\bar{x}_2$, and $\bar{x}_3$ are the means of the variables for the set of N observations. This form takes advantage of the fact that for any LS fit the constant

b_0 is always of the form:

$$b_0 = \bar{y} - \sum_{i=1}^{K} b_i \bar{x}_i, \quad \text{for } K \text{ constants fitted.}$$

Thus in our example we need to find only the coefficients b_1, b_2, and b_3.

The LS solutions for the coefficients are obtained by minimizing the quantity $[rr]$, the sum of squares of the residuals:

$$[rr] \equiv \sum_{j=1}^{N^*} (y_j - Y_j)^2$$

$$= \sum_{j=1}^{N^*} [(y_j - \bar{y}) - b_1(x_{1j} - \bar{x}_1) - b_2(x_{2j} - \bar{x}_2) - b_3(x_{3j} - \bar{x}_3)]^2$$

(For definitions of symbols, see the Glossary.)

Differentiating with respect to each b_i and setting the resultant expressions equal to zero leads to the set of three "normal equations:"

$$b_1[11] + b_2[12] + b_3[13] = [1y]$$
$$b_2[21] + b_2[22] + b_3[23] = [2y]$$
$$b_3[31] + b_2[32] + b_3[33] = [3y]$$

where

$$[11] \equiv \Sigma(x_{1j} - \bar{x}_1)^2;$$
$$[21] \equiv [12] \equiv \Sigma(x_{1j} - \bar{x}_1)(x_{2j} - \bar{x}_2);$$
$$[1y] \equiv \Sigma(x_{1j} - \bar{x}_1)(y_j - \bar{y}); \text{ etc.}$$

The solution for the b's may be obtained by the method of determinants:†

$$b_1 = \frac{\begin{vmatrix} [1y] & [12] & [13] \\ [2y] & [22] & [23] \\ [3y] & [32] & [33] \end{vmatrix}}{\begin{vmatrix} [11] & [12] & [13] \\ [21] & [22] & [23] \\ [31] & [32] & [33] \end{vmatrix}}$$

with similar expressions for b_2 and b_3.

---

* Since most of the summations are over the total number of observations from $j = 1$ to N, we will use the summation sign without indices, Σ, to indicate $\sum_{j=1}^{N}$.

† See any algebra textbook for the solution of sets of linear equations by determinants.

Designating the determinant in the denominator by Δ and expanding the numerator by the first column (expansion by minors), we have

$$b_1 = \frac{1}{\Delta}\left\{[1y]\begin{vmatrix}[22] & [23] \\ [32] & [33]\end{vmatrix} - [2y]\begin{vmatrix}[12] & [13] \\ [32] & [33]\end{vmatrix} + [3y]\begin{vmatrix}[12] & [13] \\ [22] & [23]\end{vmatrix}\right\}.$$

Making similar expansions for b_2 and b_3, we can write the solutions as:

$$b_1 = c_{11}[1y] + c_{12}[2y] + c_{13}[3y]$$
$$b_2 = c_{21}[1y] + c_{22}[2y] + c_{23}[3y]$$
$$b_3 = c_{31}[1y] + c_{32}[2y] + c_{33}[3y]$$

where, for example,

$$c_{12} = -\frac{1}{\Delta}\begin{vmatrix}[12] & [13] \\ [32] & [33]\end{vmatrix} = c_{21}.$$

The elements:

$$\begin{matrix} c_{11} & c_{12} & c_{13} \\ c_{21} & c_{22} & c_{23} \\ c_{31} & c_{32} & c_{33} \end{matrix}$$

constitute the inverse of the matrix corresponding to the determinant Δ. To simplify nomenclature, the *off-diagonal* element in the ith row and the i'th column is designated by the symbol $c_{ii'}$, and the ith *diagonal* element is designated by the symbol c_{ii}.

2B.2   Variance and Standard Errors of Coefficients

The b_i are linear functions of the y's. Note that

$$[1y] = \Sigma(x_{1j} - \bar{x}_1)(y_j - \bar{y})$$
$$= \Sigma(x_{1j} - \bar{x}_1)y_j - \bar{y}\Sigma(x_{1j} - \bar{x}_1)$$
$$= \Sigma(x_{1j} - \bar{x}_1)y_j, \quad \text{since } \Sigma(x_{1j} - \bar{x}_1) = 0.$$

Similar relations hold for $[2y]$, $[3y]$, and in general. Hence, the expression for any b_i can be written:

$$b_i = \Sigma y_j\{c_{i1}(x_{1j} - \bar{x}_1) + c_{i2}(x_{2j} - \bar{x}_2) + c_{i3}(x_{3j} - \bar{x}_3)\}.$$

For convenience let the quantity in braces be represented by k_{ij}, so that

$$b_i = \Sigma y_j\{k_{ij}\}.$$

Thus b_i is a linear function of the y_j. (This is the reason for the term linear least-squares estimates.)

The variance of any b_i is obtained directly from the equation given above. The k_{ij} are functions only of the x's, which are assumed to be known without error. The y_j are independent random variables, each with the same variance, $\sigma^2(y)$. Hence the variance of b_i can be written as

$$\text{Var } (b_i) = \sigma^2(y)\Sigma\{k_{ij}^2\}$$

Brownlee (Section 13.6) shows that

$$\Sigma\{k_{ij}^2\} = c_{ii},$$

the ith diagonal element of the inverse matrix, so that

$$\text{Var. } (b_i) = \sigma^2(y)c_{ii}$$

and the standard error of the coefficient $b_i = \sigma(c_{ii})^{1/2}$.

Exactly analogous equations relate the *estimated* variance of b_i to the *estimated* variance of y; thus the

$$\text{Est. Var. } (b_i) = s^2(y)c_{ii}.$$

2B.3   Computer Computations

The computer Linear Least-Squares Curve-Fitting Program given in the User's Manual of this book does not compute elements of the inverse matrix, $c_{ii'}$, but instead computes *multiples* of them, $c'_{ii'}$. The two are related by

$$c'_{ii} = c_{ii}[ii] \quad \text{and} \quad c'_{ii'} = c_{ii'}([ii][i'i'])^{1/2}.$$

The $c'_{ii'}$ are listed by the computer as INVERSE, C(I, I PRIME).[*] Thus for $s(b_i)$, in terms of the computer inverse elements, we have:

$$s(b_i) = s(y)\left(\frac{c'_{ii}}{[ii]}\right)^{1/2} = \text{S. E. COEF. (computer listing)},$$

where $s(y)$ is the RESIDUAL ROOT MEAN SQUARE value listed by the computer.

2B.4   Partitioning Sums of Squares

Linear LS estimation partitions the total variability in the data (expressed as a total sum of squares) into two parts:

1. The sum of squares due to the fitted equation.
2. The residual sum of squares.

The objective is to account for as much of the total variability as possible by means of the fitted equation. The partitioning is based on the identity:

$$(y_j - \bar{y}) \equiv (Y_j - \bar{y}) + (y_j - Y_j)$$

---

[*] To conform with computer printout, computer listings are capitalized.

where $Y_j = \bar{y} + \sum_{i=1}^{K} b_i(x_{ij} - \bar{x})$ is the value given for the jth observation by the fitted equation with K independent variables.

The total sum of squares is partitioned thus:

$$\sum (y_j - \bar{y})^2 = \sum (Y_j - \bar{y})^2 + \sum (y_j - Y_j)^2.$$

In the shorter nomenclature of Gauss:

$$[yy] = [YY] + [rr]$$

where (1) $[yy] \equiv \sum (y_j - \bar{y})^2$ is the TOTAL SUM OF SQUARES (computer listing),

(2) $[YY] \equiv \sum (Y_j - \bar{y})^2$ is the reduction in the total sum of squares due to the fitted equation, and

(3) $[rr] \equiv \sum (y_j - Y_j)^2$ is the RESIDUAL SUM OF SQUARES (computer listing), a measure of the squared scatter of the observed values around those calculated by the fitted equation.

2B.5 Multiple Correlation Coefficient Squared R_y^2

The MULT. CORREL. COEF. SQUARED (computer listing), R_y^2, represents the fraction of the total variation accounted for by the fitted equation:

$$R_y^2 = \frac{[YY]}{[yy]} = 1 - \frac{[rr]}{[yy]}.$$

2B.6 F-value

The F-VALUE (computer listing) used to judge the "significance" of the value of R_y^2 is also calculated from the partitioned total sum of squares:

$$\text{F-VALUE} = \frac{[YY]}{[rr]} \cdot \frac{N - K - 1}{K} = \frac{R^2}{1 - R^2} \cdot \frac{N - K - 1}{K}.$$

2B.7 Bias and Random Error

The residual sum of squares (total squared error) is also the sum of two components: (1) bias (lack of fit) and (2) random error:

$$\text{RSS} = \text{SSB} + \text{SSE}.$$

If the fitted equation contains no bias (SSB equals zero), the residual sum of squares will reflect only random error.

2B.8 Residual Mean Square

The RESIDUAL MEAN SQUARE (computer listing) with $(N - K - 1)$ degrees of freedom is:

$$\text{RMS} = \frac{[rr]}{N - K - 1} \equiv \frac{\text{RSS}}{N - K - 1}.$$

If the fitted equation contains no bias, the residual mean square is the estimated variance of y:

$$\text{RMS} = s^2(y).$$

2B.9   Residual Root Mean Square

The RESIDUAL ROOT MEAN SQUARE (computer listing) is by definition the square root of the residual mean square. Again, *if* the fitted equation contains no bias, the residual root mean square is the estimated standard deviation of y:

$$\text{RRMS} = s(y).$$

CHAPTER 3

One Independent Variable

3.1 Plotting Data and Selecting Form of Equation

Plotting is instructive when only one independent variable is thought to be influential. We can *see* whether the data fall on a straight line, show evidence of curvature, or indicate some anomaly.

When the data can be directly represented by a straight line in x and y, we are in luck. There is a simple association between the independent variable x and the dependent variable y, represented by the equation to be fitted:

$$Y = b_0 + b_1 x_1.$$

When there is curvature, we try to identify some form of equation that will fit the data better. For our purposes the fitting equations are of two types: those whose constants are linear or can be linearized by simple transformations, and those whose constants are nonlinear and cannot be easily linearized. To aid in identification, plots of some of the equation forms of each type will be given.

3.2 Plots of Linearizable Equations

The plots in Figures 3.1, 3.2, and 3.3 were made from equations which are linearizable by appropriate transformations of y and x. This system of transformations (Hald, Section 18.7) was produced by combining each of the three functions of y (y, $1/y$, $\ln y$) with each of the corresponding functions of x. Five of the combinations are shown. The asymptotes and intercepts are easily deducible by substitution in the equations.

Another set of linearizable forms is represented by polynomials with integer powers of x:

$$Y = b_0 + b_1 x_1 + b_{11} x_1{}^2 + b_{111} x_1{}^3 + \cdots.$$

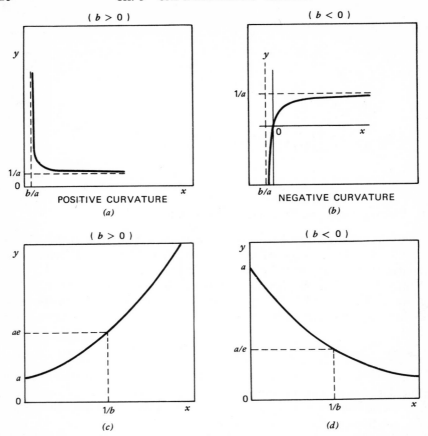

Figure 3.1   Linearizable curves. (*a*) and (*b*) Hyperbolas; $y = x/(ax - b)$. Linear form: $1/y = a - b/x$. (*c*) and (*d*) Exponential functions; $y = ae^{bx}$. Linear form: $\ln y = \ln a + bx$.

We often use a polynomial of such form—rarely going beyond the fourth power, usually not beyond the second—to represent some unknown "true" function. The higher the order of the polynomial, the more precise the data must be. Figure 3.4*a* and 3.4*b* show the diverging curves formed by the positive and negative coefficients of the quadratic term of a second-power polynomial.

Another set of useful shapes compiled by Hoerl (Section 20, page 55) is reproduced, with permission, in Figure 3.4*c*. The equation, $y = ax^b e^{cx}$, can be linearized by logs when a and x are greater than zero. Thus

$$\ln y = \ln a + b \ln x + cx$$

is linear in $\ln y$, $\ln x$, and x.

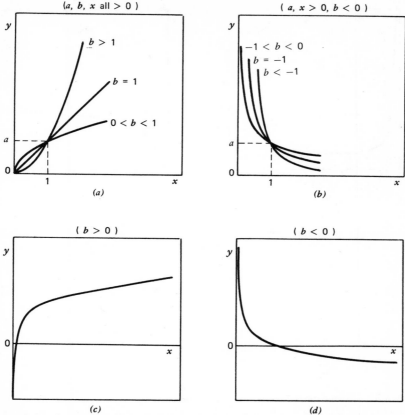

Figure 3.2 Linearizable curves. (*a*) and (*b*) Power functions; $y = ax^b$. Linear form: $\ln y = \ln a + b (\ln x)$. (*c*) and (*d*) Logarithmic functions already in linear form: $y = a + b (\ln x)$.

Although both the polynomials and Hoerl's equation may be identified by x-y plots, each term in the linearized equation must be counted as a separate independent variable in order to estimate the constants. Thus in the polynomial equation we let $x_1 = x_1'$, $x_1^2 = x_2'$, $x_1^3 = x_3'$, ..., and in Hoerl's equation $\ln y = y'$, $\ln x_1 = x_1'$, $x_1 = x_2'$ to obtain the linearized equation form: $y' = b_0 + b_1 x_1' + b_2 x_2' + b_3 x_3' + \cdots$. The added problems associated with fitting equations with two or more variables will be discussed in later chapters.

The distribution of the random errors, e_j, is changed by any nonlinear transformation of the y_j. We will often inspect the "empirical cumulative distribution of the residuals" to see whether y or its transform appears more

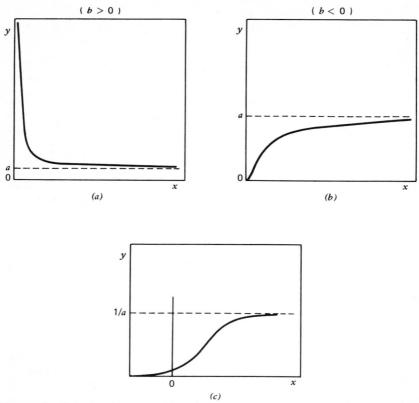

Figure 3.3 Linearizable curves. (a) and (b) Special functions; $y = ae^{b/x}$. Linear form: $\ln y = \ln a + b/x$. (c) $y = 1/(a + be^{-x})$. Linear form: $1/y = a + be^{-x}$.

nearly normal. If the transformation of y chosen for physical or algebraic reasons produces wide changes in the residuals when they are plotted against Y (fitted y-values), we may want to make some guess about how the variance of y is changing with η or with some x. We can then *weight* the y_j *inversely* by their estimated variances.

3.3   Plots of Nonlinearizable Equations

Quite a few commonly occurring shapes are not linearizable; two are shown in Figure 3.5. The S-shaped curve of Figure 3.5a may be fitted by the so-called logistic equation, but most experimenters will want mathematical help at this point. The curve in Figure 3.5b starts at zero, increases to a maximum, and then dies away to zero. The concentration of an intermediate

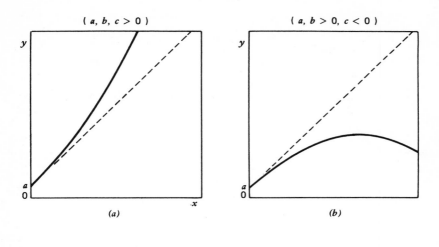

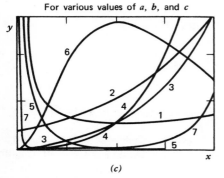

Figure 3.4 Linearizable curves. (*a*) and (*b*) Polynomial functions; $y = a + bx + cx^2$. Linear form: $y = a + bx + cx^2$. (*c*) Hoerl's special functions; $y = ax^b e^{cx}$. Linear form: $\ln y = \ln a + b (\ln x) + cx$.

	Coefficients		
Curve	*a*	*b*	*c*
1	10	−1	0.1
2	2	0	0.1
3	0.1	1	0.1
4	0.01	2	0.1
5	10	−1	−0.2
6	1	2	−0.2
7	0.1	−2	0.5

23

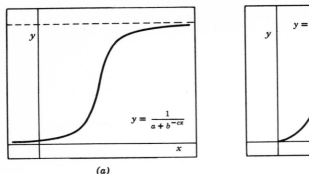 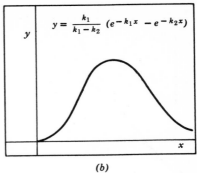

Figure 3.5   Nonlinearizable curves.

compound, when x represents time (or duration) of a chemical reaction, often follows such a curve.

A useful collection of equations of these shapes has also been made by Hoerl (Section 20, pages 56–77). We advise using the equations in this reference, but not the method of fitting, since results obtained are sensitive to the location of the coordinate points chosen for estimating the constants.

3.4 Statistical Independence and Clusters of Dependent Variable

In any fitted equation the uncertainty in the constants cannot be judged unless one knows or can deduce from the data something about the uncertainty of the y-values. It will not do to assume without check that all y-values have the same independent random variation. For example, each run may produce material which is split into several subbatches for chemical analysis. An x-y plot may then show (Figure 3.6a) several vertical clusters of points, each at a single x-value and each representing a single run. The scatter within each cluster measures only the uncontrolled variation of chemical analysis, perhaps combined with within-run product variability, and hence fails to reflect the variation from one *run* to another even at the same value of x. A statistician would say that, if only r runs were made, the true error for use in judging the precision of the two parameters of the straight line is estimable with $(r - 2)$ degrees of freedom, and not with $(N - 2)$, where N is the number of chemical analyses.

To take another very common case, suppose that a whole sequence of x-values (say, p of them) is whipped through on one batch of raw material. Additional observations may be obtained by running through the whole set of x-settings again, perhaps on other batches of raw material, q times.

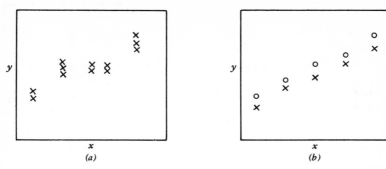

Figure 3.6 Randomization patterns. (*a*) Data at each *x*-value taken at one time. (*b*) Each series of five values taken in one run.

The p points produced by each set may lie closely on a straight line, but the different sets may lie on different lines (Figure 3.6*b*). From q such sequences, we have, at most, $(q - 1)$ degrees of freedom for judging the real error in the average values of y or the *heights* of the lines at $\bar{x}$, even though all lines are closely parallel, so that the average slope is very precisely determined.

These samples of common randomization patterns are given, not to criticize the way in which the data were taken, but to show that the data cannot be analyzed properly, or valid conclusions drawn, if the analyst has an erroneous view of how they were collected.

3.5 Allocation of Data Points

The allocation of x-values needs to be examined. All data points weigh equally in determining the average height of the line, but the slope is influenced more heavily by the extreme values of x. Also, a cluster of points is often equivalent to only one point in the determination of the slope. In Figure 3.7*a*, the cluster at low x is essentially one point, and the slope is highly dependent on the point at high x. Similarly, in Figure 3.7*b* the cluster at the mean of the x-values contributes little to determining the slope, which is influenced mainly by the two extreme values. These plots emphasize the value of designed experiments to avoid such unbalanced data.

3.6 Outliers

The data should also be examined for occasional "bad" values (wild points, mavericks, outliers). Depending on their location in "x-space," such values can affect the estimate of the average height of the line or the estimate of its slope. Figure 3.8*a* is an example of a bad value at high x which will seriously bias the estimated slope. Care is needed in this case, for we may

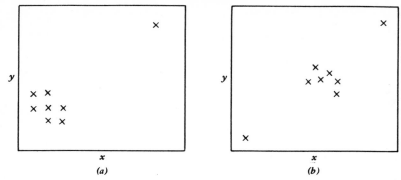

Figure 3.7   Allocation of data points.

not have a bad value; rather, the true equation may be a curve. Figure 3.8*b* is an example of a bad value at the middle of the range of x; it has little effect on the slope but will raise the height of the line, and consequently our error estimate, by a large factor.

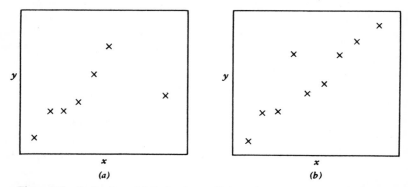

Figure 3.8   Bad values. (*a*) Bad value at high x. (*b*) Bad value at midrange of x.

3.7   Use of Computer Program

After the data have been examined and the form of the equation to be fit has been tentatively chosen, we are ready to use the computer Linear Least-Squares Program to see whether an equation of the form selected can be made to fit the data well by a suitable choice of coefficients. The key assumptions given in Chapter 2 should always be checked. Some of them may be known to hold. The others can often be supported by critical examination of the data together with the statistics produced by the computer.

When the underlying assumptions are satisfied, the computer printout gives results showing:

(a) how well the equation is fitting (MULT. CORREL. COEF. SQUARED),
(b) what the magnitudes of the intercept (COEF. B(0)) and of the coefficients (COEF. B(I)) are,
(c) how well each coefficient is estimated (S. E. COEF.), and
(d) how significant, in the statistical sense, are the coefficients taken all
together (F-VALUE).

When the "known' variance of a single observation of y is consistent with
the residual mean square value of the fitted equation with degrees of freedom
greater than 20, there is little likelihood that any large bias or lack of fit is
left in the equation. Under such circumstances we consider ourselves for-
tunate in having a "good" fit.

3.8 Study of Residuals

In the event that we have not yet obtained a "good" fit, much can be
learned from computer plots of the residuals—the differences between the
fitted or calculated value, Y, and the observed value, y. (Residuals are also
known as remainders, discrepancies, deviations, and differences.)

The computer plot of the residuals versus their corresponding equation
values will sometimes show whether there is some dependence of the magnitude of the residual on the magnitude of the equation value (Anscombe and
Tukey, pages 141–160). Four common defects which may be revealed by such
a plot are shown in Figure 3.9.

The wedge-shaped plot in Figure 3.9a indicates that the scatter of the
residuals increases with Y (fitted Y). The residuals should show roughly equal
scatter for all values of Y; otherwise the variance of the points about the
line cannot be considered constant. Using log y will often eliminate this type
of unequal scatter.

The wedge-shaped plot in Figure 3.9b indicates that the scatter of residuals
decreases as the Y-value approaches some value—in this case 100. This
suggests that a new fit should be made, but now with each point weighted by
$1/(100 - Y_j)^2$. This may give a somewhat different b-value but one with
a better fit. Such a weighting produces an equation that is nonlinear in its fitted
constants. We defer the treatment of such equations to Chapter 9.

The U-shaped scatter plot (Figure 3.9c) is the result of fitting a straight line
to data which are better represented by a simple curve.

The uneven plot (Figure 3.9d) is an example of clustered data. The relatively few observations at the lower values of Y contribute disproportionately
to the estimates made and may well determine the slope. Therefore, they
should be identified in regard to the time and conditions under which they
were obtained. If all of them are judged to be sound random data, then all

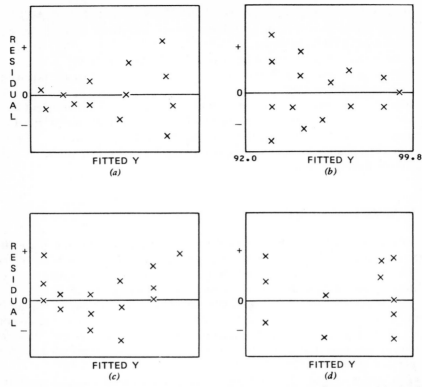

Figure 3.9 Defects revealed by residuals. (*a*) Indicates variance increases with *Y*. (*b*) Indicates variance decreases as *Y* approaches 100. (*c*) Indicates curvature. (*d*) Indicates clustered data.

should be used, but it is clear (is it not?) that we have only three well-separated levels of *x* and hence only 1 degree of freedom for judging curvature, even though we have eleven observations.

Cumulative Distribution of Residuals

The normal distribution is a two-parameter one (μ and σ are the symbols usually employed for its mean and standard deviation, respectively). The *sample* mean and standard deviation are usually denoted by $\bar{y}$ and s_y. When a *K*-parameter equation is fitted to *N* data points (which have normally distributed error), then the *N* residuals have a normal distribution with mean zero and with standard deviation $\sigma\sqrt{(N - K - 1)/(N - 1)}$. There are of course only $(N - K - 1)$ degrees of freedom in these *N* residuals.

The cumulative normal distribution can be linearized so that the height of the straight line through the data estimates the population mean, while the "slope" estimates the standard deviation. The unit for slope is the distance from the 0.1587 (or 16%) to the 0.5000 point. Alternatively, and a little more precisely, the y-distance from the 16% to the 84% point estimates 2σ.

Special grids have been prepared (the best known and in our opinion the best, is by Hazen, Whipple, and Fuller and is sold by Codex Book Co. as their No. 3227) that make it easy to plot the empirical cumulative distribution (usually abbreviated e.c.d.) for any N. The plotting points on the percentage scale are at $(j - \frac{1}{2})/N$, where $j = 1, 2, \ldots, N$, when the residuals are arranged in order of increasing magnitude.

Our purpose in plotting the cumulative distribution of the residuals is not at all to estimate a standard deviation. Rather, we do this to see whether the error structure appears to be roughly normal (Gaussian). It is, then, only the *shape* of the distribution that is of interest. We therefore suppress the actual scale of the residuals. Their actual ordered magnitudes are printed in the table just preceding the residual plot.

Since samples of the cumulative distributions will contain random errors, we need to acquire some feeling for normal departures from normality. Appendix 3A shows some actual plots of random normal deviates with standard deviation 1 and mean 0. They were taken from Rand Corporation tables, where 100,000 such deviates are given. Our sample sizes range from 8 to 384. As might be expected, samples of 8 tell us almost nothing about normality, whereas samples of 384 seem very stable except for their few lowest and highest points. Sets of 16 show shocking wobbles; sets of 32 are visibly better behaved; and sets of 64 nearly always appear straight in their central regions but fluctuate at their ends.

The commonest type of "nonnormality" is a single oversized residual. Figure 3.10 shows an actual example, which indicates too that our computer printout, although admittedly rough (it has 50 steps in each direction), is still amply fine to show curvature or other gross irregularity. Both the cumulative distribution plot and the fitted value plot are needed to identify the point as an outlier. Had the large residual occurred at an extreme Y-value, the point might have implied curvature in the true relation, or a systematic variation of variance with the value of Y. These possibilities should be checked before the point is deemed to be an outlier. Some effort should be made to find the actual cause of its excessive magnitude before it is finally discarded. If the outlier is a high-yield point or a particularly desirable property of a product, knowledge of its cause may be more valuable than all the rest of the data. The experimenter can often make a better guess than the data correlator as to where to look for causes. In many cases such a search has led to the discovery of important variables that had not previously been considered. Numerous patents have resulted from the recognition of outliers.

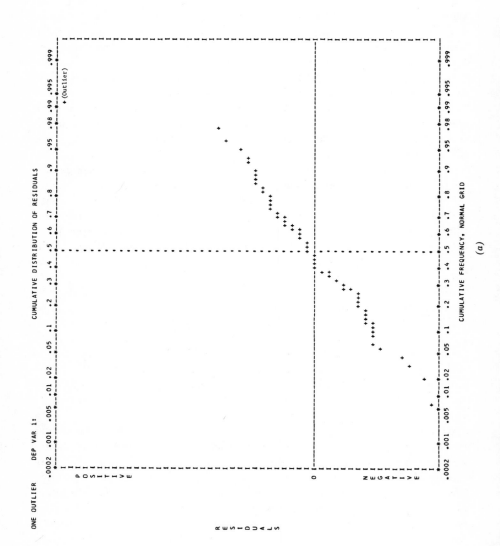

(a)

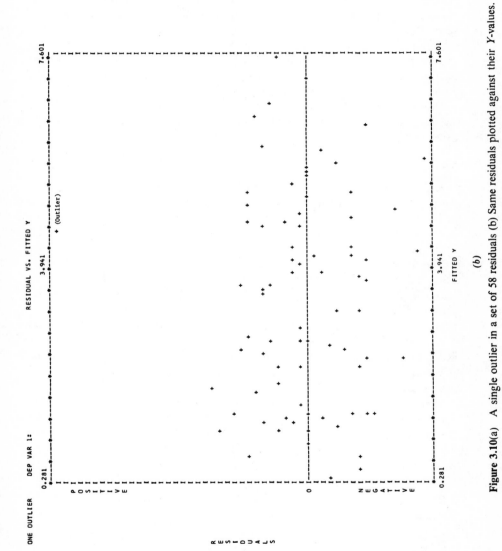

Figure 3.10(a)   A single outlier in a set of 58 residuals (b) Same residuals plotted against their Y-values.

(b)

31

The time order of the residuals is often informative in judging the quality of data. If observed values are given to the computer in the same order in which they were obtained, the computer program will list them and their residuals in that order. Such a list sometimes reveals sudden jumps, or clusters of points that may be only repeats (and not real replicates) of the same item of information. If the signs of the residuals do not appear to fluctuate from plus to minus in a random order, the independence of the observations should be measured by the runs-of-signs test (Brownlee, Section 6.3). Tables of critical values are given by Owen (Sections 12.4 and 12.5) and by Dixon and Massey (Appendix Table A10).

None of these techniques can be expected to work well with fewer than twenty observations, and all are much more efficient when the number exceeds fifty.

3.9   Dealing with Error in the Independent Variable

Least-squares analysis assumes that the independent variable is measured without error. When the x-values have considerable variance, the estimate of the coefficient (the slope) is biased toward zero. As a rule of thumb, LS analysis can be used safely if the variance of x is less than a tenth of the average scatter of the x's from their mean.

The difficult case of wide uncertainty in several x_i is not discussed in this book. (For details, see the publications by Acton and by Madansky.)

If the variance of x is too large for LS analysis, Bartlett's three-group procedure is recommended. To determine the slope of the fitted line by this procedure, the data are divided into three groups by the observed x-values. The slope of the fitted line is that of the line connecting the average points in the first and the third group. The fitted line must pass through the average point of all the data. Bartlett derives confidence intervals for the slope and height calculated in this manner.

Berkson has shown that when x is a "controlled variable," that is, when each x-value has a random component with population value zero added to a target value, standard LS methods give unbiased estimates of the equation coefficients. For details, the reader should consult Brownlee (Second Edition, page 393) and the additional references that he gives.

3.10   Example of Fitting a Straight Line to Data—One Independent Variable

A numerical example of fitting a straight line to data, using our computer program, is given in Figures 3B.1–3B.5. The example was taken from pilot-plant data. The independent variable, x, is the acid number of a chemical determined by titration, and the dependent variable, y, is its organic acid content determined by extraction and weighing. The aim was to determine how well values obtained by the relatively inexpensive titration method can serve to estimate those obtained by the more expensive extraction and weighing technique. The high values of t and of R^2, the appearance of the x-y graph, the distribution of the residuals, and the even spread of the residuals versus Y plot—all support the conclusion that x is a good estimate of y.

APPENDIX 3A

CUMULATIVE DISTRIBUTION PLOTS OF RANDOM NORMAL DEVIATES

The observed values used in these plots were taken from the Rand Corporation tables of random normal deviates; each set of z values has a standard error of 1 and a mean of 0.

Number of Observations per Set	Figures
8	3A.1–3A.2
16	3A.3–3A.4
32	3A.5–3A.6
64	3A.7–3A.13
384	3A.14

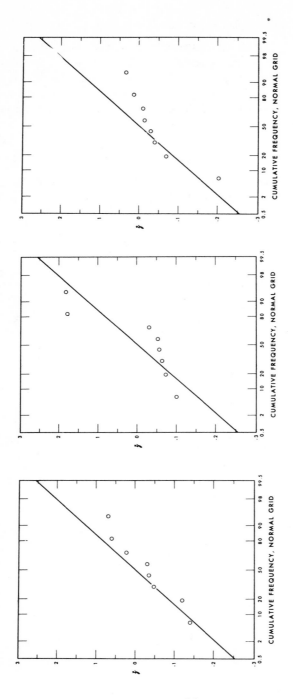

Figure 3A.1 Samples of 8 random normal deviates.

34

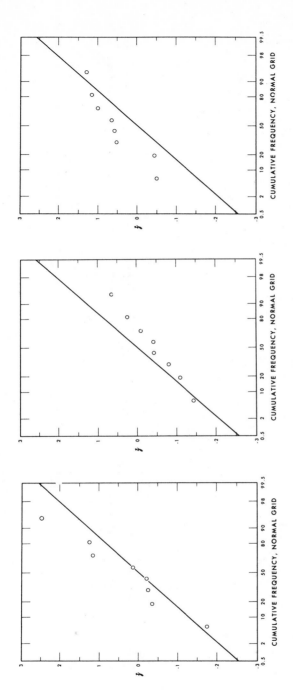

Figure 3A.2 Samples of 8 random normal deviates.

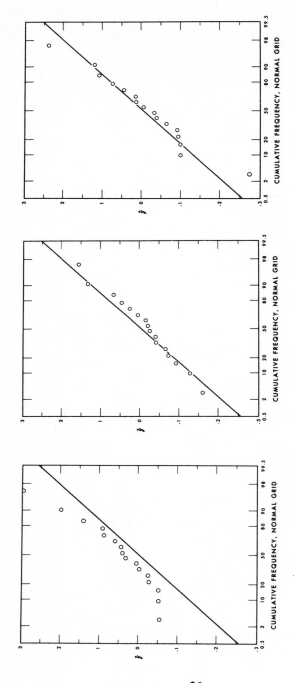

Figure 3A.3   Samples of 16 random normal deviates.

36

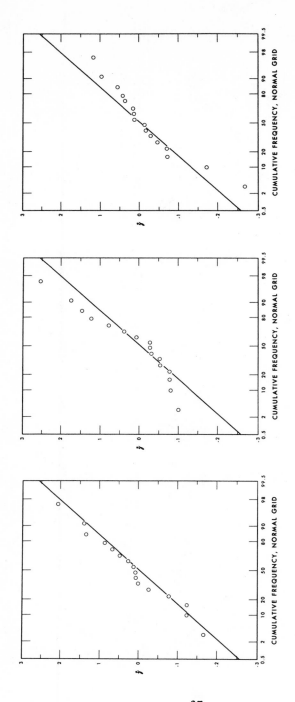

Figure 3A.4 Samples of 16 random normal deviates.

37

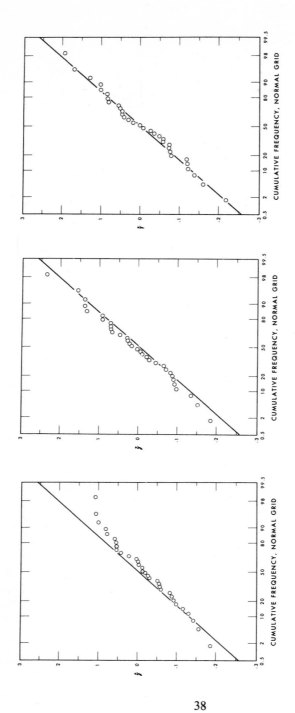

Figure 3A.5 Samples of 32 random normal deviates.

38

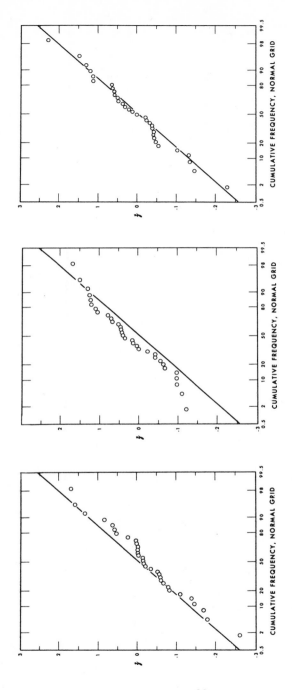

Figure 3A.6 Samples of 32 random normal deviates.

39

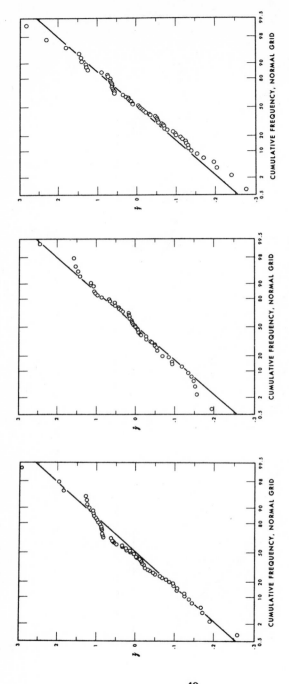

Figure 3A.7 Samples of 64 random normal deviates.

40

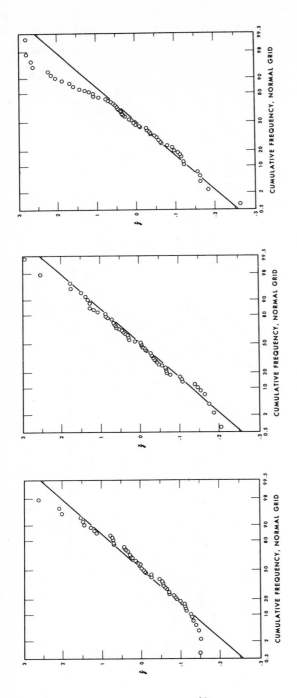

Figure 3A.8 Samples of 64 random normal deviates.

41

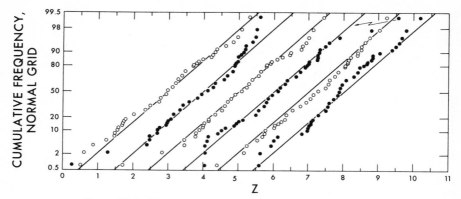

Figure 3A.9 Six samples of 64 random normal deviates.

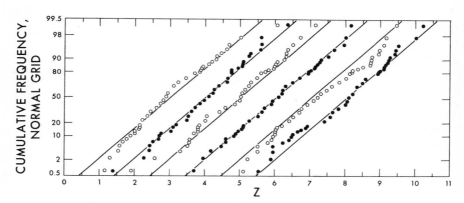

Figure 3A.10 Six samples of 64 random normal deviates.

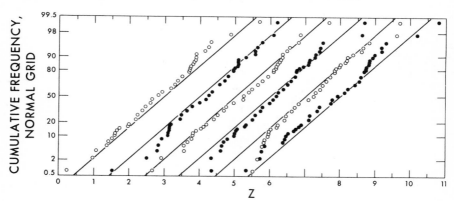

Figure 3A.11 Six samples of 64 random normal deviates.

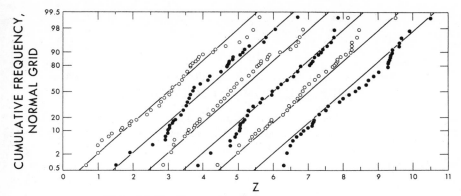

Figure 3A.12 Six samples of 64 random normal deviates.

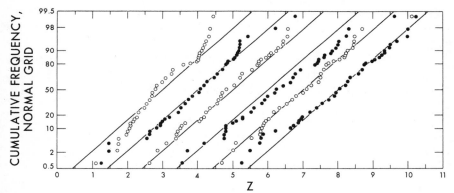

Figure 3A.13 Six samples of 64 random normal deviates.

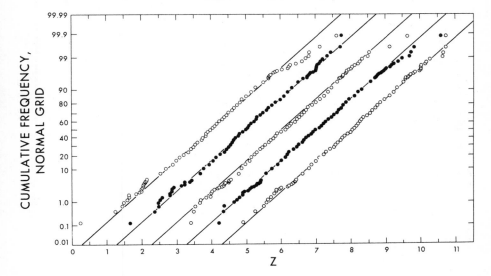

Figure 3A.14 Five samples of 384 random normal deviates.

APPENDIX 3B

EXAMPLE OF FITTING A STRAIGHT LINE TO DATA— ONE INDEPENDENT VARIABLE

The objective of this problem is to determine how well values obtained by a relatively inexpensive titration method can serve to estimate those obtained by a more expensive extraction and weighing technique. Twenty samples of a pilot-plant chemical were selected to cover the range of interest. Each sample (chosen from the twenty at random) was thoroughly mixed just before being split and analyzed by both methods.

Figure 3B.1 is a plot of the values obtained. There appears to be a reasonably good linear relationship between the acid number (x) determined by titration and the organic acid content (y) determined by extraction and weighing. The error in x is negligible.

Figures 3B.2–3B.5 are computer printouts obtained by fitting the equation $y = b_0 + b_1 x$ to the data, using the Linear Least-Squares Curve-Fitting Program. Details of the nomenclature used in the printouts are given in the Glossary.

In Figure 3B.2 the data input is listed together with the sum of each variable, the residual sums of squares and cross products, the calculated mean and root mean square of each variable, and the simple correlation coefficients. The numbers at the top of each column, such as 1-11-21 and 2-12-22, identify

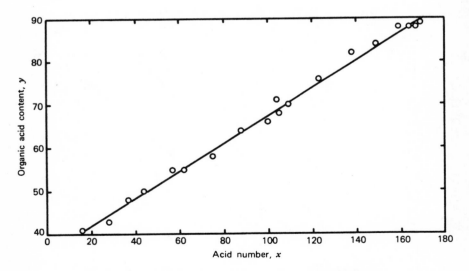

Figure 3B.1 Plot of acid number versus organic acid content.

the numbers of the variables listed in that column. In this case, they are 1 and 2. Had there been 21 variables, the 1st, 11th, and 21st variable would have been listed in that order in the first column.

Figure 3B.3 lists many of the important statistical properties of the fit. The b_1 coefficient, given in fixed decimal point form, is 0.32108. Since the standard error of the coefficient is 0.0055, the value of b_1 might well be written as 0.321 ± 0.006. The multiple correlation coefficient squared, R^2, is 0.995; this indicates that 99.5 % of the total sum of squares of y is accounted for by this equation. The residual root mean square is 1.2. Sometimes the residual column allows us to determine that the observations were not statistically independent. There were 10 positive differences where OBS. Y was larger than FITTED Y, and 10 negative differences. In all, there were 13 changes of sign. We see on page 380 of Owen that 15 changes of sign would have been too many and 6 too few. There is in this case, then, no evidence of autocorrelation.

Figure 3B.4 is a computer-made plot of the empirical cumulative distribution of the residuals. The residuals fall, as they should, approximately on a straight line, indicating that they are roughly normally distributed, with no outliers.

Figure 3B.5 is the computer plot of the residuals versus FITTED Y. Since the residuals are evenly distributed, the variance of the points about the line is roughly constant.

We have, then, a good correlation between the two methods of assaying the particular chemical tested. Which one should be used will depend on the relative costs of the two procedures.

LINEAR LEAST-SQUARES CURVE-FITTING PROGRAM

ACID NUMBER

DATA INPUT   1 INDEPENDENT VARIABLES   1 DEPENDENT VARIABLE(S)

OBSV.	SEQ.	1-11-21	2-12-22	3-13-23	4-14-24	5-15-25	6-16-26	7-17-27	8-18-28	9-19-29	10-20-30
1	0	123.000	76.000								
2	0	109.000	70.000								
3	0	62.000	55.000								
4	0	104.000	71.000								
5	0	57.000	55.000								
6	0	37.000	48.000								
7	0	44.000	50.000								
8	0	100.000	66.000								
9	0	16.000	41.000								
10	0	28.000	43.000								
11	0	138.000	82.000								
12	0	105.000	68.000								
13	0	159.000	88.000								
14	0	75.000	58.000								
15	0	88.000	64.000								
16	0	164.000	88.000								
17	0	169.000	89.000								
18	0	167.000	88.000								
19	0	149.000	84.000								
20	0	167.000	88.000								

SUMS OF VARIABLES
2.06100D+03 1.37200D+03

RESIDUAL SUMS OF SQUARES + CROSS PRODUCTS
1 4.90329D+04
2 1.57694D+04 5.09880D+03

MEANS OF VARIABLES
1.03050D+02 6.86000D+01

ROOT MEAN SQUARES OF VARIABLES
5.08004D+01 1.63816D+01

SIMPLE CORRELATION COEFFICIENTS, R(I,I PRIME)
1 1.000
2 0.997 1.000

Figure 3B.2

LINEAR LEAST-SQUARES CURVE FITTING PROGRAM

ACID NUMBER   DEP VAR 1: ORG AC   MIN Y = 4.100D 01   MAX Y = 8.900D 01   RANGE Y = 4.800D 01

RESULTANT EQUATION:
$$Y = 35.46 + 0.3216X$$
Y = ORGANIC ACID CONTENT
X = ACID NUMBER

IND.VAR(I)	NAME	COEF.B(I)	S.E. COEF.	T-VALUE	R(I)SQRD	MIN X(I)	MAX X(I)	RANGE X(I)	REL.INF.X(I)
0		3.54583D 01							
1	ACID N	3.21608D-01	5.55D-03	57.9	0.0	1.600D 01	1.690D 02	1.530D 02	1.03

NO. OF OBSERVATIONS	20
NO. OF IND. VARIABLES	1
RESIDUAL DEGREES OF FREEDOM	18
F-VALUE	3352.3
RESIDUAL ROOT MEAN SQUARE	1.22997881
RESIDUAL MEAN SQUARE	1.51284788
RESIDUAL SUM OF SQUARES	27.23126184
TOTAL SUM OF SQUARES	5098.80000000
MULT. CORREL. COEF. SQUARED	.9947

------ORDERED BY COMPUTER INPUT------

IDENT.	OBSV.	WS DISTANCE	OBS. Y	FITTED Y	RESIDUAL
AN	1	27.	76.000	75.016	0.984
AN	2	2.	70.000	70.514	-0.514
AN	3	115.	55.000	55.398	-0.398
AN	4	0.	71.000	68.906	2.094
AN	5	145.	55.000	53.790	1.210
AN	6	298.	48.000	47.358	0.642
AN	7	238.	50.000	49.609	0.391
AN	8	1.	66.000	67.619	-1.619
AN	9	518.	41.000	40.604	0.396
AN	10	385.	43.000	44.463	-1.463
AN	11	84.	82.000	79.840	2.160
AN	12	0.	68.000	69.227	-1.227
AN	13	214.	88.000	86.594	1.406
AN	14	54.	58.000	59.579	-1.579
AN	15	15.	64.000	63.760	0.240
AN	16	254.	88.000	88.202	-0.202
AN	17	297.	89.000	89.810	-0.810
AN	18	280.	88.000	89.167	-1.167
AN	19	144.	84.000	83.378	0.622
AN	20	280.	88.000	89.167	-1.167

------ORDERED BY RESIDUALS------

OBSV.	OBS. Y	FITTED Y	ORDERED RESID.	SEQ
11	82.000	79.840	2.160	1
4	71.000	68.906	2.094	2
13	88.000	86.594	1.406	3
5	55.000	53.790	1.210	4
1	76.000	75.016	0.984	5
6	48.000	47.358	0.642	6
19	84.000	83.378	0.622	7
9	41.000	40.604	0.396	8
7	50.000	49.609	0.391	9
15	64.000	63.760	0.240	10
16	88.000	88.202	-0.202	11
3	55.000	55.398	-0.398	12
2	70.000	70.514	-0.514	13
17	89.000	89.810	-0.810	14
18	88.000	89.167	-1.167	15
20	88.000	89.167	-1.167	16
12	68.000	69.227	-1.227	17
10	43.000	44.463	-1.463	18
14	58.000	59.579	-1.579	19
8	66.000	67.619	-1.619	20

Figure 3B.3

47

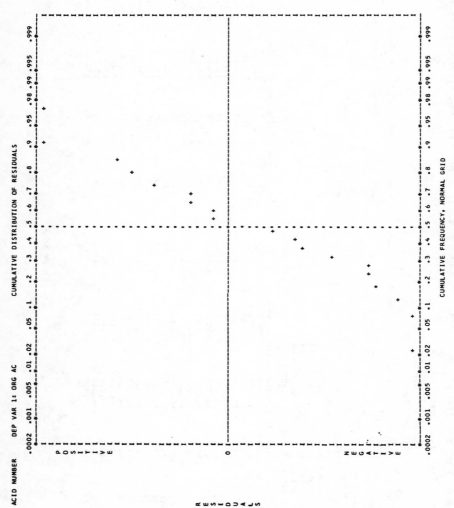

Figure 3B.4

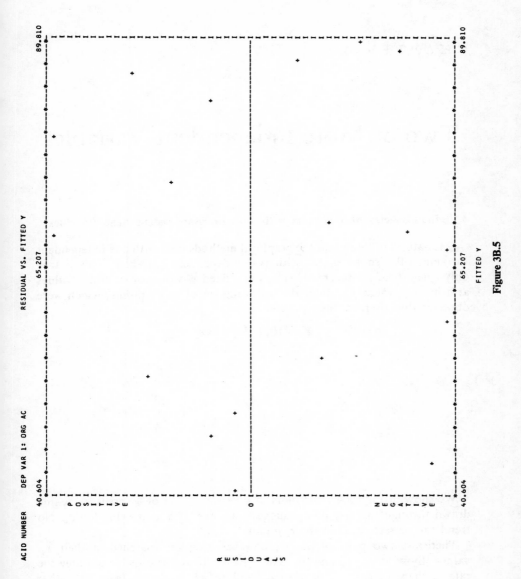

Figure 3B.5

CHAPTER 4

Two or More Independent Variables

4.1 Inadequacies of x-y Plots with Two or More Independent Variables

It is natural to think that the graphical methods used with one independent variable will also be useful with two independent variables. This is true only when there are several x_1-values at a fixed x_2-value, or several x_2-values at a fixed x_1-value. Consider the following set of seven points, which were obtained directly from the equation

$$Y = 10 + 7x_1 - 4x_2.$$

x_1	x_2	y
1	3	5
2	1	20
3	3	19
4	5	18
5	7	17
6	4	36
7	7	31

We show in Figure 4.1 the plots of y versus x_1 and of y versus x_2. A line drawn through the x_1-y points suggests a slope of 5 as a value for b_1. No trend can be seen at all in the x_2-y plot.

There are two pairs of points, however, that are matched in their x_2-values (those at $x_2 = 3$ and those at $x_2 = 7$). Thus it is possible to judge the rate of change due to x_1, *at fixed* x_2, and to get a check, from the other pair, on this rate. In this case, the same value, 7, is obtained for b_1. With this value it is now possible to estimate b_2. This set of data is exceptional, however, in having matched points.

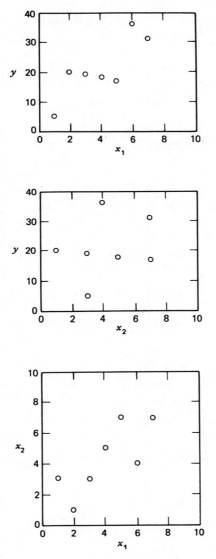

Figure 4.1 Data derived from $Y = 10 + 7x_1 - 4x_2$.

51

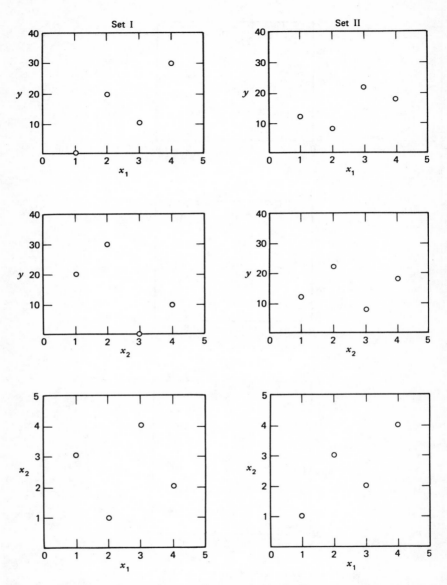

Figure 4.2 Data derived from $Y = b_0 + b_1x_1 + b_2x_2$ (Different x_1, x_2 allocations).

Let us take an even smaller set of points from an equation with different constants, where this matching is not found.

x_1	x_2	y
1	3	0
2	1	20
3	4	10
4	2	30

Here, plotting y versus x_1 (Figure 4.2, Set I), we might imagine that we see a slope, b_1, for x_1 of $+10$. Similarly b_2 might be thought to be -10. But taking another set of data points derived from the same equation, we have:

x_1	x_2	y
1	1	12
2	3	8
3	2	22
4	4	18

Plotting as before (Figure 4.2, Set II), we now find $b_1 = +5$ and $b_2 = 0$. The two sets of data appear to give different values for b_1 and b_2 because of the different allocation of the data points in "x_1-x_2" space. This is shown at the bottom of Figure 4.2, where x_2 is plotted against x_1. Clearly, plotting single independent variables versus y is worthless, or at least hazardous, when only *two* variables operate at the same time, *even though they are operating additively* with no error present. A third variable makes the situation even worse.

It is possible, and indeed advisable, to plot *contours* of constant y on an x_1-x_2 grid when only two independent variables are operating. But this expedient is not extensible to more than two independent variables and so we must draw the following conclusion: plots of x_i versus y when k is greater than 1 are rarely useful. When one independent variable dominates the situation, then of course the corresponding x_i-y plot will show this correctly. But when several x_i are influential, or when several x_i are intercorrelated, confusion is the principal product of such plots.

We must go to more generalizable analytical methods for such cases.

4.2 Equation Forms and Transformations

The general equation, with K independent variables indexed by the subscript i and N observations indexed by the subscript j, has the form:

$$Y_j = b_0 + b_1 x_{1j} + b_2 x_{2j} + \cdots + b_i x_{ij} + \cdots + b_K x_{Kj} = b_0 + \sum_{i=1}^{K} b_i x_{ij}.$$

The constants $b_0, b_1, b_2, \ldots, b_i, \ldots, b_K$ are to be estimated from the data.

A wide variety of shapes can be represented by such an equation. For example,

$$(4.1) \qquad Y = b_0 + b_1\left(\frac{x_1{}^2}{x_2}\right) + b_2 x_1 x_3 + b_3 x_1 \log x_4$$

is linear in the b's and that, as we said in Chapter 1, is all that is required for linear least-squares estimation. The computer program will operate on this equation just as well as on

$$(4.2) \qquad Y = b_0 + b_1 z_1 + b_2 z_2 + b_3 z_3,$$

since equation 4.1 is easily converted to 4.2 by setting $(x_1{}^2/x_2) = z_1$, etc.

We see, then, that many complex equations (or surfaces) are linearizable and should be linearized, when enough is known about the way in which the x_i affect the response.

It may be useful to put equation 4.1 in the form:

$$(4.3) \quad Y = b_0 + b_1 f_1(x_1, x_2, \ldots, x_K) + b_2 f_2(x_1, x_2, \ldots, x_K) + \cdots,$$

where the only restriction on the f's is that they be *known, constant-free* functions of the x_i. Thus $e^{x_1 \sin x_2}$ is acceptable, but $(x_1 - c)/(x_1 - d)$ is not, unless d is known.

If the values of y cover quite a large range, that is, if y has a natural zero and if $y_{\max}/y_{\min}$ is greater than, say, 10, it may be doubted that the residual mean square, $s^2(y)$, is constant for all values of y in the relevant range. If the error variance is not constant, we can weight the y-values. They should be weighted *inversely* as their variances (Brownlee, 1960 ed. only, Section 11.12). It can be proved (but this will not be done here) that the variance of all estimates of constants will be simultaneously minimized by using the correct weights. If guessed weights are used, we may not get minimum variance. If incorrect weights that depend, not on the observed y-values, but only on the x-values are used, we get "unbiased estimates" of the slopes, but not with minimum variance. The LINWOOD program is designed to use the weights *directly* or *inversely* as the variance of each observation. Thus, if the correct estimate of the variance is used, the estimated Residual Mean Square of the fit will have an expected value of 1.

When Var. $(x_i) \neq 0$ it can be shown that the estimate of b_i is biased downward from the true slope (Hald, pages 615–616). Since we cannot usually be certain that the variance of x_i is zero, some judgment should be made about the cases in which this variance can be considered "nearly zero" or at least "safely negligible."

Even when some x_i have considerable variances, it is possible to get an unbiased and efficient estimate of each b_i. This case is not discussed here; see Acton, Madansky, or Keeping.

4.3   Variances of Estimated Coefficients, b_i, and Fitted Values, Y_j

With two independent variables the variance of the coefficient b_1 may be calculated from:

$$\text{Est. Var. } (b_1) = \frac{s^2(y)}{[11](1 - r_{12}{}^2)},$$

where $r_{12}{}^2{}^*$ is the degree of nonorthogonality between x_1 and x_2 *in these data.* It will not have the same value in other sets, but may be only a property (an unfortunate one) of the data before us. If $r_{12}{}^2$ is large (i.e., near 1), this is very disadvantageous, since then the variances of the two b_i concerned are larger than they need be, considering only the unavoidable size of $s^2(y)$ and the spread of the x_i, [11], and [22] in these data.

An exactly analogous equation holds for the variance of each b_i in the multifactor case, that is, where there are more than two factors. We need only replace $r_{12}{}^2$ by $R_i{}^2$, the degree of nonorthogonality of the ith independent variable. Thus the amount of linear dependence of x_i on the other variables determines how much the variance of b_i is inflated. Obviously a similar statement holds for each factor, but a different $R_i{}^2$ enters for each. The STANDARD ERROR OF THE COEFFICIENT is obtained by taking the square root of Var. (b_i).

The estimated standard error of a particular Y_j, $s(Y_j)$, depends of course on the variances and covariances of all the b_i, as well as on the actual placement of the point Y_j in x-space.

The equation for computing $s(Y_j)$ is complex but computable. The values of $s^2(Y_j)$ divided by RMS are given in the Functions Related to the Variance of Fitted Y table of the computer program and will be discussed in Chapter 7. The simple bounds given below often suffice.

Minimum $s(Y)$ is always at point $x_i = \bar{x}_i$,

$$\frac{s(y)}{\sqrt{N}} < s(Y) < s(y),$$

$$\text{Average } s(Y) = \sqrt{\frac{K + 1}{N}}\, s(y).$$

---

* The term $r_{12}{}^2 = [12]^2/[11][22]$ has the form of a squared correlation coefficient. It measures the fraction of the variation, $[ii]$, in one independent variable that is accounted for by the other in an equation of the form $X_i = A + Bx_{i'}$, $i \neq i'$. It takes the value 1 for exact linear dependence and 0 when the equation gives no improvement over setting $X_i = \bar{x}_i$. When $r_{12}{}^2 = 0$, we have complete (linear) independence of x_1 and x_2 or "orthogonality." When $r_{12}{}^2 \neq 0$, we have some degree of dependence and hence a degree of nonorthogonality. The Variance-Inflation Factor $= 1/(1 - r_{12}{}^2)$.

4.4   Use of Indicator Variables for Discontinuous or Qualitative Classifications

One qualitative factor at two levels

Suppose that our data are divisible into two sets, not necessarily equal in number, and that the distinction between the two sets is not amenable to quantitative measure (e.g., day versus night, two suppliers of the same catalyst). We can use an indicator or "dummy" variable, x_1, to make the distinction, *defining* x_1 as 0 for category I, and 1 for category II. No other values of x_1 are to be allowed. Our equation is

$$Y = b_0 + b_1 x_1,$$

which looks rather familiar. For all the data in category I, Y will be b_0; for all in category II, Y will be $b_0 + b_1$. We see that b_1 is simply the difference between the two average values of y.

One qualitative two-level factor and one quantitative (continuous) factor

The equation now reads:

$$Y = b_0 + b_1 x_1 + b_2 x_2.$$

Here x_1 is defined exactly as above, but x_2 may take any value over some range. The data are now represented by two parallel lines with slope b_2, separated by the amount b_1. More than one discrete two-level factor and several continuous factors can be accommodated in this way.

Discrete factors at more than two levels

Suppose that the data produced in *three* shifts must be distinguished in our equation. Here we need *two* fitting constants, since there are 2 degrees of freedom for the three-level factor. Thus:

$$Y = b_0 + b_1 x_1 + b_2 x_2,$$

where now $x_1 = 1$ for shift II,
$\qquad\quad\; = 0$ otherwise;
$\qquad x_2 = 1$ for shift III,
$\qquad\quad\; = 0$ otherwise.

In a simple table of translation we have then

	I	II	III
x_1	0	1	0
x_2	0	0	1

and by direct substitution

$$Y_I = b_0,$$

$$Y_{II} = b_0 + b_1,$$

$$Y_{III} = b_0 + b_2.$$

Other allocations of the indicator variables to the levels of the discontinuous variable are permissible, but the one shown above gives simple meanings to the three constants.

Discrete factors interacting with continuous factors

If the linear effect (i.e., the slope) of some continuous factor, x_2, is not the same at the two levels of a discontinuous factor, x_1, we can represent this as a product term, $x_1 x_2$:

$$Y = b_0 + b_1 x_1 + b_2 x_2 + b_{12} x_1 x_2.$$

Again $x_1 = 0$ or 1, and x_2 is continuous.

Details of specifying the allocation of indicator variables are discussed in Chapter 8 and illustrated in Figures 8A.1 and 8A.11.

4.5 Allocation of Data in Factor Space

The deliberate allocation of the levels of factors so as to minimize bias and random error is the subject of the branch of statistics called "design of experiments." Statistically designed multifactor experiments are becoming fairly common in some branches of science and engineering research, but the data analyst rarely sees the results of a single balanced experiment. In the first place the analysis and interpretation of such experiments is usually easy and is familiar to the experimenter. A single set of balanced data rarely brings the research work to a satisfactory conclusion, and therefore more experimental work is usually done to clear up ambiguities, to extend the range of some factors, and to include new factors. After all this is done, the entire set is never balanced, and the more general methods of least squares, not analysis of variance, become necessary.

Nearly all of the devices proposed in this work are responses to the fact that most multifactor data are not taken—and often cannot be taken—in balanced arrays. The disposition of the data points in "factor space" then becomes a matter for study. Failure to examine the actual disposition may result in severe misinterpretation.

Statisticians regularly recommend that data would be better if taken in balanced arrays, in accordance with their requirements as to randomization. Doing this is sometimes acceptable but often not practicable. We take the position here that the data have been taken and must be analyzed as they stand. Recommendations to take more data will be more acceptable if preceded by a conscientious study of all that has been done so far. In this way the consumer will see what his data show and what he should do if they do not show enough.

It seems to have escaped the notice of some statisticians that a very large part—perhaps nine tenths—of scientific and engineering research work is carried through to some kind of successful result without any aid from experts in experiment design. There are several reasons for this. Many scientists take their data varying one factor at a time, and so have little trouble analyzing it. Contrariwise, sometimes when an experiment has not come out well, the able researcher will vary three or even eight conditions simultaneously in directions that make sense to him. If the new experiment works, he will hurry on, not going back to find out which of the conditions that he varied was decisive. Moreover some technical work is closely linked to theory and to a wide background of experience. Crucial experiments then permit valid decisions about the part of the theory that is under test.

For most researchers, however, the time comes when they must review and summarize what they have found, perhaps over a long period of time. At this point a mass of data must be studied; it is almost never balanced, is rarely homogeneous, and seldom contains real replicates. It is about these cases and some of their common defects that we write.

Linear dependences among the x_i

One of the x_i, say x_1, may be linearly related to one or more of the other factors. There will then be a linear relationship among some of the x_i, perhaps disturbed by random fluctuations, of the form:

$$x_1 = c_1 + c_2 x_2 + c_3 x_3 + \cdots.$$

In studying two variables we saw that the slopes b_1 and b_2 cannot be estimated *at all* from B_1 and B_2, where these are the constants in the two equations $Y_I = B_I + B_1 x_1$ and $Y_{II} = B_{II} + B_2 x_2$. Even if we look at them two at a time, we can hope for little or no illumination concerning the true linear dependencies among the x_i. We must, in fact, go through the fitting of a full linear equation, using each x_i in turn as the *dependent* variable, in order to detect such relationships.

It is often possible to cut down the full equation to find that only a few of the other factors are linearly related to a particular x_i. It may even happen that only *one* factor is left that is strongly linearly correlated with our present one. But this will not always be the case.

As we have indicated previously, the *variance* of b_i in the equation relating Y to x_i is *inflated* by the factor $1/(1 - R_i^2)$. Thus, if any R_i^2 is as large as 0.9, the variance of the corresponding b_i is inflated by a factor of at least 10.

The outermost points in data space

A process that is well described by a simple equation when running within some standard range of practice may follow a different equation, or be much less stable, when operating outside this range. Therefore we require some means of detecting extreme conditions in multifactor situations. We propose a (squared) measure of "distance from standard" that appears to take satisfactory account of the differing influences of the several factors. It is automatically computed for each data point and is printed along with the fitted values and the residuals. See WSSD in Figures 6A.2, 6B.2, 7A.2, and 7A.9–7A.11 for examples.

When there are no heavy dependencies among the x_i, it is not difficult to locate the points in factor space that are farthest away from the middle of the data. Once again, it is not safe to look at the factors singly, or even in pairs, and to stop there.

Nested data

Plant and pilot-plant data are often taken in naturally grouped sets. Hard-to-change conditions are held constant or nearly constant for considerable periods of time, while those easy to change are varied. If this is done *and recorded*, the analysis of the data, although not completed in a single computer run, is fairly straightforward.

We can expect the quickly changed factors to produce changes in results with small variance. The levels of the hard-to-change factors are run through more slowly and might therefore be expected to reflect weather and other changes with larger random components and therefore greater variance. It will be necessary to run *two* fittings: one on the nested data "within plots" and one "among plots." (The term plots comes from the first appearance of such groupings in agricultural data.)

After the within-plot analysis has been made, we must correct or adjust all plot averages to the same values of the within-plot factors. These revised averages are then treated as new data for the among-plot analysis. All the warnings, pitfalls, and devices indicated above may be expected to apply twice over to "split plot" or "nested" data.

If equations are fitted to such data without taking into account that they are nested, the wrong factors may appear to be significant. The residual mean square will be an unevenly pooled average of the two basic variances; this gives us *too many* whole-plot factors showing "significant" effects, and *too few* (or not any) subplot factors showing "significant" effects.

If the data are recorded helter-skelter, with no record of the time order in which they were taken or of what "nesting" has taken place, then the facts about nesting become more difficult to disentangle. In Chapter 5 we treat a case of nesting by direct inspection, since the situation is quite simple. In Chapter 8 a more complex case is studied, and more general rules are given.

CHAPTER 5

Fitting an Equation in Three Independent Variables

5.1 Introduction

In looking for a multifactor example using real data, we were struck by the fact that no clean-cut, undesigned data were on hand. By "clean-cut" we mean "meeting all the necessary statistical requirements without any complicating circumstances." When we tried to analyze textbook examples, we found unforeseen complications in each. One of the more interesting problems is presented here, not only as an example of fitting an equation with three independent variables, but also to point out some of the precautions that are necessary.

The example is from Brownlee (Section 13.12). It is also presented in Draper and Smith (Chapter 6). The data, given in Table 5.1, represent 21 successive days of operation of a plant oxidizing ammonia to nitric acid. Factor x_1 is the flow of air to the plant. Factor x_2 is the temperature of the cooling water entering the countercurrent nitric oxide absorption tower. Factor x_3 is the concentration of nitric acid in the absorbing liquid (coded for ease of hand calculation by subtracting 50 and then multiplying by 10). The response, y, is 10 times the percentage of the ingoing ammonia that is lost as unabsorbed nitric oxides; it is an indirect measure of the yield of nitric acid.

5.2 First Trials

Figures 5.1–5.3 show summary computer printouts after fitting the equation

$$Y = b_0 + b_1x_1 + b_2x_2 + b_3x_3.$$

Although the standard statistics—the t-values, the F-value, and the multiple correlation coefficient squared—all seem acceptably significant, the

TABLE 5.1

DATA FROM OPERATION OF A PLANT FOR THE OXIDATION OF AMMONIA
TO NITRIC ACID

Observation Number	Air Flow, x_1	Cooling Water Inlet Temperature, x_2	Acid Concentration, x_3	Stack Loss, y
1	80	27	89	42
2	80	27	88	37
3	75	25	90	37
4	62	24	87	28
5	62	22	87	18
6	62	23	87	18
7	62	24	93	19
8	62	24	93	20
9	58	23	87	15
10	58	18	80	14
11	58	18	89	14
12	58	17	88	13
13	58	18	82	11
14	58	19	93	12
15	50	18	89	8
16	50	18	86	7
17	50	19	72	8
18	50	19	79	8
19	50	20	80	9
20	56	20	82	15
21	70	20	91	15

normal plot (Figure 5.2) indicates one very low point (*21*), and the residuals versus fitted *Y* plot (Figure 5.3) is irregular in an unfamiliar way. There appear to be five points (*1, 2, 3, 4,* and *21*) whose residuals are widely separated from the others, *and* the remaining sixteen residuals seem to show some sort of subsidiary trends in sets.

5.3 Possible Causes of Disturbance

In reflecting on possible causes of—or at least names for—these oddities, we list four conceivable sorts of disturbance:

1. A small minority of points, perhaps only one or two, may be entirely erroneous.

LINEAR LEAST-SQUARES CURVE FITTING PROGRAM

STACK LOSS. PASS 1   DEP VAR 1: S.LOSS   MIN Y = 7.000D+00   MAX Y = 4.20CD+01   RANGE Y = 3.500D+01

Y = B(0) + B(1)X1 + B(2)X2 + B(3)X3
Y = STACK LOSS
X1 = AIR FLOW
X2 = COOLING WATER TEMPERATURE
X3 = NITRIC ACID CONCENTRATION
ALL DATA USED

IND.VAR(I)	NAME	(CEF.B(I)	S.E. COEF.	T-VALUE	R(I)SQRD	MIN X(I)	MAX X(I)	RANGE X(I)	REL.INF.X(I)
0		-3.951970+01							
1	A.FLOW	7.915400-01	1.250-01	5.3	0.6559	5.000D+01	8.000D+01	3.000D+01	0.61
2	W.TEMP	1.295290+00	3.680-01	3.5	0.6113	1.700D+01	2.700D+01	1.000D+01	0.37
3	A.CONC	-1.521230-01	1.560-01	1.0	0.2501	7.200D+01	9.300D+01	2.100D+01	0.09

NO. OF OBSERVATIONS   21
NO. OF IND. VARIABLES   3
RESIDUAL DEGREES OF FREEDOM   17
F-VALUE   59.9
RESIDUAL ROOT MEAN SQUARE   3.24336392
RESIDUAL MEAN SQUARE   10.51940951
RESIDUAL SUM OF SQUARES   178.82996160
TOTAL SUM OF SQUARES   2065.23809524
MULT. CORREL. COEF. SQUARED   .9136

---ORDERED BY COMPUTER INPUT---

IDENT.	OBSV.	WSS DISTANCE	OBS. Y	FITTED Y	RESIDUAL
S LOSS	1	24.	42.000	38.765	3.235
S LOSS	2	24.	37.000	38.917	-1.917
S LOSS	3	13.	37.000	32.444	4.556
S LOSS	4	0.	28.000	22.302	5.698
S LOSS	5	0.	18.000	19.712	-1.712
S LOSS	6	1.	18.000	21.007	-3.007
S LOSS	7	2.	19.000	21.389	-2.389
S LOSS	8	2.	20.000	21.389	-1.389
S LOSS	9	1.	15.000	18.144	-3.144
S LOSS	10	2.	14.000	12.733	1.267
S LOSS	11	3.	14.000	11.364	2.636
S LOSS	12	3.	13.000	10.221	2.779
S LOSS	13	2.	11.000	12.429	-1.429
S LOSS	14	1.	12.000	12.050	-0.050
S LOSS	15	7.	8.000	5.639	2.361
S LOSS	16	6.	7.000	6.095	0.905
S LOSS	17	6.	8.000	9.520	-1.520
S LOSS	18	6.	8.000	8.455	-0.455
S LOSS	19	6.	8.000	9.598	-0.598
S LOSS	20	1.	15.000	13.598	1.412
S LOSS	21	5.	15.000	22.238	-7.238

---ORDERED BY RESIDUALS---

OBSV.	OBS. Y	FITTED Y	ORDERED RESID.	STUD.RESID.	SEQ
4	28.000	22.302	5.698	1.9	1
3	37.000	32.444	4.556	1.5	2
1	42.000	38.765	3.235	1.2	3
12	13.000	10.221	2.779	1.0	4
11	14.000	11.364	2.636	0.9	5
15	8.000	5.639	2.361	0.8	6
20	15.000	13.588	1.412	0.5	7
10	14.000	12.733	1.267	0.4	8
16	7.000	6.095	0.905	0.3	9
14	12.000	12.050	-0.050	-0.0	10
18	8.000	8.455	-0.455	-0.2	11
19	8.000	9.598	-0.598	-0.2	12
8	20.000	21.389	-1.389	-0.5	13
13	11.000	12.429	-1.429	-0.5	14
17	8.000	9.520	-1.520	-0.6	15
5	18.000	19.712	-1.712	-0.5	16
2	37.000	38.917	-1.917	-0.7	17
7	19.000	21.389	-2.389	-0.8	18
6	18.000	21.007	-3.007	-1.0	19
9	15.000	18.144	-3.144	-1.0	20
21	15.000	22.238	-7.238	-2.6	21

Figure 5.1

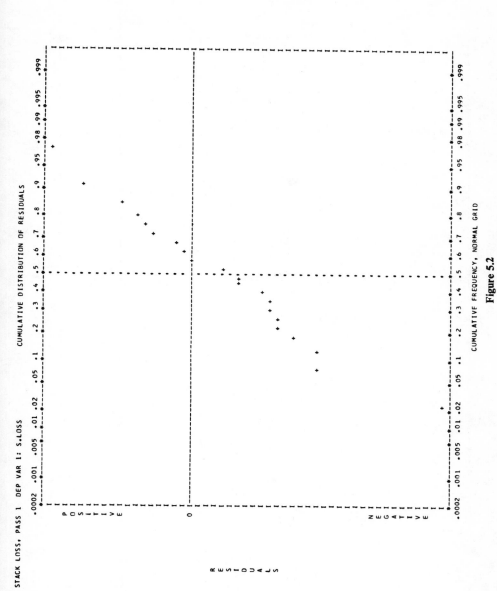

Figure 5.2

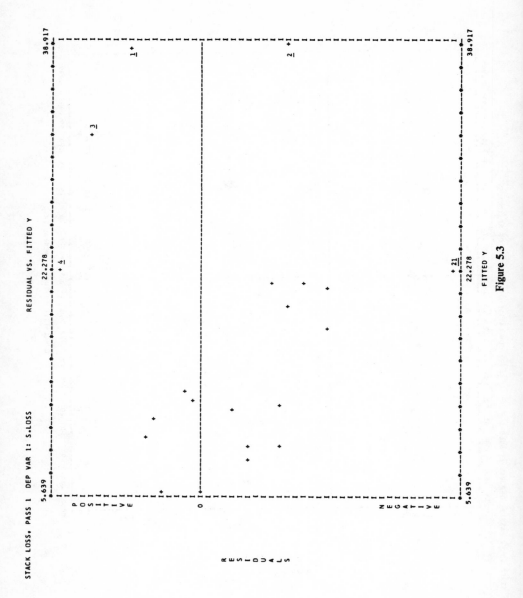

Figure 5.3

64

2. We may not have the right functional form. The large residuals may be reflecting "function bias."

3. The random error may be quite irregular, not at all "normal."

4. Some points may reflect changing conditions, soon after large changes in some influential factor. The system may not yet have come to equilibrium under the new conditions.

We focus our attention on possibilities 1 and 2. We ask first whether perhaps observation *21* alone was disturbing the equation, and, second, whether there might not be a trend in error increasing with the true magnitude of the response. The latter possibility could be accommodated by using $\log y$ instead of y as the dependent variable.

In Pass 2 we drop observation *21*. Figures 5.4–5.6 show the results. We see from the t-values, the F-value, the multiple correlation coefficient squared (R_y^2), and the residual mean square (RMS) that we have a noticeably better fit. But the normal plot of residuals is now skew, and the residuals versus fitted Y plot shows the same undesirable trends in sixteen points that it did in Pass 1. Point *21* was the last piece of data to be taken and reflects a jump in x_1 (from 56 to 70) that might signal lack of equilibrium, but removal of *21* has not sufficed to make everything satisfactory.

5.4   Outlier or Logged Response or Squared Independent Variable

Figures 5.7–5.9 show the computer summary and plots from Pass 3, using $\log y$ instead of y. We have no better fit than with y, but the normal plot is greatly improved and the queer trends in the residuals have largely disappeared.

Since x_3 has shown negligible influence in the three passes just made, we drop it and run through the same alternatives (the full set of data, dropping point *21*, using $\log y$ for response). The first six lines of Table 5.2 summarize our work so far.

We make two more passes (7 and 8 in Table 5.2) so that all eight alternatives are before us. These may be viewed as a 2^3 factorial experiment:*

Factor A has two levels: observation *21* in or out,

Factor B has two levels: y or $\log y$ used as response,

Factor C has two levels: x_3 in or out.

Thus each equation studied can be specified by a simple symbol, for example, *abc* means "*21* out, $\log y$ used, x_3 out." We resist the temptation to carry

---

* The language and nomenclature of 2^K factorial designs will be used in more detail later. Here we only require representation of an equation by indicating the presence or absence of a small letter, one for each change in the equation. The reader can easily see this by study of columns 2–5 in Table 5.2.

LINEAR LEAST-SQUARES CURVE FITTING PROGRAM

STACK LOSS, PASS 2   DEF VAR 1: S.LOSS   MIN Y = 7.000D+00   MAX Y = 4.200D+01   RANGE Y = 3.500D+01

```
Y = B(0) + B(1)X1 + B(2)X2 + B(3)X3
Y  = STACK LOSS
X1 = AIR FLOW
X2 = COOLING WATER TEMPERATURE
X3 = NITRIC ACID CONCENTRATION
OBSERVATION 21 OMITTED
```

INC.VAR(I)	NAME	COEF.B(I)	S.E.COEF.	T-VALUE	R(I)SQRD	MIN X(I)	MAX X(I)	RANGE X(I)	REL.INF.X(I)
0		-4.37040D+01							
1	A.FLOW	8.869108D-01	1.19D-01	7.5	0.7052	5.000D+01	8.000D+01	3.000D+01	0.76
2	W.TEMP	8.16620D-01	3.25D-01	2.5	0.6853	1.700D+01	2.700D+01	1.000D+01	0.23
3	A.CONC	-1.07141D-01	1.25D-01	0.9	0.2276	7.200D+01	9.300D+01	2.100D+01	0.06

```
NC. OF OBSERVATIONS            20
NC. OF IND. VARIABLES           3
RESIDUAL DEGREES OF FREEDOM    16
F-VALUE                        98.8
RESIDUAL ROOT MEAN SQUARE        2.56920122
RESIDUAL MEAN SQUARE             6.60079490
RESIDUAL SUM OF SQUARES        105.61271844
TOTAL SUM OF SQUARES          2062.55000000
MULT. CORREL. COEFF. SQUARED     .9468
```

-------- ORDERED BY COMPUTER INPUT --------

IDENT.	OBSV.	WSS DISTANCE	OBS. Y	FITTED Y	RESIDUAL
S LOSS	1	52.	42.000	39.538	2.062
S LOSS	2	52.	37.000	40.045	-3.045
S LOSS	3	29.	37.000	33.752	3.248
S LOSS	4	1.	28.000	21.698	6.302
S LOSS	5	1.	18.000	20.065	-2.065
S LOSS	6	1.	18.000	20.882	-2.882
S LOSS	7	1.	19.000	21.055	-2.055
S LOSS	8	1.	20.000	21.055	-1.055
S LOSS	9	1.	15.000	17.325	-2.325
S LOSS	10	2.	14.000	13.992	0.008
S LOSS	11	2.	14.000	13.028	0.972
S LOSS	12	2.	13.000	12.318	0.682
S LOSS	13	1.	11.000	13.778	-2.778
S LOSS	14	1.	12.000	13.416	-1.416
S LOSS	15	13.	8.000	5.915	2.085
S LOSS	16	13.	7.000	6.236	0.764
S LOSS	17	13.	8.000	8.553	-0.553
S LOSS	18	12.	8.000	7.803	0.197
S LOSS	19	12.	9.000	8.512	0.488
S LOSS	20	2.	15.000	13.633	1.367

-------- ORDERED BY RESIDUALS --------

OBSV.	OBS. Y	FITTED Y	ORDERED RESID.	STUD.RESID.	SEQ
4	28.000	21.698	6.302	2.6	1
3	37.000	33.752	3.248	1.4	2
15	8.000	5.915	2.085	0.9	3
1	42.000	39.938	2.062	1.0	4
20	15.000	13.633	1.367	0.6	5
11	14.000	13.028	0.972	0.4	6
16	7.000	6.236	0.764	0.3	7
12	13.000	12.318	0.682	0.3	8
19	9.000	8.512	0.488	0.2	9
18	8.000	7.803	0.197	0.1	10
10	14.000	13.992	0.008	0.0	11
17	8.000	8.553	-0.553	-0.3	12
8	20.000	21.055	-1.055	-0.5	13
14	12.000	13.416	-1.416	-0.6	14
7	19.000	21.055	-2.055	-0.9	15
5	18.000	20.065	-2.065	-0.8	16
9	15.000	17.325	-2.325	-1.0	17
13	11.000	13.778	-2.778	-1.2	18
6	18.000	20.882	-2.882	-1.2	19
2	37.000	40.045	-3.045	-1.5	20

Figure 5.4

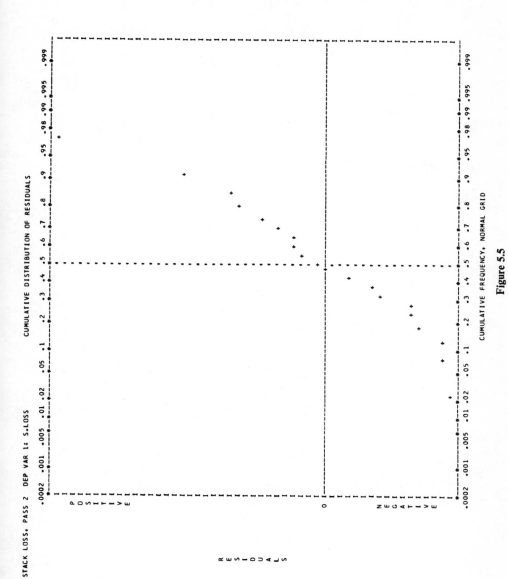

Figure 5.5

67

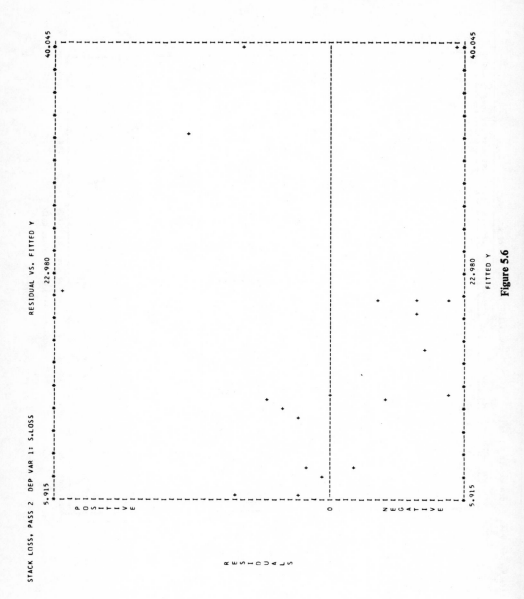

Figure 5.6

LINEAR LEAST-SQUARES CURVE FITTING PROGRAM

STACK LOSS, PASS 3   DEP VAR 1: LOG Y   MIN Y = 8.451D-01   MAX Y = 1.623D+00   RANGE Y = 7.782D-01

```
LOG Y = B(0) + B(1)X1 + B(2)X2 + B(3)X3
Y  = STACK LOSS
X1 = AIR FLOW
X2 = COOLING WATER TEMPERATURE
X3 = NITRIC ACID CONCENTRATION
ALL DATA USED
```

IND.VAR(I)	NAME	CCEF.B(I)	S.E. CCEF.	T-VALUE	R(I)SQRD	MIN X(I)	MAX X(I)	RANGE X(I)	REL.INF.X(I)
0		-4.12028D-01							
1	A.FLOW	1.50113D-02	3.19D-03	4.7	0.6559	5.000D+01	8.000D+01	3.000D+01	0.58
2	W.TEMP	2.75624D-02	8.70D-03	3.2	0.6113	1.700D+01	2.700D+01	1.000D+01	0.35
3	A.CONC	1.24238D-03	3.70D-03	0.3	0.2501	7.200D+01	9.300D+01	2.100D+01	0.03

```
NO. OF OBSERVATIONS          21
NO. OF IND. VARIABLES         3
RESIDUAL DEGREES OF FREEDOM  17
F-VALUE                      52.9
RESIDUAL ROOT MEAN SQUARE    0.0766956
RESIDUAL MEAN SQUARE         0.00588213
RESIDUAL SUM OF SQUARES      0.09999625
TOTAL SUM OF SQUARES         1.0239056
MULT. CORREL. COEF. SQUARED  .9033
```

			ORDERED BY COMPUTER INPUT			
IDENT.	OBSV.	WSS DISTANCE	OBS. Y	FITTED Y	RESIDUAL
S LOSS	1	19.	1.623	1.644	-0.021
S LOSS	2	19.	1.568	1.643	-0.074
S LOSS	3	10.	1.568	1.515	0.053
S LOSS	4	1.	1.447	1.298	0.155
S LOSS	5	0.	1.255	1.233	0.022
S LOSS	6	1.	1.261	1.261	-0.006
S LOSS	7	1.	1.279	1.296	-0.017
S LOSS	8	1.	1.301	1.296	0.005
S LOSS	9	1.	1.176	1.201	-0.025
S LOSS	10	1.	1.145	1.054	0.092
S LOSS	11	1.	1.146	1.065	0.081
S LOSS	12	2.	1.114	1.037	0.077
S LOSS	13	1.	1.041	1.057	-0.015
S LOSS	14	1.	1.079	1.098	-0.019
S LOSS	15	5.	0.903	0.945	-0.042
S LOSS	16	5.	0.845	0.942	-0.057
S LOSS	17	5.	0.903	0.952	-0.049
S LOSS	18	5.	0.903	0.960	-0.057
S LOSS	19	4.	0.954	0.989	-0.035
S LOSS	20	1.	1.176	1.082	0.054
S LOSS	21	4.	1.176	1.303	-0.127

		ORDERED BY RESIDUALS			
OBSV.	OBS. Y	FITTED Y	ORDERFD RESID.	STUD.RESID.	SEQ
4	1.447	1.288	0.159	2.2	1
20	1.176	1.082	0.094	1.3	2
10	1.146	1.054	0.092	1.3	3
11	1.146	1.065	0.081	1.1	4
12	1.114	1.037	0.077	1.1	5
3	1.568	1.515	0.053	0.8	6
5	1.255	1.233	0.022	0.3	7
8	1.301	1.296	0.005	0.1	8
6	1.261	1.261	-0.005	-0.1	9
13	1.041	1.057	-0.015	-0.2	10
7	1.279	1.296	-0.017	-0.3	11
14	1.079	1.098	-0.019	-0.3	12
1	1.623	1.644	-0.021	-0.3	13
9	1.176	1.201	-0.025	-0.3	14
19	0.954	0.989	-0.035	-0.5	15
15	0.903	0.945	-0.042	-0.6	16
17	0.903	0.952	-0.049	-0.8	17
18	0.903	0.960	-0.057	-0.8	18
2	1.568	1.643	-0.074	-1.2	19
16	0.845	0.942	-0.097	-1.4	20
21	1.176	1.303	-0.127	-2.0	21

Figure 5.7

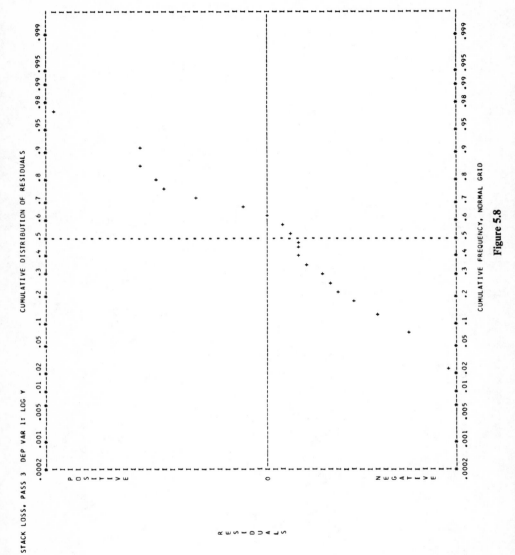

STACK LOSS, PASS 3   DEP VAR 1: LOG Y   CUMULATIVE DISTRIBUTION OF RESIDUALS

Figure 5.8

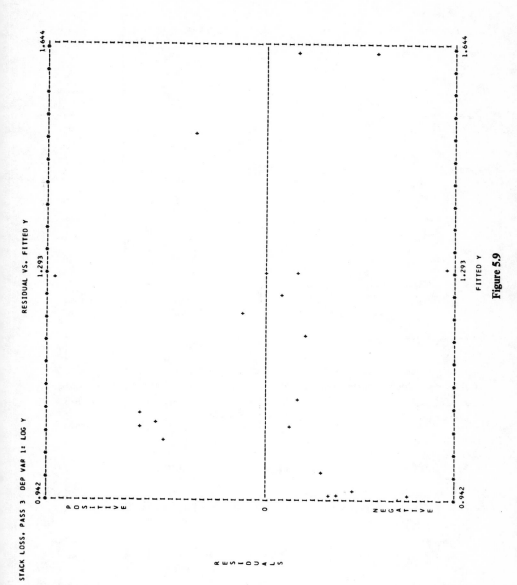

Figure 5.9

71

TABLE 5.2

EXAMINATION OF ALL ALTERNATIVES FOR SEVERAL INDEPENDENT REVISIONS

		Action			Result				
Pass	Factorial Symbol	Obs. 21 (A)	Response (B)	x_3 (C)	F-Value	$R_y{}^2$	RMS	Normal Plot	Residual vs. Fitted Y Plot
1	(1)	in	y	in	60	0.91	10.5	21 low	1, 2, 3, 4 outside points
2	a	out	y	in	99	0.95	6.6	Curved	1, 2, 3, 4 outside points
3	b	in	$\log y$	in	55	0.90	0.0059	O.K.	O.K.
4	c	in	y	out	90	0.91	10.5	21 low	1, 2, 3, 4 outside points
5	ac	out	y	out	150	0.95	6.5	Curved	1, 2, 3, 4 outside points
6	bc	in	$\log y$	out	83	0.90	0.0056	21 high	O.K.
7	ab	out	$\log y$	in	66	0.92	0.0048	4 high	
8	abc	out	$\log y$	out	102	0.92	0.0046	4 high	
			Observations 1, 3, 4, and 21 Omitted						
9	(1)	out	y	in	169	0.975	1.6	20 low	Curvature?
10	a	out	y	out	250	0.973	1.6	O.K.	Curvature?
11	b	out	$\log y$	in	53	0.92	0.0032	20 high	Curvature?
12	ab	out	$\log y$	out	83	0.92	0.0031	20 high 16, 2 low	Curvature?

through the estimation of average factorial effects on F, $R_y{}^2$, etc. It is clear that ac, which means "use y as response and drop _21_ and x_3," gives highest F and $R_y{}^2$. Its nearest competitor is abc, which directs us to use $\log y$ as response, while dropping observation _21_ and x_3.

5.5   Random Error Estimated from Near Neighbors

The results of Passes 1–8 supply clear directives as to factor x_3 and point _21_. Omitting x_3 gives no worse fit; dropping _21_ produces a notable improvement. But the evidence has not sufficed to indicate clearly a choice between y and $\log y$.

If $\log y$ is the right response, we expect the error to increase with y; in fact, the "per cent error in y" should be constant. If y is the correct response for our form of equation in the x_i, then the average error should be the same at all y. Now that x_3 has been dropped, we see that there are three sets of near replicates in the data as shown at the top of page 73.

If we dare assume that these three sets of consecutive duplicates are measuring random error among days, then two facts strike the eye. In the first

Observations	Residual SS	Degrees of Freedom	Mean Squares	Std. Dev.	Fitted Y-Value	Per Cent Error, s/Y
5, 6, 7, 8	2.8	3	0.93	1.0	21	5
10, 11, 12, 13, 14	6.8	4	1.70	1.3	12	11
15, 16, 17, 18, 19	2.0	4	0.50	0.7	8	9
Pooled	11.6	11	1.05	1.02		

place there is no evidence of trend in magnitude of error with Y. Second, the random error is much smaller than the value $s = 2.6$ of Pass 5 had led us to suppose. The trend in the estimated standard deviation, s (and hence in s^2), is not firmly ruled out, since we have only 11 degrees of freedom in all. But the new mean square of 1.05 gives definite information about lack of fit of our equation to the 20 data points.

$$\text{Mean square for lack of fit} = \frac{\text{RSS (Pass 5)} - \text{RSS (new replicates)}}{\text{df (Pass 5)} - \text{df (new replicates)}}$$

$$= (110.5 - 11.6)/(17 - 11) = 16.48,$$

$$F_{11}^{6} \text{ (lack of fit)} = \frac{\text{(Mean square for lack of fit)}}{\text{(Mean square error)}}$$

$$= 16.48/1.05$$

$$= 15.7,$$

$$F_{11}^{6} (0.0005) = 10.6.$$

We do not know yet whether this lack of fit is due to our having chosen a wrong equation form or to the presence of some "bad" data points.

5.6 Independence of Observations

Examining the second possibility first, we take a longer look at the data in the order of appearance of the points. It is distressing to see y drop from 42 to 37 with no change in x_1 or x_2 (points *1* and *2*). It is also upsetting to see y hold at 37 (point *3*) while x_1 drops from 80 to 75. Some light begins to appear on scanning the next five points, since x_1 is steady and y drops (in two days) to 18 and then holds fairly steady for four days. *There appears to be a lag of about one day in the response to a change in x_1.* Figure 5.10 shows all 21 points plotted in time order versus y, with x_1 and x_2 labeled for each point.

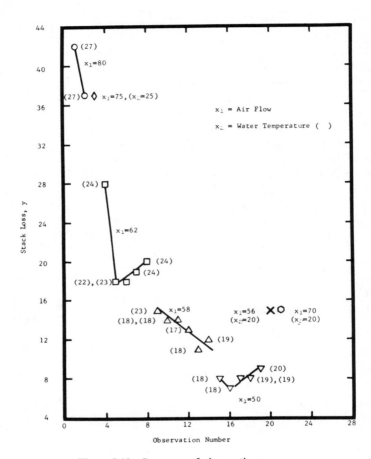

Figure 5.10   Sequence of observations.

We conclude that when x_1 was above 60 the plant required about a day to come to equilibrium ("line-out" is the term used by plant operators). We decide, therefore, to drop points *1, 3, 4,* and *21* permanently, since they correspond to transient states that cannot possibly be represented by the same equation as the "lined-out" points. We now return to the fray to see whether we can find an acceptable equation with satisfactory residual plots.

5.7 Systematic Examination of Alternatives Discovered Sequentially

Although we usually get our notions one at a time, it does not follow that we will find the best equation(s) by accepting or rejecting each notion after examining it *once*. It will often be possible to broaden our experience by looking at each new idea in the full context of all its predecessors.

Now that we have decided to drop four points because they are not representative, we must reconsider our equation for the remaining seventeen points. In strict analogy with factorial experimentation, we examine all the contingencies that we can *in a block* instead of one at a time. Thus, if we want to see the effects of logging y *and* of dropping x_3, we must make all four equation fits.

The lower panel of Table 5.2 gives the key statistics for each of the four equations under study. Now we see, as we did not before, that there is some hint of curvature in the residual versus fitted Y plots.

The term $b_{11}x_1^2$ is the natural one to choose, since x_1 is by far the more influential factor. But the curvature might be better represented by $b_{22}x_2^2$, by $b_{12}x_1x_2$, or by some combination of all three quadratic terms. Furthermore, we do not really know whether x_3 is uninfluential or has just been lying doggo beneath these quadratic terms. Finally, the dilemma of y versus log y is still with us.

We propose then 2^5 trials, each with a different equation, to test all combinations of x_1^2, x_2^2, x_1x_2, x_3, and log y, keeping x_1 and x_2 in all equations. We will call these alternatives A, B, C, D, and E, respectively. Table 5.3 shows the residual sums of squares (RSS) in standard order. Those for log y obviously cannot be compared directly with those for y, but each set of sixteen can be internally compared. A more objective way of comparing large sets of RSS will be offered in Chapter 6; entirely by luck it is not needed here.

We note three equations—(1), b, and d—which give visibly poorer fits than the remainder, so we drop them forthwith. But there is something odd about the remaining thirteen equations in each set of sixteen: they all give very closely the same residual sums of squares. We would expect a drop of about 1.05 (the mean-square value from duplicates) when fitting an extra constant even to random responses, but nothing of the kind occurs. All thirteen values for y lie between 16.20 and 16.59. The values for log y are similarly restricted.

In the interests of simplicity we choose, as our best equation, a, that is,

$$Y = b_0 + b_1x_1 + b_2x_2 + b_{11}x_1^2,$$

TABLE 5.3

RESIDUAL SUMS OF SQUARES FOR 32 EQUATIONS

Observations 1, 3, 4, and 21 omitted, basic equation $= b_0 + b_1 x_1 + b_2 x_2$

$a = x_1^2$ in
$b = x_2^2$ in
$c = x_1 x_2$ in

$d = x_3$ in
$e = \log y$ in place of y
[x_1 and x_2 are always in]

Pass	Factorial Symbol	RSS	Pass	Factorial Symbol	RSS
10	(1)	22.26	28	e	0.0434
13	a	16.44	29	ae	0.0211
14	b	18.34	30	be	0.0272
15	ab	16.43	31	abe	0.0209
16	c	16.59	32	ce	0.0201
17	ac	16.44	33	ace	0.0200
18	bc	16.59	34	bce	0.0201
19	abc	16.42	35	$abce$	0.0200
20	d	20.40	36	de	0.0420
21	ad	16.29	37	ade	0.0204
22	bd	16.29	38	bdc	0.0204
23	abd	16.20	39	$abde$	0.0204
24	cd	16.21	40	cde	0.0199
25	acd	16.21	41	$acde$	0.0199
26	bcd	16.18	42	$bcde$	0.0199
27	$abcd$	16.17	43	$abcde$	0.0199

since it gives us as good a fit as any of its competitors and has only one extra term. It should be noted that equation c (adding only the term $b_{12} x_1 x_2$) is almost as good.

Although equation a fits the present data far better than our old equation fitted the complete data, it is not a very nice fit anyway, as Figure 5.11 shows. A *residual* is attached to each point as it is located in x_1-x_2 space. We see immediately that the five largest residuals (both positive and negative) are close together in the middle of this plot. The largest residuals remain in these positions for all the other equations of the set tested. We conclude that the equation form we are using cannot accommodate this sort of wrinkle.

Nor do we think it worthwhile to press further. The data have been handled rather more rudely than is permissible without consultation with those who produced them. But the plant is long since shut down, and the process surely obsolete. Our methodological points have been made.

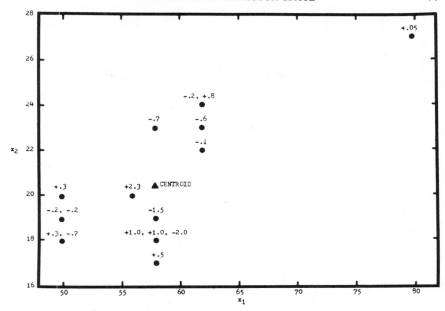

Figure 5.11 Stack loss residuals from $Y = b_0 + b_1 x_1 + b_2 x_2 + b_{11} x_1^2$. Observations *1, 3, 4*, and *21* removed.

5.8 Remote Points in Factor Space

Figure 5.11 shows that point *2* is remote from the rest of the data. This remains true however x_1 and x_2 are scaled. (As already indicated in Chapter 4, it will be necessary in more complicated cases to weight each x_i by its importance, but this step is not needed here.)

A remote observation need not have a large y-residual. On the contrary, it will more often have a very small one, as in the present case. But point *2* was the first of the seventeen observations and may be suspected of controlling our estimates, b_1, b_2, and b_{11}. The only way to see if this is happening is to drop the point and repeat the whole computation. We have done this and find that b_1 and b_2 are only slightly changed while b_{11} drops to negligibility. The variances of the two b_i increase greatly. The reader is spared the documentation, but our conclusion is that point *2* is a valid one and should be retained.

5.9 Equation Using "Lined-Out" Data

We give the usual computer summary for our final equation in Figures 5.12–5.14 and a final summary in Table 5.4. Although we see nothing

LINEAR LEAST-SQUARES CURVE FITTING PROGRAM

STACK LCSS PASS 13   DEP VAR 1: S+LCSS   MIN Y = 7.000D+00   MAX Y = 3.70CD+01   RANGE Y = 3.000D+01

Y = E(C) + B(1)X1 + B(2)X2 + B(3)X1SQRD
Y = STACK LOSS
X1 = AIR FLOW - MEAN VALUE OF AIR FLCW(57.7)
X2 = COOLING WATER TEMPERATURE - MEAN COOLING WATER TEMPERATURE(20.4)
1ST DAY OBSERVATIONS WITH FLCW GREATER THAN 60 OMITTED(1,3,4,21)

IND.VAR(I)	NAME	COEF.B(I)	S.E. COEF.	T-VALUE	R(I)SQRD	MIN X(I)	MAX X(I)	RANGE X(I)	REL.INF.X(I)
0		1.40685D+01							
1	A.FLCW	7.17651C-01	6.410D-02	11.2	0.6469	-7.7C0D+CC	2.230D+01	3.000D+01	0.72
2	W.TEMP	5.27804D-01	1.500-C1	3.5	0.5750	-3.400D+00	6.600D+00	1.000D+01	0.18
3	FLCWSQ	6.81831D-03	3.180D-03	2.1	0.4351	9.CC0D-02	4.973D+02	4.972D+02	0.11

NO. OF OBSERVATIONS   17
NC. OF IND. VARIABLES   3
RESIDUAL DEGREES CF FREEDCM   13
F-VALUE   210.8
RESIDUAL RCCT MEAN SQUARE   1.12455407
RESIDUAL MEAN SQUARE   1.26462187
RESIDUAL SUM OF SQUARES   1t.44008427
TOTAL SUM CF SQUARES   816.23529412
MULT. CORREL. COEF. SQUARED   .5755

ORDERED BY COMPUTER INPUT

IDENT.	OBSV.	WSS DISTANCE	CBS. Y	FITTED Y	RESIDUAL
S LCSS	2	218.	37.000	36.547	0.053
S LOSS	5	8.	18.000	18.125	-0.125
S LOSS	6	9.	18.000	18.653	-0.653
S LOSS	7	10.	19.000	19.181	-C.181
S LOSS	8	10.	20.000	19.181	0.819
S LOSS	9	2.	15.000	15.657	-0.657
S LOSS	10	1.	14.003	13.018	0.982
S LOSS	11	1.	14.030	13.018	0.982
S LCSS	12	3.	13.000	12.490	C.510
S LCSS	13	1.	11.030	13.018	-2.018
S LCSS	14	1.	12.000	13.546	-1.546
S LCSS	15	26.	8.000	7.680	0.320
S LOSS	16	26.	7.000	8.209	-0.680
S LCSS	17	25.	8.000	8.208	-C.2C8
S LOSS	18	25.	3.300	7.680	-0.208
S LOSS	19	25.	9.000	8.735	0.265
S LCSS	20	1.	15.000	12.657	2.343

ORDERED BY RESIDUALS

OBSV.	OBS. Y	FITTED Y	ORDERED RESID.	STUD.RESID.	SEQ
20	15.000	12.657	2.343	2.2	1
10	14.0C0	13.018	0.982	1.0	2
11	14.000	13.018	0.982	1.0	3
8	20.000	19.181	0.819	0.8	4
12	13.000	12.490	0.510	0.5	5
15	8.000	7.680	0.320	0.3	6
19	5.0C0	8.735	0.265	0.3	7
2	37.000	36.947	0.053	0.6	8
5	18.000	18.125	-0.125	-0.1	9
7	19.0C0	19.181	-0.181	-0.2	10
9	8.000	8.208	-0.208	-0.2	11
16	8.0C0	8.208	-0.208	-0.2	12
6	18.000	18.653	-0.653	-0.6	13
17	15.000	15.657	-0.657	-0.7	14
18	7.000	7.680	-0.680	-0.7	15
14	12.000	13.54t	-1.546	-1.5	16
13	11.0C0	13.018	-2.018	-2.0	17

Figure 5.12

78

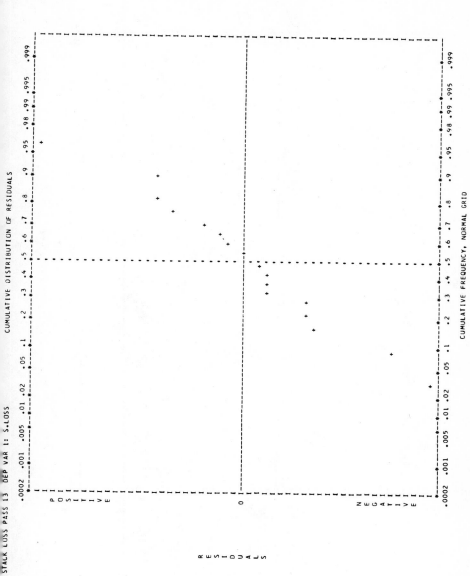

STACK LOSS PASS 13 DEP VAR 1: S.LOSS

CUMULATIVE DISTRIBUTION OF RESIDUALS

CUMULATIVE FREQUENCY, NORMAL GRID

Figure 5.13

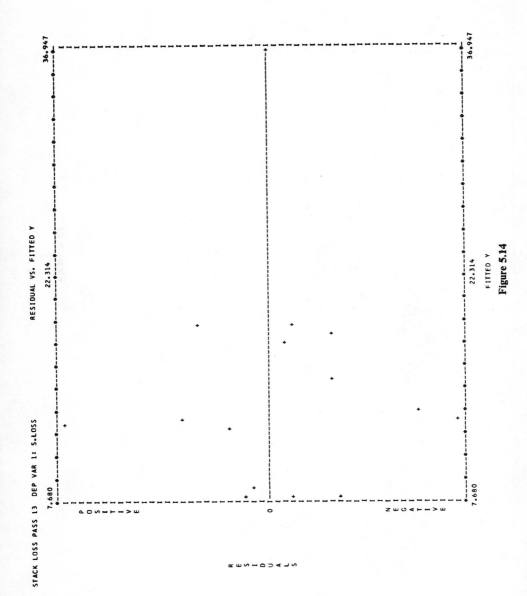

Figure 5.14

80

drastically wrong in any of the tables or plots, there is still some uncomfortable suggestion of lack of fit at the center of the data. We are now so near the ultimate error mean square of 1.05, however, that there is little hope of producing a notably better fit. (Our duplicates produce an expected residual sum of squares of 13 × 1.05 or 13.65. The current total RSS is 16.44; only 2.78, then, remains to be accounted for.)

TABLE 5.4
SUMMARY OF FINAL EQUATION OF STACK LOSS PROBLEM

$$Y = -15.6 - 0.06x_1 + 0.53x_2 + 0.0068x_1{}^2$$

Number of observations	17
Number of variables	3
F-value	211
Multiple correlation coefficient squared	0.98
Residual mean square	1.26
Normal plot	O.K.
Residual versus fitted Y plot	O.K.

i	Term	t_i	$R_i{}^2$	RI
1	x_1'	11.2	0.65	0.72
2	x_2'	3.5	0.57	0.18
3	$x_1'^2$	2.1	0.43	0.11

5.10 Conclusions on Stack Loss Problem

We conclude:

1. Of the twenty-one successive points given, four (*1, 3, 4, 21*) represent transitional states.

2. The remaining seventeen points are well fitted by either of the equations

$$Y_1 = 14.1 + 0.72x_1' + 0.53x_2' + 0.0068x_1'^2$$

$$= -15.4 - 0.07x_1 + 0.53x_2 + 0.0068x_1{}^2$$

or

$$Y_2 = 14.1 + 0.71x_1' + 0.51x_2' + 0.0254x_1'x_2',$$

where $x_1' = x_1 - 57.7$ and $x_2' = x_2 - 20.4$.

3. The residual error has a root-mean-square value of 1.1, with 14 degrees of freedom, inside the region of x_1-x_2 space spanned by the data (shown in Figure 5.11).

4. There is still some lack of fit in the five neighboring points *10*, *11*, *13*, *14*, *20*.

We return to these data in Chapter 7.

5. Retrospectively, we could have come to our final answer immediately after Passes 1 and 2 ($N = 21$, $K = 3, 2$)! The largest residuals (from the wrong equation, look you) are for points *1*, *3*, *4*, and *21*. If we had only had the perspicacity to ask then and there, "What is identifiable about these particular points?" we might have found our answer. They are at the highest Y-values and were taken immediately after changes in plant conditions. We have preferred to give a more nearly historical account of our trials and attendant fumblings in order to show the steps involved.

5.11   General Conclusions

1. Clues found in sequence should be tested all together. This often means that 2^L computer passes—or some balanced fraction thereof—should be made for L clues.

2. A better bound on random error can sometimes be obtained from near replicates (especially if these are not consecutive in time) than is given by the usual residual mean square from an empirical fitting equation.

3. Thorough search need not end with a single "best" equation.

4. The conclusions just given would be more persuasive if a larger set of data had been used in the example. In such larger sets, only drastic improvements should be accepted as proven.

CHAPTER 6

Selection of Independent Variables

6.1 Introduction

Assumptions

In the preceding chapters we have dealt with independent variables known to be influential. More strictly, we have been assuming that:

a. All independent variables are known and are used in the fitting equation.*
b. The form of the equation is known.
c. The fitted equation is linear in its $(K + 1)$ unknown constants.
d. The fitted equation is to be used only on a system that operates like the one producing the data, especially with regard to its correlation structure.

We now consider a modification of assumption a. We have a list of variables that quite surely include all the important factors, but we are not sure that every factor listed is really needed in the equation representing the data. We rely as before on others for assurance that the factors and their function forms [$x_1 = 1/T$, $x_2 = \log$ (space velocity), $x_3 = x_1 x_2$, etc.] are correct.

Obvious imperfections

Obvious imperfections in the data should of course be rectified before an elaborate search is made for the best equation or equations. But it is not possible to give a standard order of procedure that can be followed in all cases. Some defects will be visible at the very beginning; others will appear only after lengthy analysis. We mention here three sorts of defects, one familiar, two less well known.

---

* Some obviously uninfluential variables may have been dropped.

83

Extremely wild points are easy to spot. Often they are seen to represent conditions that are physically impossible. Many computer-oriented editing programs designed to spot such bad values are available. We do not recommend any "statistical" criterion for data deletion. The points that we are thinking of are those that would be removed as "impossible" by 99% of the technical experts in the field under study.

Since start-up difficulties are not rare in plants, initial runs or even early sets of data points are often not representable by the same equation as data collected subsequently. A plant that takes a long time to come to equilibrium after any large change in some factor (or factors) is really afflicted with repeated start-up difficulties. Chapter 5 has shown an example of this situation, and the reader has seen how tedious the spotting and elimination of such points may be.

Another common type of defect in plant data is "nesting." Some conditions (factor levels) are changed rarely; others, more frequently. It is unlikely that data taken under these conditions have uncorrelated errors. We will discuss a clear case of this defect in Chapter 8. Here it is only necessary to point out that plots of each x_i versus the residuals or versus the components and components-plus-residuals discussed in Chapter 7 will help to spot nesting or to rule it out. The general method of equation choice proposed below may yield entirely wrong answers if the data are nested and if this fact is ignored.

2^K possible equations from K candidate variables

When there are K potential independent variables, there are 2^K possible equations. There will be one equation with none of the independent variables in it ($Y = \bar{y}$). There will be K equations each using one variable ($Y = b_0 + b_i x_i$). There will be $K(K - 1)/2$ equations using two variables, etc. Often there are several equations that represent the data equally well. Sometimes the simplest of these will be preferred. In other cases an equation with more factors will be chosen in the hope that it will avoid or decrease bias in the estimates of some constants.

On stepwise regression

One class of strategies for searching the 2^K possible equations is called by the general name "stepwise regression." There are two basic versions, forward and backward. In the forward version (Ralston and Wilf, Section 17*) independent variables are introduced one at a time. At each stage the variable which produces the greatest reduction in the residual sum of squares and whose $[b_i/s(b_i)]^2$ exceeds a preselected value is brought into the equation. The backward version begins with the full regression equation and eliminates the least important variables one at a time, using similar criteria.

* By M. A. Efroymson

Stepwise regression can lead to confusing results, however, when the independent variables are highly correlated (see the publications by Hamaker and by Oosterhoff). Moreover, it is based on the implicit assumption that there is *one* best equation and that stepwise regression will find it. In our experience there are often better equations with different sets of independent variables that are overlooked by this procedure.

F-test

The *F*-test is widely used to judge the need for adding to the equation a *set* of higher-order terms in the basic variables [see Brownlee, Section 13.8, and Davies (1958, Section 11B.32)]. The same procedure might be used to determine whether adding or deleting basic variables substantially affects the fit. But, as with stepwise regression, this search for essential variables is risky when the independent variables are correlated, because the result is not unique. Adding or deleting associated variables in different orders can lead to different equations.

All 2^K equations and fractions

The most comprehensive approach is to fit all 2^K equations and then to compare them. This is not a new idea, and even before the widespread use of electronic computers several authors published such results computed by hand (e.g., Hald, Section 20.3, and Moriceau). Improvements in programming now permit the computation of 2^{12} (4096) residual sums of squares in less than 1 second of IBM 3033 computer time *and* of 2^{18} (262, 144) sums of squares in about 3 minutes with no increase in computer core requirements.

There is, then, little point in looking for more efficient search methods when $K \leqslant 18$. By brute-strength application of our routines, it is possible to extend K to $(18 + Q)$ factors by doing 2^Q sets of size 2^{18} to exhaust all possibilities for small values of Q.

However, for searches with a large number of variables, two alternatives are recommended:

1. A $t_{K,i}$-directed search.
2. Fractional replication of the 2^K possible equations.

Neither is guaranteed to find all of the better-fitting equations, but both have worked well on all cases tried so far. The method of searching for influential subsets proposed by Furnival and Wilson is rapid and allows a selected number of the better equations to be found at each level of p terms ($p < K$). This method is particularly useful when there is a large number of variables to be searched. Search time has recently been greatly reduced by not considering any combinations of a given level of p variables unless the lowest C_p value is less or equal to the C_p value of the full equation.

6.2   Total Squared Error as a Criterion for Goodness of Fit—C_p

Definition

When a large number of alternative equations are being considered, it is imperative that some simple criterion of goodness (or badness) of fit be chosen to characterize each equation. We recommend a measure of "total squared error" given by Mallows. This statistic, called C_p, measures the sum of the *squared biases* plus the *squared random errors* in Y at all N data points. It is a simple function of the residual sum of squares from each fitting equation.

Derivation of the C_p statistic

The total squared error (bias plus random) for N data points, using a fitted equation with p terms, is

$$\sum_{j=1}^{N} (\nu_j - \eta_j)^2 + \sum_{j=1}^{N} \text{Var.} \left\{ Y_{pj} \right\}$$

where
$\nu_j = \nu(x_{1j}, x_{2j}, \ldots)$, expected value from *true* equation,

$\eta_j = \beta_0 + \sum_{i=1}^{k} \beta_i x_{ij}$, expected value from the fitting equation being used,

$(\nu_j - \eta_j) =$ bias at the jth data point,

$p = k + 1$ when β_0 is present,

$\quad = k$ when β_0 is absent.

For convenience, let SSB_p stand for $\sum_{j=1}^{N} (\nu_j - \eta_j)^2$ and define a quantity, Γ_p, the standardized total squared error, by

$$(6.1) \qquad \Gamma_p = \frac{\text{SSB}_p}{\sigma^2} + \frac{1}{\sigma^2} \sum_{j=1}^{N} \text{Var.} (Y_{pj}).$$

It can be shown that:

$$(6.2) \qquad \sum_{j=1}^{N} \text{Var.} (Y_{pj}) = p\sigma^2.$$

Combining these equations gives:

$$(6.3) \qquad \Gamma_p = \frac{\text{SSB}_p}{\sigma^2} + p.$$

Now the residual sum of squares, RSS_p, from a p-term equation has the expectation:

$$E\{\text{RSS}_p\} = \text{SSB}_p + (N - p)\sigma^2.$$

Solving for SSB_p gives:

$$SSB_p = E\{RSS_p\} - (N - p)\sigma^2;$$

and so

$$\Gamma_p = \frac{E\{RSS_p\}}{\sigma^2} - (N - p) + p$$

$$= \frac{E\{RSS_p\}}{\sigma^2} - (N - 2p).$$

With a good estimate of σ^2 (call it s^2), the statistic C_p is an estimate of Γ_p.

(6.4) $$C_p = \frac{RSS_p}{s^2} - (N - 2p).$$

When a p-term equation has negligible bias, $SSB_p \approx 0$, $E\{RSS_p\} = (N-p)\sigma^2$, and

$$E\{C_p \mid SSB_p = 0\} = \frac{(N - p)\sigma^2}{\sigma^2} - (N - 2p) = p.$$

Mallows' graphical method of comparing fitted equations

If the C_p's for a number of equations are plotted against p, those for equations with small bias will tend to cluster about the line $C_p = p$ (Figure 6.1, point A), while those for equations with substantial bias will fall above the line (Figure 6.1, point B).* Although point B in Figure 6.1 is above the line $C_p = p$, it is still below point A and consequently represents an equation with slightly lower total error. Thus, adding terms may reduce bias, but at the expense of increasing the total variance of prediction for the N points and, consequently, the average variance per point. If an equation is needed for interpolation in the region of factor space spanned by the data points, it may pay to drop a few terms in order to obtain a simpler equation. In other words, it may pay to accept some bias in order to get a lower average error of prediction.

We need an unbiased estimate of σ^2 to calculate C_p. Often the mean square from the complete equation will serve this purpose. This forces C_p to be $p_{\max}$ or $(K + 1)$ for the complete equation. Substituting this estimate for σ^2 in the C_p calculation assumes that the complete equation has been carefully chosen to give negligible bias. A step is made in Chapter 7 toward finding a less biased estimator of σ^2 than s^2 from the complete equation.

---

* We see from the form of equation 6.4 that C_p may decrease by as much as 2 units as p decreases by 1 because of dropping a term from an equation. (The RSS cannot decrease on dropping a term; s^2 and N are fixed.)

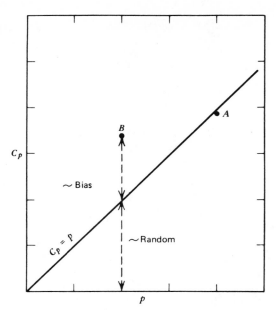

Figure 6.1 C_p plot.

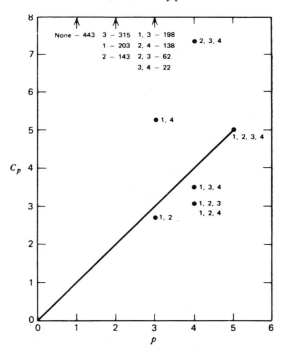

Figure 6.2 C_p plot for Hald's data.

88

6.3 A Four-Variable Example

Hald's book (Section 20.3) contains an interesting example involving four independent variables. The data, given in Appendix 6A.1, concern the heat evolved during the hardening of certain cements, as it depends on the percentage of four constituents. The equation is

$$Y = b_0 + b_1x_1 + b_2x_2 + b_3x_3 + b_4x_4.$$

Statistics for the complete equation and C_p's for all sixteen equations, using the four variables in all combinations, are given in Table 6.1.

It is easy to see from the listing (and from Figure 6.2) which are the

TABLE 6.1
STATISTICS FOR HALD'S DATA

$N = 13$
(b_0 included in all equations)

Term	Complete Equation				Simple Correlation Coefficients, r_{ii}'					
	b_i	$s(b_i)$	t_i	R_i^2		x_1	x_2	x_3	x_4	y
Constant	62.40									
x_1	1.55	0.74	2.08	0.97	x_1	1.0				
x_2	0.51	0.72	0.70	0.99	x_2	0.23	1.0			
x_3	0.10	0.75	0.14	0.98	x_3	-0.82	-0.14	1.0		
x_4	-0.14	0.71	0.20	0.99	x_4	-0.24	-0.97	0.03	1.0	
F-value	111.4				y	0.73	0.82	-0.53	-0.82	1.0
R_y^2	0.98									
RSS	47.9									
$s^2(y)$	5.98									

C_p's for All Equations

Variables in Equation	p	C_p	Variables in Equation	p	C_p
None	1	443.2	x_4	2	138.8
x_1	2	202.7	x_1, x_4	3	5.5
x_2	2	142.6	x_2, x_4	3	138.3
x_1, x_2	3	2.7	x_1, x_2, x_4	4	3.0
x_3	2	315.3	x_3, x_4	3	22.4
x_1, x_3	3	198.2	x_1, x_3, x_4	4	3.5
x_2, x_3	3	62.5	x_2, x_3, x_4	4	7.3
x_1, x_2, x_3	4	3.0	x_1, x_2, x_3, x_4	5	5.0

best-fitting equations and that should end our story. But we can hardly help noticing that, when factor x_1 is omitted, C_p is always greater than 7. It seems that we have computed at least half of our C_p's only to discard them. Table 6.1 shows that t_1 is 2.08, the largest of the four. If we had decided to retain factor x_1 in all equations, we would have computed only eight C_p's:

Variables in Equation		C_p
1	(0)	202.7
1	2	2.7*
1	3	198.2
1	23	3.0*
1	4	5.5
1	24	3.0*
1	34	3.5*
1	234	5.0

In this case, t_i is an excellent indicator; all four best C_p's are spotted in seven trials. We note that, if we had added the next largest t_i (only 0.7!), we would have gone directly to equation 12 and its neighbors 123 and 124, finding the winners immediately.

Even though the sixteen C_p-values constitute a 2^4 arrangement, there is little point in putting them through the usual Yates computation (see Davies, 1960, Sections 7.45 and 7.61, and Daniel, 1976, Section 3.3) to find "effects of entering factors" and their interactions. Inspection of the first four values in Table 6.1 shows immediately that there is no hope of getting a simple factorial representation in terms of average effects of entering factors. Similarly, a short look at the eight C_p's including x_1—just given—tells us that factors 2, 3, and 4 do not have simple effects on C_p. We have all the information we need in the eight values as they stand.

Here is our conclusion, to be documented further, about interpreting sets of 2^K C_p-values and their fractions 2^{K-Q}. We do full factorial sets of C_p when we can, for the sake of completeness, not hoping to discover factors whose entry always produces the same decrease in C_p. When we must do some fraction of the full set, we do not use the usual fractions because these are bound to omit influential factors in half of their runs. The useful fractions are always in corners of the full experiment-space. *Clues to the location of these corners are always given by the factors that have largest t_i-values in the full equation.*

Hamaker used two standard "stepwise regression" methods on these data but found that the forward and backward procedures selected different variables. These data, which are only a part of a much larger set, are discussed again in Chapter 9.

6.4 Disposition of Data in Factor Space

The disposition of data in factor space is the prime cause of confusion in selecting candidate equations. Consider an artificial problem involving two independent variables with observations taken at four points in factor space such that the coefficient of determination, R_i^2, for x_1 given x_2 (or x_2 given x_1) is 0.90 (with only two independent variables $R_1^2 = R_2^2 = r_{12}^2$, the square of the simple correlation coefficient). The observations, four values at each point in factor space for a total of sixteen, are shown in Figure 6.3. They were generated from the equation

$$y_j = 1 + x_{1j} + x_{2j} + e_j$$
$$e_j = \text{random standard normal error.}$$

A data analyst, faced with these data and technical information indicating that the form

$$Y = b_0 + b_1 x_1 + b_2 x_2$$

is a reasonably complete equation, would fit all four possible equations to see

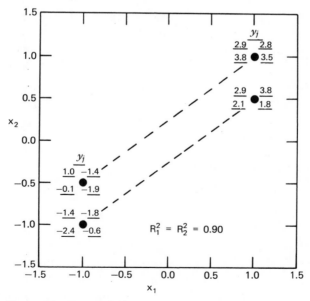

Figure 6.3 Two-variable example: $y_j = 1 + x_{1j} + x_{2j} + e_j$.

whether a shorter equation would do. Table 6.2 shows the results of fitting these equations:

$$Y_1 = b_0 + b_1 x_1 + b_2 x_2$$
$$Y_2 = b_0' + b_1' x_1$$
$$Y_3 = b_0'' + b_2'' x_2$$
$$Y_4 = b_0'''.$$

(Note that x_1 and x_2 have mean values of zero in the example. Thus $b_0 = \bar{y}$ for all four equations.)

The three equations using either one or both of the independent variables are far better than the equation with neither and have nearly equivalent residual mean squares (RMS) and multiple correlation coefficients, R_y^2. However the coefficients of x_1 and x_2 differ considerably among the three equations:

$$Y_1 = 0.94 + 0.85x_1 + 1.55x_2$$
$$Y_2 = 0.94 + 2.01x_1$$
$$Y_3 = 0.94 \qquad\qquad + 2.57x_2$$

because the b's in the shorter equations are biased estimates. (A short proof of the algebraic relationship between the values of the coefficients of the full and shorter equations is given in the paper by Gorman and Toman, 1966.) Thus if there is a high linear association among the effective independent variables (as indicated by a high R_i^2), we can expect wide variations in the coefficients when an effective independent variable is omitted.

In this example, the ratio of the standard errors of the coefficients of the full equation to those for the equations with one variable is 3.16. The reason for this is again the high linear dependence between x_1 and x_2. For the full equation the standard error of b_1 can be written directly as a function of R_1^2:

$$\text{S.E. } b_1 = \frac{\text{RMS}}{\sqrt{(1 - R_1^2) \sum_j (x_{1j} - \bar{x}_1)^2}}.$$

For the equation with x_1 only (x_2 omitted):

$$\text{S.E. } b_1 = \frac{\text{RMS}'}{\sqrt{\sum_j (x_{1j} - \bar{x}_1)^2}}.$$

The ratio of the two is proportional to $1/\sqrt{1 - R_1^2}$. In the example, this equals $1/\sqrt{1 - 0.9} = 3.16$. The corresponding t-tests fluctuate even more because they reflect changes in both the b's and the RMS's.

If estimates of the effects of x_1 and x_2 are wanted, only the complete equation gives unbiased estimates of the coefficients with proper standard errors.

TABLE 6.2
FITTED EQUATIONS OF THE TWO-VARIABLE EXAMPLE

Term	x_1 and x_2			x_1 only			x_2 only			Neither x_1 nor x_2		
	Coefficient	S.E. Coefficient	t-Value	Coefficient	S.E. Coefficient	t-Value	Coefficient	S.E. Coefficient	t-Value	Coefficient	S.E. Coefficient	t-Value
Constant	0.94			0.94			0.94			0.94		
x_1	0.85	0.70	1.2	2.01	0.24	8.5	—	—	—	—	—	—
x_2	1.55	0.88	1.8	—	—	—	2.57	0.28	9.1	—	—	—
RMS	0.78			0.89			0.80			5.15		
R_y^2	0.87			0.84			0.85			—		
p	3			2			2			1		
C_p	3.0			4.1			2.5			85.7		

A data analyst seeking these estimates would have only the information in Table 6.2 to help decide which equation to use. The need for the complete equation is indicated by the nearly equivalent fits of the three equations to the data and the wide fluctuations in the b's and their t-tests as one or the other variable is dropped.

On the other hand, if a simple interpolation formula for estimating responses is wanted, all three equations should be considered. If predicted values are to be calculated only within the area outlined by the four points in factor space (Figure 6.3) then the equations with only x_1 or x_2 might be suitable. In this

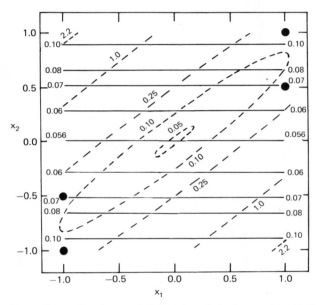

Figure 6.4 Comparison of variance of Y for: $Y = b_0 + b_1x_1 + b_2x_2$ (dashed line) and $Y = b_0'' + b_2''x_2$ (solid line).

region the variance of predicted values calculated from the full equation and, for example, from the equation with x_2 only, are in fair agreement. This is shown in Figure 6.4 where contours of constant variance for the complete equation are shown as dashed ellipses and those for the equation with x_2 only as solid horizontal lines. However, outside the region where the data were taken the shorter equation gives a false sense of security; for example, in the neighborhood of the point $x_1 = -1$ and $x_2 = 1$ (Figure 6.4) the ratio of the proper variance, given by the complete equation, to that given by the shorter equation is $2.2/0.10 = 22$.

Use of the shorter equation also gives biased estimates of Y. Figure 6.5 is a contour plot of the expected value of the difference between Y calculated from the complete equation and Y calculated from the equation using x_2 only. In the region where the data were taken the expected value of the bias ranges from 0 to a maximum of ± 0.4. Hence if the shorter equations are used as a matter of convenience, their region of applicability must be clearly specified in terms of both x_1 and x_2.

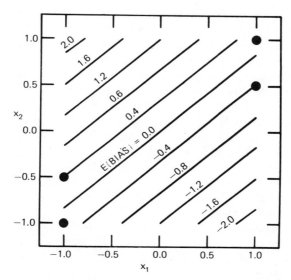

Figure 6.5 Expected value of bias for $Y = b_0'' + b_2'' x_2$ when the true equation is $\eta = 1 + x_1 + x_2$.

6.5 A Six-Variable Example

a. Search of all 2^K equations

The data that we shall consider [previously discussed by Gorman and Toman (1966)] are measurements of the "rate of rutting" of 31 experimental asphalt pavements, each prepared under different conditions specified by the values of five of the independent variables. An indicator variable is used to separate the results for 16 pavements tested in one set of runs from those for the 15 tested in the second set. The following equation was first used to fit the data:

$$Y = b_0 + b_1 x_1 + b_2 x_2 + b_3 x_3 + b_4 x_4 + b_5 x_5 + b_6 x_6$$

where Y = log (change of rut depth in inches/million wheel passes),

   x_1 = log (viscosity of asphalt),

   x_2 = percentage of asphalt in surface course,

   x_3 = percentage of asphalt in base course,

   x_4 = indicator variable to separate two sets of runs,

   x_5 = percentage of fines in surface course, and

   x_6 = percentage of voids in surface course.

The original data are given in Figure 6B.1.

The objectives of the analysis were (1) to estimate the effects of the six individual factors, thus providing information for the design of an experiment, and (2) to find a simple interpolation formula for interim use. We will use the smallest equation that shows no serious lack of fit for the first objective, but will find that there is no "simple" equation which will satisfy the second objective.

The familiar statistics b_i, $s(b_i)$, t_i, R_i^2 and $r_{ii'}$ are given in Table 6.3, along with the complete set of sixty-four C_p-values for all subsets of the six factors. Using the usual 2^K factorial nomenclature, we indicate the presence of a factor by lower-case letters a, b, c, d, e, and f for $x_1, \ldots, x_6$, respectively, and so identify each equation by a "word" of six letters or less. Figure 6.6 shows all C_p less than 100 plotted versus p. It is obvious that a and b are present in all equations with C_p less than 18.

The list of C_p's contains a set of eight that are smallest (starred in the table). Once found, they are seen to have a simple structure: each contains abf plus a combination of c, d, and e. Figure 6.7, which shows these eight plotted against their respective p-values on a larger scale, makes it clear that $abdf$ is the best-fitting equation. Inclusion of d improves the fit over abf for all combinations of c and e. We see further that inclusion of c or e or both only adds random error and reduces bias negligibly.

It is well to compare the coefficients of all the better equations in order to judge the consequences of dropping other factors. Table 6.4 shows all b_i for the better-fitting equations. We see that these effects are about the same for the four equations that contain a, b, d, and f (i.e., x_1, x_2, x_4, and x_6). But when d is dropped, b_1 changes considerably. The factor x_1 is then measuring $(x_1 + 1.007x_4)$, since x_1 and x_4 have nearly the same spread and also high correlation ([44]/[11] = 1.14; $r_{14} = 0.94$).

Since equation $abdf$ fits best, the harried data analyst might simply advise the experimenters to use it for prediction or extrapolation. Unfortunately, the equation has a serious shortcoming: x_4 is not reproducible. As you will recall, x_4 is the indicator variable that separates the two sets of runs. We do not know what caused the two runs to respond differently. Hence, we cannot set factor x_4 at either of its former levels. The data do show that there was a

TABLE 6.3
STATISTICS FOR SIX-VARIABLE EXAMPLE
$N = 31$
(b_0 included in all equations)

Term	\multicolumn{4}{Complete Equation}				Simple Correlation Coefficients, $r_{ii'}$						
	b_i	$s(b_i)$	t_i	$R_i{}^2$		a	b	c	d	e	f	y
Constant	−2.64	—	—	—								
x_a	−0.51	0.073	7.0	0.91	a	1.0						
x_b	0.50	0.12	4.3	0.32	b	−0.06	1.0					
x_c	0.10	0.14	0.7	0.32	c	−0.22	−0.26	1.0				
x_d	−0.13	0.064	2.1	0.90	d	0.93	0.01	−0.12	1.0			
x_e	0.02	0.034	0.6	0.31	e	−0.30	−0.14	0.44	−0.23	1.0		
x_f	0.14	0.047	2.9	0.44	f	0.46	−0.43	−0.03	0.40	0.12	1.0	
F-value	140				y	−0.97	0.15	0.19	−0.93	0.32	−0.41	1.0
$R_y{}^2$	0.972											
RSS	0.3070											
$s^2(y)$	0.01279											

C_p's for All Equations

Variables in Equation	p	C_p	Variables in Equation	p	C_p	Variables in Equation	p	C_p	Variables in Equation	p	C_p
(1)—none	1	835.6	e	2	748.8	f	2	692	ef	3	575.5
a	2	20.43	ae	3	21.68	af	3	20.00	aef	4	21.87
b	2	817.0	be	3	716.1	bf	3	693.7	bef	4	577.4
ab	3	13.98	abe	4	14.29	abf	4	5.67*	$abef$	5	7.45*
c	2	806.7	ce	3	748.3	cf	3	666.8	cef	4	577.3
ac	3	21.90	ace	4	22.53	acf	4	21.37	$acef$	5	22.84
bc	3	770.0	bce	4	709.0	bcf	4	668.0	$bcef$	5	579.1
abc	4	15.95	$abce$	5	16.17	$abcf$	5	7.56*	$abcef$	6	9.43*
d	2	92.0	de	3	84.2	df	3	92.8	def	4	82.7
ad	3	20.53	ade	4	21.46	adf	4	20.61	$adef$	5	22.24
bd	3	70.23	bde	4	57.56	bdf	4	70.21	$bdef$	5	59.09
abd	4	12.22	$abde$	5	11.32	$abdf$	5	4.26*	$abdef$	6	5.51*
cd	3	88.6	cde	4	85.1	cdf	4	89.36	$cdef$	5	83.9
acd	4	22.42	$acde$	5	22.86	$acdf$	5	22.36	$acdef$	6	23.65
bcd	4	58.38	$bcde$	5	53.92	$bcdf$	5	56.83	$bcdef$	6	54.38
$abcd$	5	13.51	$abcde$	6	13.24	$abcdf$	6	5.31*	$abcdef$	7	7.00*

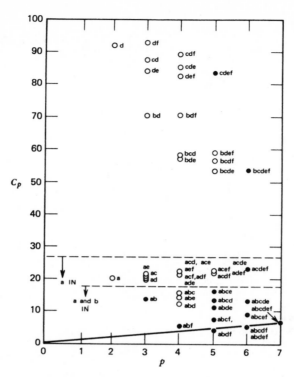

Figure 6.6 C_p plot for six-variable example.

TABLE 6.4
COMPARISON OF EQUATIONS FOR SIX-VARIABLE EXAMPLE

	Equation									
	abf		*abdf*		*abcdf*		*abdef*		*abcdef*	
Term	b_i	(t_i)	b_i	(t_i)	b_i	(t_i)	b_i	(t_i)	b_i	(t_i)
Constant	−1.57	(—)	−1.85	(—)	−2.69	(—)	−2.05	(—)	−2.64	(—)
$x_1 = a$	−0.66	(25.5)	−0.55	(8.4)	−0.52	(7.4)	−0.52	(7.5)	−0.51	(7.0)
$x_2 = b$	0.43	(3.9)	0.46	(4.3)	0.50	(4.4)	0.47	(4.4)	0.50	(4.3)
$x_3 = c$	—	(—)	—	(—)	0.12	(1.0)	—	(—)	0.10	(0.7)
$x_4 = d$	—	(—)	−0.11	(1.9)	−0.13	(2.0)	−0.12	(2.0)	−0.13	(2.1)
$x_5 = e$	—	(—)	—	(—)	—	(—)	0.03	(0.9)	0.02	(0.6)
$x_6 = f$	0.15	(3.1)	0.14	(3.2)	0.14	(3.2)	0.13	(2.8)	0.14	(2.9)
s	0.112		0.111		0.111		0.112		0.113	
C_p	5.67		4.26		5.31		5.51		7.00	

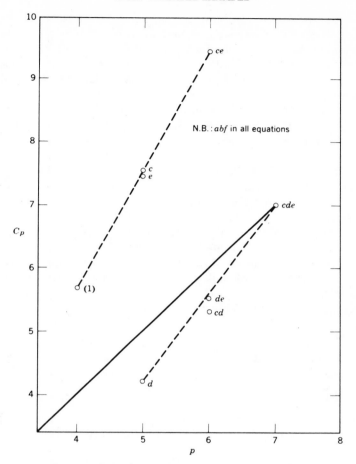

Figure 6.7 Plot of eight smallest C_p-values (six-variable example).

considerable difference between the log rutting rates in the two sets of data that is not accounted for by the change in asphalt viscosity. It is not possible to judge whether this discrepancy is a result of some difference between batches of asphalt, a change in tires used on the test machine, a change in some other raw material, or a combination of these. It will not do, either, to christen this variation a random one. All we can do at this time is to record the difference and study the data to see whether it is a consistent one.

In order to challenge the form of equation being used, we added the five cross-product terms—$x_i x_4$ with $i = 1, 2, 3, 5, 6$—in a separate computer pass, not shown. We found all t-values for the five new terms to be less than 1.0, while all the linear terms maintained their former importance. We also made separate least-squares fits for each of the two sets of data. Again we found no large differences in effects or in residual mean squares. Thus, we are left with *abdf* as our best equation for summarizing these data.

As estimates of the true effects of x_1, x_2, and x_6, we would recommend those given by the equation *abdf*. The usual statistics are all shown in the computer printout reproduced in Figure 6B.5. But the experimenter cannot safely use this equation for prediction, since he cannot set the level of x_4 in future data.

In regard to factors x_3 and x_5, which have been dropped, we must also hedge a little. If these factors had been varied over realistic ranges in these tests, we could boast that our estimated values, b_1, b_2, and b_6, have some degree of validity, since they hold over quite a range of experience. (In actual fact, wider variations of x_3 and x_5 were possible and, for x_3, profitable.)

The indefensible blocking of the present data serves only to provide us with a cautionary tale. It is not only post hoc that we would recommend *proper* blocking of any future runs. Since the test machine holds only sixteen pavement segments, it is obvious that future blocks will be at most of that size. Every factor must be varied in each block. The rules for doing this are not given here but are well known to design statisticians.

b. $t_{K,i}$-Directed search

Ordering the six (in general K) factors by their decreasing t-values (called $t_{K,i}$ in order to distinguish them from the t-values found in subsidiary equations, we have the results shown in the table. Computing directly the six C_p-values for equations 1, 12, 126, 1246, 12346, and 123456, we obtain the values given to the right in the same table. This method immediately gives us, at least in this example, the winner and two competitors (indicated by **).

i	Added Factor	$t_{k,i}$	Cumulative p	C_p
1	a	7.0	2	20.4
2	b	4.3	3	14.0
6	f	2.9	4	5.7**
4	d	2.1	5	4.3**
3	c	0.7	6	5.3**
5	e	0.6	7	7.0

Since $t_i^2 s^2$ measures the increase in the *fitting* sum of squares due to x_i, and since the *residual* sum of squares determines C_p, there is a simple identity relating the $t_{K,i}$ for factor x_i in the K-constant equation to the C_p (called now

$C_{K,-i}$), measuring the total squared error made by the equation *with* x_i *removed*:

$$C_{K,-i} = t^2_{K,i} + K - 1,$$

since $C_{K+1} = K + 1$.

Thus, if a $t_{K,i}$ as large as 3 appears, the $C_{K,-i}$ found on dropping the corresponding factor will be $(K + 8)$. The C_p's related to this one by further deletions of factors cannot be more than 2 units below $(K + 8)$ for each factor dropped. It will only be possible to arrive at a C_p roughly equal to p along this route by dropping at least 8 factors. If the uninfluential factors being dropped produce their expected decrease in the residual sum of squares of s^2 (rather than the minimum possible decrease of zero), then 16 factors will have to be dropped to negate the effect on C_p of dropping a factor with a $t_{K,i}$ of 3.

By similar reasoning, a factor with a $t_{K,i}$ of 2 can be omitted from a well-fitting equation only if from three to six other factors are dropped along with it to bring C_p down to p or less. We will see several cases (not chosen for this reason) in which factors with $t_{K,i}$-values as small as 1 have remained in all acceptable equations.

c. Fractional replicates

The fractions of 2^K needed here are not the usual ones, since our aim is so different from that of the factorial experimenter. We are looking for subsets of small C_p-values. We are not looking for the factors whose presence uniformly reduces C_p by a constant amount over the whole system. There are, usually, no such factors. Indeed every correlation between two factors appears in the 2^K set of C_p as a two-factor interaction. It may well be that once we have included a controlling factor another set of less influential factors operates nearly additively. This subset will not usually be discovered by a balanced fraction, however, since even in the simplest case (one dominant factor) only half of the fraction gives information about the operation of the other factors in the favorable context. More commonly two or more factors are present in *all* favorable equations. We are then painted into a quadrant, an octant, or an n-tant of the full set of 2^K possibilities and correspondingly into a *fraction* of the balanced fraction usually chosen. It is only in this favorable "corner" that a fractional replicate may show which secondary factors are helpful, and which are not.

In the present case factors a and b (x_1 and x_2) are visibly important ($t_{K,i} = 7.0, 4.3$). As a further check, the C_p's for the equations dropping a, b, or both are calculated and compared with the value for the full equation:

	p	C_p
$cdef$	5	83.9
$acdef$	6	23.6
$bcdef$	6	54.4
$abcdef$	7	7.0

From these results it is plausible to conclude that only equations containing both a and b will be satisfactory.

Consider equation $acdef$ with $C_p = 23.6$. The only way in which an equation with a lower C_p and p can be obtained is by eliminating variables that contribute little to the fit. Such eliminations will on the average reduce C_p by one unit (and cannot possibly reduce it by more than two units) per variable dropped. If all five variables are dropped from equation $acdef$, the value of C_p cannot be reduced by more than ten units, or to 13.6, which still exceeds 7.0, the value for the full equation. Hence, no equation containing a but not b can have a C_p less than 7.0.

There are 2^4 or 16 equations involving c, d, e, and f in all combinations, with a and b always included. For illustrative purposes we select a half-replicate of these, using the identifying letters of the equations as factorial treatment combinations. We choose the defining contrast $I + CDEF$.

		p	C_p
ab		3	13.98
ab	cf	5	7.56
ab	df	5	4.26
ab	cd	5	13.51
ab	ef	5	7.45
ab	ce	5	16.17
ab	de	5	11.32
ab	$cdef$	7	7.0

The usual Yates calculation (Davies, 1960, Section 7.45, and Daniel, 1976, Section 3.3) is performed on the C_p's, as shown in the table. Effects C, D, E, etc., show how the presence of the corresponding variable affects the value of C_p. [The pattern, which shows clearly in the eight ordered C_p's in the table, is repeated in the half-replicate not used. The effects found also agree closely. This is a rather stringent test of our "factorial assumption," since the half used first contains the smallest C_p (4.26), whereas the other half does not.]

Effect F predominates and on the average reduces C_p by 7.2 units. One more check on the importance of F is made by calculating C_p for $abcde$; we get 13.2 (Table 6.4).

		p	C_p	I	II	III	III/4	Effects
ab		3	13.98	21.54	39.31	81.25	—	I + CDEF
ab	cf	5	7.56	17.77	41.94	7.23	1.8	C
ab	df	5	4.26	23.62	2.83	−9.07	−2.3	D
ab	cd	5	13.51	18.32	4.40	2.63	0.7	CD + EF
ab	ef	5	7.45	−6.42	−3.77	2.63	0.7	E
ab	ce	5	16.17	9.25	−5.30	1.57	0.4	CE + DF
ab	de	5	11.32	8.72	15.67	−1.53	−0.4	DE + CF
ab	cdef	7	7.00	−4.32	−13.04	−28.71	−7.2	F

The average effect of inserting d (x_4) into equations containing ab is to reduce C_p by 2.3 units. If d were without effect in reducing the residual sum of squares at all, it would *increase* C_p by 1 unit. We have little doubt, then, that d is also a useful term. This assumption is tested in both directions by first dropping d from our best equation, $abdf$, to get abf ($C_p = 5.7$), and then by looking at the set of four equations that contain $abdf$ along with c and/or e:

		C_p	Effect	
abdf	(1)	4.26		
	c	5.31	+1.27	C
	e	5.51	+1.47	E
	ce	7.00	+0.22	CE

Summarizing, we have found that $abdf$ must be in any well-fitting equation, and that c and e then only add variance (the expected increase in C_p for a factor having no effect on bias is 1.0).

The purpose of factorial exploration is to locate variables common to all the better equations. The $t_{K.i}$-values for the coefficients in the complete equation are good first indicators, but variables picked in this way must be checked by wider sampling to make sure that they are essential. The remaining useful factors are found by using full factorials or fractional sampling plans. This approach has been used successfully on problems with as many as twenty-five variables.

In this discussion, we have used very small fractional replicates for expository purposes. In real life fractional replicates will of course always be much larger and involve much larger sets of factors. Appendix 6C gives three sets of fractions that may be found useful. It is perhaps needless (and certainly redundant) to remind the user of this appendix that these fractions are recommended for use on the set of *less* influential factors, with the more surely influential ones firmly fixed in all equations to be tested.

APPENDIX 6A

COMPUTER PRINTOUTS OF FOUR-VARIABLE EXAMPLE (page 106)

Pass	Figures
1	6A.1 and 6A.2
2	6A.3

APPENDIX 6B

COMPUTER PRINTOUTS OF SIX-VARIABLE EXAMPLE (page 109)

Pass	Figures
1	6B.1–6B.4
2	6B.5–6B.7

APPENDIX 6C

FRACTIONAL REPLICATION FOR 2^K EQUATIONS

Three sets of fractions of 2^K are given below. The first set is of "Resolution V." This means that all simple nonadditivities (usually called two-factor interactions and abbreviated 2fi) are estimable. Thus, if factors A and B (or x_1 and x_2) have small effects in reducing the total squared error, C_p, the data analyst may want to know whether their simultaneous presence in the fitting equation has a large effect. If this is the case, it is called an "AB interaction." It can be detected only by fitting all four equations:

Factors in Equation	Literal Designation	Equation
None	(1)	$Y = \bar{y}$
x_1	a	$Y = \bar{y} + b_1(x_1 - \bar{x}_1)$
x_2	b	$Y = \bar{y} + b_2(x_2 - \bar{x}_2)$
x_1, x_2	ab	$Y = \bar{y} + b_1(x_1 - \bar{x}_1) + b_2(x_2 - \bar{x}_2)$

Of course we do not want to double the number of equations for each new factor. The device of "Resolution V" fractional replication permits estimation of all 2fi *provided* there are no combinations of *three* factors that produce unexpectedly large reductions in C_p. If 3fi are expected, we will require a "Resolution VII" plan. See Table 6C.3.

The plans given in Table 6C.1 are adapted from the National Bureau of Standards booklet, "Fractional Factorial Experimental Designs for Factors at Two Levels," *Applied Mathematics Series* 48 (reprinted with corrections in 1962). Some effort has been made to select the simplest set of generators to ease transcription and multiplication. All relevant information is given for identifying each 2fi.

The resulting 2^{K-Q} values of C_p must be put through the usual Yate's algorithm for estimation of the "effects and interactions" of the various factors on the goodness of fit (see Davies, 1960, Sections 7.45 and 7.61, and Daniel, 1976, Section 3.3). Thus for the 2^{5-1} the standard order, bringing in

(*continued on page 118*)

TABLE 6C.1
"Resolution V" Plans for 2^K Equations (K Factors)

N.B.S. page	K	Q	2^{K-Q} =N	Equation Generators	Key Interactions in Alias Subgroup
5	5	1	16	ae,be,ce,de	-ABCD(E)
7	6	1	32	af,bf,cf,df,ef	+ABCDE(F)
8	7	1	64	ag,bg,cg,dg,eg,fg	-ABCDEF(G) *
22	8	2	64	agh,bgh,cg,dh,eg,fh	-ABCE(G), -ABDF(H)
24	9	2	128	aj,bj,chj,dh,ehj,fh,ghj	-ABCDG(J),CDEFG(H) **
36	10	3	128	ahk, bjk, ck, dhjk, eh,fhj,gj	-ABCD(K), -ADEF(H), -BDFG(J) ***
46	11	4	128	ahkl,bhj,chk,dhjl, ehl, fhjk, ghjkl	-ADEG(L),-ACFG(K), -BDFG(J), ABCDEFG(H) -BCFH(L), ACEH(J)
57	12	5	128	ahkl,bhjl,chjm,dhkm, ehklm,fhjlm,ghjk	-ABEF(L),-CDEF(M), -BCFG(J),-ADEG(K), ABCDEFG(H),-CDEF(M), ABGH(M), CDGH(L)

* actually of Resolution VII

** actually of Resolution VI

*** GH+JK, GJ+HK, GK+HJ estimable

LINEAR LEAST-SQUARES CURVE-FITTING PROGRAM

4 VARIABLES PASS 1

DATA INPUT   4 INDEPENDENT VARIABLES   1 DEPENDENT VARIABLE(S)

OBSV.	SEQ.	1-11-21	2-12-22	3-13-23	4-14-24	5-15-25	6-16-26	7-17-27	8-18-28	9-19-29	10-20-30
1	0	7.000	26.000	6.000	60.000	78.500					
2	0	1.000	29.000	15.000	52.000	74.300					
3	0	11.000	56.000	8.000	20.000	104.300					
4	0	11.000	31.000	8.000	47.000	87.600					
5	0	7.000	52.000	6.000	33.000	95.900					
6	0	11.000	55.000	9.000	22.000	109.200					
7	0	3.000	71.000	17.000	6.000	102.700					
8	0	1.000	31.000	22.000	44.000	72.500					
9	0	2.000	54.000	18.000	22.000	93.100					
10	0	21.000	47.000	4.000	26.000	115.900					
11	0	1.000	40.000	23.000	34.000	83.800					
12	0	11.000	66.000	9.000	12.000	113.300					
13	0	10.000	68.000	8.000	12.000	109.400					

DATA TRANSFORMATIONS

POSITION	CODE	OPERATION
1		NONE
2		NONE
3		NONE
4		NONE
5		NONE

CONSTANT LOCATION	OMIT	VARIABLE	NAME
1	0	1	C3A
2	0	2	C3S
3	0	3	CAF
4	0	4	C2S
5	0	5	HEAT

MEANS OF VARIABLES
7.46154D+00 4.81538D+01 1.17692D+01 3.00000D+01 9.54231D+01

ROOT MEAN SQUARES OF VARIABLES
5.88239D+00 1.55609D+01 6.40513D+00 1.67382D+01 1.50437D+01

SIMPLE CORRELATION COEFFICIENTS, R(I,I PRIME)

1	1.000				
2	0.229	1.000			
3	-0.824	-0.139	1.000		
4	-0.245	-0.973	0.030	1.000	
5	0.731	0.816	-0.535	-0.821	1.000

Figure 6A.1

LINEAR LEAST-SQUARES CURVE FITTING PROGRAM

4 VARIABLES PASS 1   DEP VAR 1: HEAT

MIN Y = 7.250D+01   MAX Y = 1.155D+02   RANGE Y = 4.340D+01

$Y = B(0) + B(1)X1 + B(2)X2 + B(3)X3 + B(4)X4$
Y = CUMULATIVE HEAT OF HARDENING AFTER 180 DAYS, CALORIES/GRAM OF CEMENT
X1 = PCT. TRICALCIUM ALUMINATE, C3A
X2 = PCT. TRICALCIUM SILICATE, C3S
X3 = PCT. CALCIUM ALUMIUM FERRATE, CAF
X4 = PCT. DICALCIUM SILICATE, C2S
NOTE: CLINKER COMPOSITION CALCULATED FROM ANALYSIS OF OXIDES.

IND.VAR(I)	NAME	COEF.B(I)	S.E. COEF.	T-VALUE	R(II)SQRD	MIN X(I)	MAX X(I)	RANGE X(I)	REL.INF.X(I)
0		6.24054D+01							
1	C3A	1.55110D+00	7.45D-01	2.1	0.9740	1.000D+00	2.100D+01	2.000D+01	0.71
2	C3S	5.10168D-01	7.24D-01	0.7	0.9961	2.600D+01	7.100D+01	4.500D+01	0.53
3	CAF	1.01990D-01	7.55D-01	0.1	0.9787	4.000D+00	2.300D+01	1.900D+01	0.04
4	C2S	-1.44061D-01	7.05D-01	0.2	0.9965	6.000D+00	6.000D+01	5.400D+01	0.18

NO. OF OBSERVATIONS   13
NO. OF IND. VARIABLES   4
RESIDUAL DEGREES OF FREEDOM   8
F-VALUE   111.5
RESIDUAL ROOT MEAN SQUARE   2.446C075E
RESIDUAL MEAN SQUARE   5.98295492
RESIDUAL SUM OF SQUARES   47.86363935
TOTAL SUM OF SQUARES   2715.76307652
MULT. CORREL. COEF. SQUARED   .9824

IDENT.	OBSV.	WSS DISTANCE	OBS. Y	FITTED Y	RESIDUAL		OBSV.	OBS. Y	FITTED Y	ORDERED RESID.	STUD.RESID.	SEQ
		ORDERED BY COMPUTER INPUT						ORDERED BY RESIDUALS				
C 22	1	25.	78.530	78.495	0.005		6	105.200	105.275	3.925	1.7	1
E 23	2	34.	74.300	72.789	1.511		11	83.800	81.809	1.991	1.1	2
M 52	3	8.	104.333	105.971	-1.671		2	74.300	72.789	1.511	0.8	3
E 88	4	19.	87.600	89.327	-1.727		9	87.600	91.722	1.378	0.7	4
N 96	5	1.	95.900	95.649	0.251		12	113.300	112.327	0.973	0.5	5
T 65	6	7.	109.203	105.275	3.925		10	115.900	115.618	0.282	0.2	6
94	7	33.	102.700	104.149	-1.449		5	95.900	95.649	0.251	0.1	7
S 24	8	30.	72.503	75.675	-3.175		1	105.200	78.495	0.005	0.0	8
A 89	9	14.	93.100	91.722	1.378		7	102.700	104.149	-1.449	-0.7	9
M 90	10	74.	115.900	115.618	0.282		3	104.300	105.971	-1.671	-1.1	10
D 25	11	20.	83.800	81.809	1.991		4	87.600	89.327	-1.727	-0.8	11
L 55	12	20.	113.300	112.327	0.973		13	109.400	111.694	-2.294	-1.1	12
E 91	13	21.	109.400	111.694	-2.294		8	72.500	75.675	-3.175	-1.7	13

Figure 6A.2

107

LINEAR LEAST-SQUARES CURVE FITTING PROGRAM

4 VARIABLES PASS 2 DEP VAR 1: HEAT

MIN Y = 7.250D+01   MAX Y = 1.015D+02   RANGE Y = 4.340D+01

Y = E(0) + E(1)X1 + E(2)X2
Y = CUMULATIVE HEAT OF HARDENING AFTER 180 DAYS, CALORIES/GRAM OF CEMENT
X1 = PCT. TRICALCIUM ALUMINATE, C3A
X2 = PCT. TRICALCIUM SILICATE, C3S
NOTE: CLINKER COMPOSITION CALCULATED FROM ANALYSIS OF OXIDES.

IND.VAR(I)	NAME	CCEF.B(I)	S.E. COEF.	T-VALUE	R(I)SQRD	MIN X(I)	MAX X(I)	RANGE X(I)	REL.INF.X(I)
0		5.25973D+01							
1	C3A	1.46831D+00	1.21D-01	12.1	0.0522	1.000D+00	2.100D+01	2.000D+01	0.68
2	C3S	6.62250C-01	4.59D-02	14.4	0.0522	2.600D+01	7.100D+01	4.500D+01	0.69

NO. OF OBSERVATIONS   13
NO. OF IND. VARIABLES   2
RESIDUAL DEGREES OF FREEDOM   10
F-VALUE   229.5
RESIDUAL ROOT MEAN SQUARE   2.40633504
RESIDUAL MEAN SQUARE   5.79044832
RESIDUAL SUM OF SQUARES   57.90448218
TOTAL SUM OF SQUARES   2715.76307692
MULT. CORREL. COEF. SQUARED   .9767

-----------ORDERED BY COMPUTER INPUT-----------

IDENT.	OBSV.	WSS DISTANCE	OBS. Y	FITTED Y	RESIDUAL
C 22	1	37.	78.500	80.074	-1.574
E 23	2	43.	74.300	73.251	1.049
M 92	3	9.	104.300	105.815	-1.515
E 88	4	27.	87.600	89.258	-1.658
N 56	5	1.	95.900	97.293	-1.393
T 85	6	8.	109.200	105.152	4.048
94	7	47.	102.700	104.002	-1.302
S 24	8	38.	72.500	74.575	-2.075
A 89	9	14.	93.100	91.275	1.825
M 50	10	68.	115.900	114.538	1.362
P 25	11	21.	83.800	80.536	3.264
L 55	12	29.	113.300	112.437	0.863
E 91	13	32.	109.400	112.293	-2.893

-----------ORDERED BY RESIDUALS-----------

OBSV.	OBS. Y	FITTED Y	ORDERED RESID.	STUD.RESID.	SEQ
6	105.200	105.152	4.048	1.8	1
11	83.800	80.536	3.264	1.5	2
9	93.100	91.275	1.825	0.8	3
10	115.900	114.538	1.362	0.8	4
2	74.300	73.251	1.049	0.5	5
12	113.300	112.437	0.863	0.4	6
7	102.700	104.002	-1.302	-0.7	7
5	95.900	97.293	-1.393	-0.6	8
3	104.300	105.815	-1.515	-0.7	9
1	78.500	80.074	-1.574	-0.8	10
4	87.600	89.258	-1.658	-0.8	11
8	72.500	74.575	-2.075	-1.0	12
13	109.400	112.293	-2.893	-1.4	13

Figure 6A.3

LINEAR LEAST-SQUARES CURVE-FITTING PROGRAM

6 VARIABLES PASS 1

DATA INPUT   6 INDEPENDENT VARIABLES   1 DEPENDENT VARIABLE(S)

OBSV.	SEQ.	1-11-21	2-12-22	3-13-23	4-14-24	5-15-25	6-16-26	7-17-27	8-18-28	9-19-29	10-20-30
1	0	2.800	4.680	4.870	-1.000	8.400	4.916	6.750			
2	0	1.400	5.190	4.500	-1.000	6.500	4.563	13.000			
3	0	1.400	4.820	4.730	-1.000	7.900	5.321	14.750			
4	0	3.300	4.850	4.760	-1.000	8.300	4.865	12.600			
5	0	1.700	4.860	4.950	-1.000	8.400	3.776	8.250			
6	0	2.900	5.160	4.450	-1.000	7.400	4.397	10.670			
7	0	3.700	4.820	5.050	-1.000	6.800	4.867	7.280			
8	0	1.700	4.860	4.700	-1.000	8.600	4.828	12.670			
9	0	0.920	4.780	4.840	-1.000	6.700	4.865	12.580			
10	0	0.680	5.160	4.760	-1.000	7.700	4.034	20.600			
11	0	6.000	4.570	4.820	-1.000	7.400	5.450	3.580			
12	0	4.300	4.610	4.650	-1.000	6.700	4.853	7.000			
13	0	0.600	5.070	5.100	-1.000	7.500	4.257	26.200			
14	0	1.800	4.660	5.090	-1.000	8.200	5.144	11.670			
15	0	6.000	5.420	4.410	-1.000	5.800	3.718	7.670			
16	0	4.400	5.010	4.740	-1.000	7.100	4.715	12.250			
17	0	88.000	4.970	4.660	1.000	6.500	4.625	0.760			
18	0	62.000	5.010	4.720	1.000	8.000	4.977	1.350			
19	0	50.000	4.960	4.900	1.000	6.800	4.322	1.440			
20	0	58.000	5.200	4.700	1.000	8.200	5.087	1.600			
21	0	90.000	4.800	4.600	1.000	6.600	5.971	1.100			
22	0	66.000	4.980	4.690	1.000	6.400	4.647	0.850			
23	0	140.000	5.350	4.760	1.000	7.300	5.115	1.200			
24	0	240.000	5.040	4.800	1.000	7.800	5.939	0.560			
25	0	420.000	4.800	4.800	1.000	7.400	5.916	0.720			
26	0	500.000	4.830	4.600	1.000	6.700	5.471	0.470			
27	0	180.000	4.660	4.720	1.000	7.200	4.602	0.330			
28	0	270.000	4.670	4.500	1.000	6.300	5.043	0.260			
29	0	170.000	4.720	4.700	1.000	6.800	5.075	0.760			
30	0	98.000	5.000	5.070	1.000	7.200	4.334	0.800			
31	0	35.000	4.700	4.800	1.000	7.700	5.705	2.000			

DATA TRANSFORMATIONS

POSITION	CODE	OPERATION	CONSTANT LOCATION	OMIT	VARIABLE	NAME
1	2	COMMON LOG	1	0	1	A
2	0	NONE	2	0	2	B
3	0	NONE	3	0	3	C
4	0	NONE	4	0	4	D
5	0	NONE	5	0	5	E
6	0	NONE	6	0	6	F
7	2	COMMON LOG	7	0	7	LOG Y

Figure 6B.1

DATA AFTER TRANSFORMATIONS   THE FITTED EQUATION HAS 6 INDEPENDENT VARIABLES, 1 DEPENDENT VARIABLE(S)

OBSV.	A 1-11-21	B 2-12-22	C 3-13-23	D 4-14-24	E 5-15-25	F 6-16-26	LOG Y 7-17-27	8-18-28	9-19-29	10-20-30
1	4.47158D-01	4.68000D+00	4.87000D+00	-1.00000D+00	8.40000D+00	4.91600D+00	8.29304D-01			
2	1.46128D-01	5.15000D+00	4.50000D+00	-1.00000D+00	6.50000D+00	4.55280D+00	1.11394D+00			
3	1.46128D-01	4.82000D+00	4.73000D+00	-1.00000D+00	7.90000D+00	5.32C80D+00	1.16879D+00			
4	5.18514D-01	4.85000D+00	4.76000D+00	-1.00000D+00	8.30000D+00	4.86500D+00	1.10037D+00			
5	2.30449D-01	4.86000D+00	4.95000D+00	-1.00000D+00	8.40000D+00	3.77560D+00	9.16454D-01			
6	4.62398D-01	5.16000D+00	4.45000D+00	-1.00000D+00	7.40000D+00	4.39720D+00	1.02816D+00			
7	5.68202D-01	4.82CCD+00	5.05000D+00	-1.00000D+00	6.80000D+00	4.86700D+00	8.62310D-01			
8	2.30449D-01	4.86000D+00	4.70000D+00	-1.00000D+00	8.60000D+00	4.82840D+00	1.10278D+00			
9	-3.62122D-02	4.78000D+00	4.84000D+00	-1.30000D+00	6.70000D+00	4.86540D+00	1.09568D+00			
10	-1.67491D-01	5.1CCCD+00	4.76000D+00	-1.00000C+00	7.70000D+00	4.03430D+00	1.31387D+00			
11	7.78151D-01	4.57000D+00	4.82000D+00	-1.00000D+00	7.40000D+00	5.45050D+00	5.53883D-01			
12	6.33468D-01	4.C1CCD+00	4.65000D+00	-1.00000D+00	6.70000D+00	4.85260D+00	8.45098D-01			
13	-2.21849D-01	5.07000D+00	5.10000D+00	-1.00000D+00	7.50000D+00	4.25740D+00	1.41830D+00			
14	2.55273D-01	4.66000D+C0	5.09000D+00	-1.00000D+00	8.20000D+00	5.14280D+C0	1.06707D+00			
15	7.78151D-01	4.82C0D+00	4.41000D+00	-1.00000D+00	7.10000D+00	3.71800D+00	8.84795D-01			
16	6.43453D-01	5.01000D+00	4.74000D+00	-1.00000D+00	5.50000D+00	4.71460D+00	1.08814D+00			
17	1.94448D+C0	4.97000D+30	4.66000D+30	-1.00000D+00	5.50000D+00	4.62550D+00	1.30334D-01			
18	1.79239D+00	5.01000D+00	4.72000D+00	1.00000D+00	6.80000D+00	4.32220D+00	1.58362D-01			
19	1.69897D+00	4.96000D+C0	4.90000D+00	1.00000D+00	8.20000D+00	4.97690D+00	2.04120D-01			
20	1.76343D+00	5.2CCCD+00	4.70000D+00	1.00000D+00	8.20000D+00	5.087A0D+00	4.13927D-02			
21	1.95424D+00	4.80000D+00	4.60000D+00	1.00000D+00	6.CC0D+00	5.97C80D+00	-7.05811D-02			
22	1.81554D+C0	4.58000D+C0	4.90000D+00	1.00000D+00	6.40000D+00	4.64660D+00	-7.91812D-02			
23	2.14613D+00	4.76000D+00	4.76000D+00	1.00000D+00	7.30000D+00	5.11540D+00	-2.51812D-01			
24	2.38021D+00	4.80400D+00	4.80000D+00	1.00000D+00	7.80000D+00	5.93940D+00	-1.42668D-01			
25	2.62325D+00	4.8CCCD+00	4.60000D+00	1.00000D+00	7.40000D+00	5.91630D+00	-1.42668D-01			
26	2.69897D+00	4.83000D+00	4.60000D+00	1.00000D+00	7.C000D+00	5.47090D+00	-3.27902D-01			
27	2.25527D+00	4.66000D+00	4.72000D+00	1.00000D+00	7.20000D+00	4.6C1C0D+00	-4.81486D-01			
28	2.43136D+00	4.67C00D+00	4.67000D+00	1.00000D+00	6.30000D+00	5.04320D+00	-5.85027D-01			
29	2.23045D+00	4.72000D+00	4.72000D+00	1.00000D+00	6.80000D+00	5.07480D+00	-1.19186D-01			
30	1.99123D+00	5.00000D+00	5.07000D+00	1.00000D+00	7.20000D+00	4.33360D+00	-9.69100D-02			
31	1.54407D+00	4.7C000D+C0	4.80000D+00	1.00000D+00	7.70000D+00	5.70510D+00	3.01030D-01			

MEANS OF VARIABLES
1.18343D+00 4.91CCCD+00 4.75613D+00 -3.22581D-02 7.30000D+00 4.88386D+00 4.87498D-01

ROOT MEAN SQUARES OF VARIABLES
9.44940D-01 2.17286C-01 1.76450D-01 1.01600D+00 7.2E526D-01 5.7EC12D-01 6.C7145D-01

SIMPLE CORRELATION COEFFICIENTS, R(I,I PRIME)

1	1.000						
2	-0.058	1.000					
3	-0.217	-0.263	1.000				
4	0.939	0.012	-0.120	1.000			
5	-0.303	-0.138	0.437	-0.234	1.000		
6	0.469	-0.432	0.027	0.405	0.116	1.000	
7	-0.972	0.154	0.189	-0.929	0.320	-0.410	1.000

Figure 6B.1 (continued)

110

Y = E(0) + B(1)X1 + E(2)X2 + E(3)X3 + B(4)X4 + B(5)X5 + B(6)X6
Y = LOG(RATE OF CHANGE CF RUT CEPTH IN INCHES / MILLION WHEEL PASSES)
X1 = A, LOG(VISCOSITY OF ASPHALT)
X2 = F, PER CENT ASPHALT IN SURFACE COURSE
X3 = C, PER CENT ASPHALT IN EASE COURSE
X4 = D, INDICATOR VARIABLE TO SEPARATE TWO SETS OF RUNS
X5 = E, PER CENT FINES IN SURFACE COURSE
X6 = F, PER CENT VOIDS IN SURFACE COURSE

IND.VAR(I)	NAME	COEF.+B(I)	S.E. COEF.	T-VALUE	R(I)SQRD	MIN X(I)	MAX X(I)	RANGE X(I)	REL.INF.X(I)
0	A	-2.64431D+00							
1	A	-5.123C7D-01	7.21D-02	7.0	0.9105	-2.218D-01	2.6990+00	2.9210+00	0.75
2	B	4.97988C-01	1.15D-01	4.3	0.3210	4.570D+00	5.4200+CC	8.5200-01	0.21
3	C	1.C1C54D-01	1.42D-01	0.7	0.3191	4.410D+00	5.100D+00	6.9000-01	0.03
4	D	-1.34394D-01	6.39D-02	2.1	0.8989	-1.0000+00	1.0000+00	2.0000+00	0.13
5	E	1.885E2D-02	3.42C-02	0.6	0.3150	5.8C00+CC	8.6000+CC	2.8000+00	0.03
6	F	1.374F3D-01	4.79D-02	2.9	0.4439	3.718D+00	5.9710+00	2.2530+00	0.15

NO. OF OBSERVATIONS   31
NO. OF IND. VARIABLES   6
RESIDUAL DEGREES OF FREEDOM   24
F-VALUE   140.1
RESIDUAL ROOT MEAN SOLARE   0.1131021C
RESIDUAL MEAN SQUARE   0.0127920E
RESIDUAL SUM OF SQUARES   0.30701002
TOTAL SUM OF SQUARES   11.05E1E356
MULT. COPREL. COEF. SCUFRED   .9722

-----ORDERED BY COMPUTER INPUT-----

IDENT.	OBSV.	WSS DISTANCE	OBS. Y	FITTED Y	RESIDUAL
108	1	14.	0.929	0.918	-0.088
109	2	25.	1.114	1.204	-0.090
110	3	24.	1.159	1.174	-0.305
111	4	11.	1.100	0.996	0.155
112	5	22.	0.916	0.970	-0.053
113	6	14.	1.028	1.016	0.012
114	7	5.	0.862	0.507	-0.044
115	8	20.	1.103	1.093	0.010
116	9	32.	1.110	1.173	-0.074
117	10	41.	1.314	1.327	-0.013
118	11	7.	0.551	0.743	-0.189
119	12	5.	0.845	0.724	0.121
120	13	43.	1.418	1.371	0.047
121	14	20.	1.067	1.056	0.011
122	15	12.	0.885	0.656	0.029
123	16	8.	1.088	0.916	-0.172
124	17	8.	-0.119	-0.072	-0.047
125	18	5.	0.130	0.108	0.022
126	19	5.	0.158	0.121	0.121
127	20	10.	0.204	0.235	-0.030
129	21	10.	0.041	0.016	-0.031
130	22	16.	-0.071	0.001	-0.071
131	23	24.	-0.079	-0.106	-0.023
132	24	33.	-0.143	-0.042	-0.927
133	25	46.	-0.252	-0.297	-0.210
134	26	49.	-0.328	-0.415	0.154
136	27	26.	-0.481	-0.370	-0.111
137	28	35.	-0.565	-0.434	-0.151
138	29	25.	-0.119	-0.272	-0.153
139	30	16.	-0.097	-0.067	-0.030
140	31	6.	0.301	0.184	0.117

---ORDERED BY RESIDUAL---

OBSV.	OBS. Y	FITTED Y	CRDEREC RESID.	STUD.RESID.	SEQ
16	1.088	0.946	0.172	1.6	1
4	1.100	-0.297	0.155	1.5	2
25	-C.143	-0.272	0.154	1.5	3
29	-0.119	-0.272	0.153	1.4	4
19	-0.053	0.158	0.155	1.4	5
12	0.845	0.724	0.121	1.2	6
31	0.645	0.724	0.121	1.2	7
26	-C.328	-0.184	0.117	1.2	8
13	1.093	-0.415	0.088	0.9	9
15	1.418	1.371	0.047	0.5	10
21	0.041	0.856	0.029	0.4	11
18	0.485	0.019	0.022	0.2	12
6	0.130	0.108	0.012	0.2	13
14	1.067	1.056	0.011	0.1	14
3	1.103	1.028	0.011	0.1	15
16	1.169	1.093	0.010	0.1	16
8	1.174	1.327	-0.005	-0.0	17
5	-C.057	3.106	-0.013	-0.1	18
30	-C.067	-C.030	-0.027	-0.3	19
23	0.079	3.106	-0.031	-0.3	20
20	0.204	0.235	-0.031	-0.3	21
7	0.862	-0.077	-0.044	-0.5	22
17	-0.119	0.907	-0.047	-0.6	23
9	-0.071	0.001	-0.047	-0.7	24
1	1.100	0.918	-0.053	-0.7	25
28	0.829	3.918	-0.074	-0.8	26
27	-C.001	1.204	-0.084	-0.9	27
2	1.114	-0.370	-0.090	-0.9	28
24	-0.481	-0.434	-0.111	-1.1	29
11	C.554	-0.189	-0.189	-1.9	30
31	-C.252	-0.042	-0.210	-2.2	31

Figure 6B.2

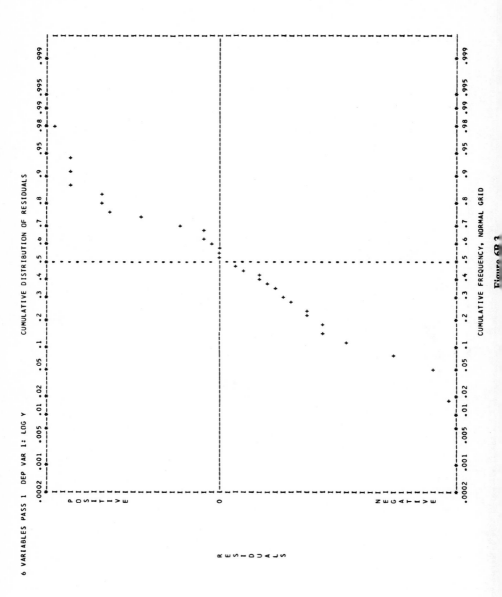

Figure 6B.3

112

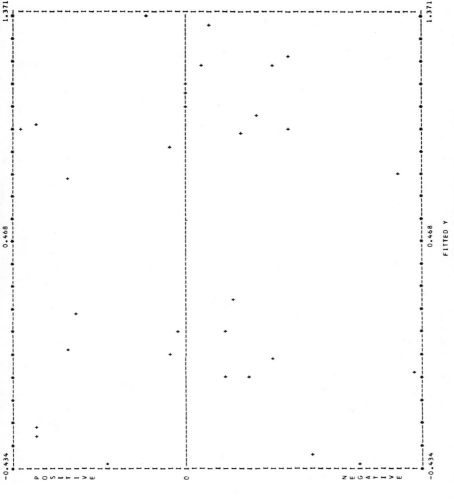

6 VARIABLES PASS 1 DEP VAR 1: LOG Y   RESIDUAL VS. FITTED Y

FITTED Y

Figure 6B.4

LINEAR LEAST-SQUARES CURVE FITTING PROGRAM

6 VARIABLES PASS 2   DEP VAR 1: LOG Y   MIN Y = -5.850D-01   MAX Y = 1.418D 00   RANGE Y = 2.003D 00

Y = B(0) + B(1)X1 + B(2)X2 + B(3)X4 + B(4)X6
Y = LOG(RATE OF CHANGE OF RUT DEPTH IN INCHES / MILLION WHEEL PASSES)
X1 = A, LOG(VISCOSITY OF ASPHALT)
X2 = B, PER CENT ASPHALT IN SURFACE COURSE
X4 = D, INDICATOR VARIABLE TO SEPARATE TWO SETS OF RUNS
X6 = F, PER CENT VOIDS IN SURFACE COURSE

IND.VAR(I)	NAME	COEF.B(I)	S.E. COEF.	T-VALUE	R(I)SQRD	MIN X(I)	MAX X(I)	RANGE X(I)	REL.INF.X(I)
0		-1.85226D 00							
1	A	-5.74751D-01	6.55D-02	8.4	0.8918	-2.218D-01	2.699D 00	2.921D 00	0.80
2	B	4.64864D-01	1.07D-01	4.3	0.2331	4.570D 00	5.420D 00	8.500D-01	0.20
3	D	-1.11184D-01	5.94D-02	1.9	0.8863	-1.000D 00	1.000D 00	2.000D 00	0.11
4	F	1.43649D-01	4.49D-02	3.2	0.3847	3.718D 00	5.971D 00	2.253D 00	0.16

NO. OF OBSERVATIONS   31
NO. OF IND. VARIABLES   4
RESIDUAL DEGREES OF FREEDOM   26
F-VALUE   216.0
RESIDUAL ROOT MEAN SQUARE   0.11148083
RESIDUAL MEAN SQUARE   0.01242798
RESIDUAL SUM OF SQUARES   0.32312739
TOTAL SUM OF SQUARES   11.05875356
MULT. CORREL. COEFF. SQUARED   .9708

COMPONENT EFFECT OF EACH VARIABLE ON EACH OBSERVATION (IN UNITS OF Y)
(VARIABLES ORDERED BY THEIR RELATIVE INFLUENCE --- OBSERVATIONS ORDERED BY INFLUENCE OF MOST INFLUENTIAL VARIABLE)

SEQ.	OBSV.	VARIABLES			
		1 A	2 B	4 F	3 D
1	13	0.77	0.07	-0.09	-0.11
2	10	0.74	0.12	-0.12	-0.11
3	9	0.67	-0.06	-0.00	-0.11
4	2	0.57	0.13	-0.05	-0.11
5	3	0.57	-0.04	0.06	-0.11
6	5	0.52	-0.02	-0.16	-0.11
7	8	0.52	-0.02	-0.01	-0.11
8	14	0.51	-0.12	0.04	-0.11
9	1	0.40	-0.11	-0.00	-0.11
10	6	0.39	-0.12	-0.07	-0.11
11	4	0.36	-0.03	-0.03	-0.11
12	7	0.34	-0.04	-0.00	-0.11
13	12	0.30	-0.14	-0.00	-0.11
14	16	0.30	0.05	-0.02	-0.11
15	11	0.22	-0.16	0.08	-0.11
16	15	0.22	0.24	-0.17	-0.11
17	31	-0.20	-0.10	-0.12	-0.11
18	19	-0.28	0.02	-0.08	-0.11
19	20	-0.32	0.13	-0.03	-0.11
20	18	-0.33	0.05	-0.01	-0.11
21	22	-0.35	0.03	-0.03	-0.11
22	17	-0.42	0.03	-0.04	-0.11
23	21	-0.42	-0.05	0.16	-0.11
24	30	-0.44	0.04	-0.04	-0.11
25	23	-0.53	0.20	0.03	-0.11
26	29	-0.57	-0.09	0.03	-0.11
27	27	-0.59	-0.12	-0.04	-0.11
28	24	-0.66	-0.06	0.15	-0.11
29	28	-0.68	-0.11	0.02	-0.11
30	25	-0.79	-0.05	0.15	-0.11
31	26	-0.83	-0.04	0.08	-0.11

Figure 6B.5

114

LINEAR LEAST-SQUARES CURVE FITTING PROGRAM

6 VARIABLES   PASS 2   DEP VAR 1: LOG Y

IDENT.	OBSV.	WSS DISTANCE	OBS. Y	FITTED Y	RESIDUAL	OBSV.	OBS. Y	FITTED Y	ORDERED RESID.	STUD. RESID.	SEQ
108	1	15.	0.829	0.896	-0.067	25	-0.143	-0.318	0.176	1.8	1
109	2	28.	1.114	1.247	-0.133	16	1.088	0.913	0.175	1.7	2
110	3	27.	1.169	1.184	-0.015	4	1.100	0.929	0.172	1.6	3
111	4	12.	1.100	0.929	0.172	29	-0.119	-0.261	0.142	1.4	4
112	5	25.	0.915	0.934	-0.018	19	0.158	0.033	0.125	1.2	5
113	6	15.	1.028	1.036	-0.008	31	0.301	0.196	0.105	1.1	6
114	7	10.	0.862	0.888	-0.026	12	0.845	0.752	0.093	0.9	7
115	8	23.	1.103	1.086	0.017	26	-0.328	-0.410	0.082	0.8	8
117	9	37.	1.100	1.200	-0.100	13	1.418	1.349	0.069	0.7	9
118	10	47.	1.314	1.329	-0.015	14	1.067	1.024	0.043	0.4	10
119	11	7.	0.554	0.740	-0.186	18	0.130	0.099	0.031	0.3	11
120	12	10.	0.845	0.752	0.093	30	-0.097	-0.107	0.017	0.2	12
121	13	50.	1.418	1.349	0.069	8	1.103	1.086	0.010	0.1	13
122	14	23.	1.067	1.024	0.043	15	0.885	0.887	-0.002	-0.0	14
123	15	12.	0.885	0.887	-0.002	23	0.079	0.083	-0.004	-0.0	15
124	16	6.	1.088	0.913	0.175	6	1.028	1.036	-0.008	-0.1	16
125	17	15.	-0.119	-0.053	-0.066	21	0.041	0.056	-0.014	-0.1	17
126	18	10.	0.130	0.099	0.031	10	1.314	1.329	-0.015	-0.1	18
127	19	8.	0.158	0.033	0.125	3	1.169	1.184	-0.015	-0.1	19
129	20	11.	0.204	0.219	-0.015	20	0.204	0.219	-0.015	-0.1	20
130	21	16.	0.041	0.056	-0.014	5	0.915	0.934	-0.018	-0.2	21
131	22	11.	-0.071	0.023	-0.094	7	0.862	0.888	-0.026	-0.2	22
132	23	27.	0.079	0.083	-0.004	17	-0.119	-0.053	-0.066	-0.6	23
133	24	38.	-0.252	-0.070	-0.181	1	0.829	0.896	-0.067	-0.6	24
134	25	53.	-0.143	-0.318	0.176	22	-0.071	0.023	-0.094	-0.9	25
135	26	57.	-0.328	-0.410	0.082	9	1.100	1.200	-0.100	-1.0	26
136	27	30.	-0.481	-0.371	-0.111	27	-0.481	-0.371	-0.111	-1.1	27
137	28	40.	-0.585	-0.399	-0.186	2	1.114	1.247	-0.133	-1.3	28
138	29	28.	-0.119	-0.261	0.142	24	-0.252	-0.070	-0.181	-1.9	29
139	30	17.	-0.097	-0.107	0.010	28	-0.585	-0.399	-0.186	-1.8	30
140	31	6.	0.301	0.196	0.105	11	0.554	0.740	-0.186	-1.9	31

Figure 6B.5 (*continued*)

115

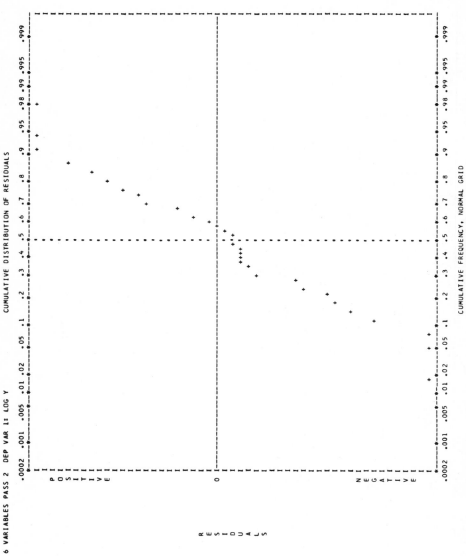

Figure 6B.6

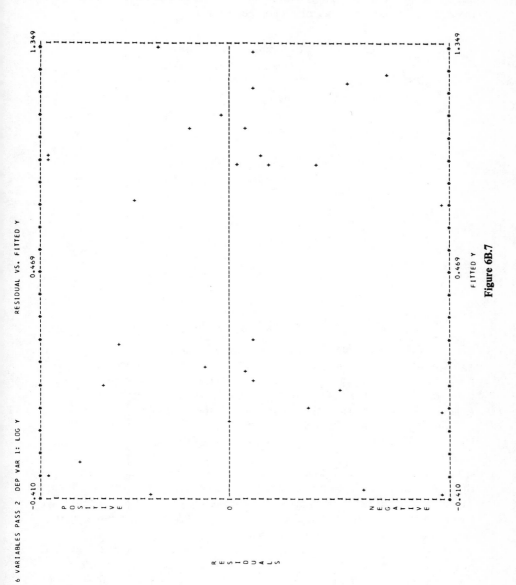

Figure 6B.7

one live letter (underlined in the table) at a time, is given at the left below. At the right the relevant aliases of each effect are listed.

	Equation (1)	Aliases
gen.	ae	A
gen.	be	B
	ab	$AB - CDE$
gen.	ce	C
	ac	$AC - BDE$
	bc	$BC - ADE$
	$abce$	$ABC - DE$
gen.	de	D
	ad	$AD - BCE$
	bd	$BD - ACE$
	$abde$	$ABD - CE$
	cd	$CE - ABE$
	$acde$	$ACD - BE$
	$bcde$	$BCD - AE$
	$abcd$	$ABCD - E$

As a further example, for 12 factors in 128 equations, we generate the 128 C_p-values to be computed from the 7 generators of lower-case letters given at the bottom of Table 6C.1. These are then put through Yate's algorithm to obtain the "effects on C_p." All the largest effects, *when negative* (since we want to consider the equations with the smallest C_p), are inspected for their usefulness. A "half-normal plot" [Daniel, 1959] of the full set of effects and interactions may be useful in choosing the factors with influential separate or combined effects.

It will be necessary to know the two-factor aliases of each effect, since these are the most likely interpretations of any large interaction contrasts. These aliases are determined from the entries in the column of capitalized letters at the right of Table 6C.1. The reader will see, for example, that if the contrast labeled $ABCDEFG$ in terms of the live letters $A, \ldots, G$ came out large, this would mean that H has a large effect. Similarly, if CDE came out large, this would probably mean FM, its two-factor interaction alias; thus factors F and M do not operate additively on C_p. The analyst would then have to put the contrasts measuring F, $M (= CDEF)$ and $FM (= CDE)$ through the *reverse* Yates algorithm [Yates] to determine what combination of f and m gives the lowest C_p. We see that negative interactions are favorable, and that positive interactions indicate that the two factors together do not produce as much decrease in C_p as their individual effects led us to expect.

When it appears that these sets of 2^{K-Q} equations are too large, then, with greater risk, we can try a fractional replicate of "Resolution IV," which inevitably measures *sums* of two or more 2fi. A list of these is given in Table 6C.2, with sympathy for those who must use them. The detailed listing of 2fi aliases for the 2^{16-11} is also given.

TABLE 6C.2
"Resolution IV" Plans for 2^K Equations (K Factors)

K	Q	2^{K-Q} = N	Equation Generators	Key Interactions in Alias Subgroup
4	1	8	ad,bd,cd	ABCD
5	1	16	ae,be,ce,de	-ABCD(E)*
6	2	16	adf,bdf,cd,ef	ABC(D),ABE(F)
7	3	16	adfg,bdf,cdg,efg	ABC(D),ABE(F),ACE(G)
8	4	16	adfg,bdfh,cdgh,efgh	ABC(D),ABE(F),ACE(G), ABG(H)
9-15		32	Drop letters from end of generators below	Drop generators containing letters not used.
16	11	32	adfgklnq, bdfhkmoq, cdghlmpq, efghnopq, jklmnopq	See next page for aliasing of 2fi.
32	26	64	adfg klnq ruwx b'c'e'h' bdfh kmoq suwy b'd'f'h' cdgh lmpq tuxy c'd'g'h' efgh nopq vwxy e'f'g'h' jklm nopq a'b'c'd' e'f'g'h' rstu vwxy a'b'c'd' e'f'g'h'	**

* actually of Resolution V
** letters r to h' are in correspondence with letters a to q:
      ABCD EFGH JKLM NOPQ
      RSTU VWXY A'B'C'D' E'F'G'H'

THE 15 STRINGS OF 2*fi* IN THE SYMMETRICAL 2^{16-11}

AB	*CD*	*EF*	*GH*	*JK*	*LM*	*NO*	*PQ*
AC	*BD*	*EG*	*FH*	*JL*	*KM*	*NP*	*OQ*
AD	*BC*	*EH*	*FG*	*JM*	*KL*	*NQ*	*OP*
AE	*BF*	*CG*	*DH*	*JN*	*KO*	*LP*	*MQ*
AF	*BE*	*CH*	*DG*	*JO*	*KN*	*LQ*	*MP*
AG	*BH*	*CE*	*DF*	*JP*	*KQ*	*LN*	*MO*
AH	*BG*	*CF*	*DE*	*JQ*	*KP*	*LO*	*MN*
AJ	*BK*	*CL*	*DM*	*EN*	*FO*	*GP*	*HQ*
AK	*BJ*	*CM*	*DL*	*EO*	*FN*	*GQ*	*HP*
AL	*BM*	*CJ*	*DK*	*EP*	*FQ*	*GN*	*HO*
AM	*BL*	*CK*	*DJ*	*EQ*	*FP*	*GO*	*HN*
AN	*BO*	*CP*	*DQ*	*EJ*	*FK*	*GL*	*HM*
AO	*BN*	*CQ*	*DP*	*EK*	*FJ*	*GM*	*HL*
AP	*BQ*	*CN*	*DO*	*EL*	*FM*	*GJ*	*HK*
AQ	*BP*	*CO*	*DN*	*EM*	*FL*	*GK*	*HJ*

TABLE 6C.3
"RESOLUTION VII" PLANS FOR 2^K EQUATIONS (K FACTORS)

K	Q	2^{K-Q} = N	Equation Generators	Key Interactions in Alias Subgroup
8	1	128	ah,bh,ch,dh,eh,fh,gh	ABCD EFG(H)
9	1	256	aj,bj,cj,dj,ej,fj, gj,hj	-ABCD EFGH(J)
11	2	512	akl,bkl,ckl,dkl, el,fl,gk,hk,jk	-ABCDEF(L), ABCD GHJ(K)
13	3	1024	almn,blm,clm,dln,eln, fl, gmn, hmn, jm,kn	-ABCDEF(L), -ABCGHJ(M), -ADEGHK(N)

120

CHAPTER 7

Some Consequences of the Disposition of the Data Points

7.1 Introduction

Most of the operations that have been proposed so far are *global* in that they describe the whole set of data without dissection. The statistics t_i, F, R_y^2, C_p, s^2, and even the b_i are global in this sense. Five new suggestions are made that deal with the *interior* analysis of the data. They consider the arrangement of the data points taking into account their relative positions and influences.

A small set of petroleum refining unit data ($N = 36$, $K = 10$) is studied first, both to extend the earlier work and to introduce the new suggestions. Two examples, one from refining and the other from marketing, are then considered ($N = 82$, $K = 4$ and $N = 124$, $K = 11$) each using one of the new techniques to show the various effects of the distribution of the data in x-space. In one example we see unusual "far-out" observations; in the other we detect a difference in level and a difference in slope signaled by the inner and outer observations. Later the new suggestions are applied to some sets used in earlier chapters.

Points taken under extreme conditions have great influence on the co-efficients found and on their apparent correlations. On the other hand, runs made under nearly the same conditions may be used to estimate random error relatively free from function bias. Means are given for spotting remote and neighboring points. In the case examined first, the outlying points show that we have curvature in one direction. For the same case we find twenty-four pairs of near neighbors that give a value just matching the residual root mean square from the fitting equation. There is, then, no evidence of system-atic lack of fit of the equation proposed.

When an equation has been chosen, it is often informative to the research

121

worker or engineer to see in detail the effect of each variable on each observation. In this way the operation of trade-offs and reinforcements can be seen
and studied. A table of component effects is given, which shows in an orderly
way how each variable contributes to the fitted value of each observation.

Component and component-plus-residual plots (defined in Sections 7.3 and
7.4) for each independent variable show how well each term in the equation is
fitting the data. If there are consistent patterns, alternative forms of the equation may be suggested. From these plots we can observe the distribution of
the observations over the range of each independent variable and estimate the
influence of each observation on each component of the equation.

One of the most rigorous tests that can be applied to the "final" fitted equation is cross verification with a second sample of data taken at a different time.
This is important if we expect the fitted equation to have broad application.
As a supplement to the comparison of the global statistics of the second set of
data with the first set, component and component-plus-residual plots are often
revealing. The *component* is retained from the equation fitted with the first
set of data, while the *residual* is calculated from the second set of data. From
these plots we can see which of the observations in the second set of data fit
well and which do not. Perhaps some variable was changed in the second set
of data that was not varied in the first set. In the ten-variable example we see
how the technique can be used to compare the fit of the equations with and
without three "start-up" observations.

7.2 Description of Example with Ten Independent Variables

The data shown in Figure 7A.1, which were first discussed by Gorman
and Toman (1966), were used to derive an interpolation formula for describ-
ing the operation of a petroleum refining unit. The points are given in
chronological order. The month and day are listed in the identification
column, and each response is the result of a week's operation under the
conditions indicated. There are two 2-week gaps (points *7–9* and *15–16*) and
two 3-week gaps (*17–19* and *35–36*). Points numbered *8* and *18* are, for
unknown reasons, missing.

The data have been used by many authors for many purposes [Hocking and
Leslie (1967), La Motte and Hocking (1970), Hoerl and Kennard (1970),
Lindley and Smith (1972), Mallows (1973), Mayer and Wilke (1973), McDonald
and Galarneu (1975), and Denby and Mallows (1977)]. All have assumed this
to be a particularly well-behaved set of data. None of the global techniques
proposed by these authors brought to light the influence of one observation
which is discussed in Section 7.4. We too missed it in the first edition of this
book.

First steps

Technical knowledge of the process indicated that the ten independent variables should be entered linearly without any square or cross-product terms, and that the dependent variable should be entered as a common log function. The results of fitting this full equation are shown in Figures 7A.1–7A.16 are summarized in the first line of Table 7.1.

The plots and standard statistics appear acceptable. However, we note that the largest residual is caused by the first day's observation.

TABLE 7.1
SUMMARY OF COMPUTER PASSES: TEN-VARIABLE EXAMPLE

| | | Conditions | | | Results | |
| | | Variables | | | | |
Pass	Figures	Out	New In	Observa-tions Out	RMS	C_p, Drop Variables
1	7A.1 –7A.16				0.0122	4, 9
2	7A.17–7A.19		x_3^2, point 33		0.0108	7, (9 or point 33)
3	7A.20–7A.21	7, 9	x_3^2, point 33		0.0107	
4	7A.22–7A.24		x_3^2, point 33, 1, 19, 20		0.0090	4, 9, x_3^2
5	7A.25–7A.26	4, 9, x_3^2	point 33, 1, 19, 20		0.0081	
6	7A.27–7A.30		x_3^2, point 33, start, start (3)		0.0086	4, 9, x_3^2, start (3)
7	7A.31–7A.33	4, 9, x_3^2	point 33, start		0.0076	
8	7A.34–7A.36	4, 9, x_3^2, point 33, start		1, 19, 20, 33	0.0081	
9	7A.37–7A.48	4, 9, x_3^2, point 33, start		33	0.0123	

7.3 Interior Analysis I. "Component-Effects" Table

Now that the equation appears to satisfy the *global* requirements, let us make an *interior analysis*. We wish to *expand* the equation, to see in detail what its various additive parts are doing. One way to gain this insight is to display and study the numerical value of each summand of the current fitting equation:

$$Y_j - \bar{y} = \sum_{i=1}^{K} b_i(x_{ij} - \bar{x}_i) = \sum_{i=1}^{K} C_{ij},$$

as it appears at each of the data points.

The individual terms, $C_{ij} = b_i(x_{ij} - \bar{x}_i)$, may be dubbed the *component effects* of x_i on Y_j. In Figure 7A.5 they are ordered by their decreasing overall relative influences in columns and by decreasing effects of the most influential variable in rows.

This ordering is chosen so that the largest effects will be easily visible. The table may be expected to have smaller values in the later columns and in the middle rows. Simple correlations among the x_i become apparent in a more informative way. Thus, in the present example, x_5 is clearly correlated with x_6, and this correlation is important since the effect of x_5 is often nearly as large as that of x_6. On the other hand, the correlation of x_6 with x_1, although formally even larger, is of less importance, as the column of C_{1j} shows. Only in observations *4, 9, 19,* and *20* does variable x_1 have effects greater than 0.1.

7.4   Interior Analysis II.   Component and Component-Plus-Residual Plots

The component-effects table shows the influence of each observation on each component of the fitted equation. However, in larger problems with many observations it is often difficult to see the distribution of observations within each variable. Plots help. The component effect, $b_i(x_{ij} - \bar{x}_i)$, at each observation (indexed by j) for each variable (indexed by i) can readily be seen when plotted against each variable. Adding the residual of each observation to its component effect allows us to gain some insight into patterns in the residuals, and into their dependence on each component of the fitted equation. They are called component and component-plus-residual plots [see Wood (1973)].

Plots somewhat similar to the component-plus-residual portion of the plots have been discussed by M. Ezekiel (1924), C. I. Bliss (1970), and W. A. Larsen and S. J. McCleary (1972). The latter, without the component portion of the plot for reference, have called theirs partial residual plots. However, if there are two or more influential independent variables, the residuals will not be a least-squares fit of any one variable. Without the component for reference, it is impossible to judge exactly where the fitted line would lie. Hence abnormal patterns which sometimes suggest another form of the equation may be missed.

Figures 7A.7–7A.16 are component and component-plus-residual plots of the ten independent variables of Pass 1. The component effect is denoted by a C at each observation. A $+$ symbol denotes the component-plus-residual for each observation. A $*$ symbol is used when the two are indistinguishable. Each is plotted versus its corresponding independent variable. Since all the components of these plots are in the same units and on the same scale of the dependent variable (requested by a 4 in column 40 of the LINWOOD control card), the slopes of the component lines can be compared.

In the plots of x_1 and x_2, Figures 7A.7 and 7A.8, the spread of the residuals about each component line appears to be fairly uniform. Observation 1 with the largest residual is identified to show its location.

In the plot of x_3, Figure 7A.9, we can see the observations that suggest curvature, possibly in the form of a squared term. Excluding observation 1 would make the curvature more pronounced.

In the plot of x_4, Figure 7A.10, we see that observation 1 may well be causing the slight slope. Without this point, x_4 may not be considered an influential variable.

In the plots of x_5, x_6, x_7, and x_8, Figures 7A.11–7A.14, the spread of the residuals about the component line again appears to be fairly uniform. Variable x_6 has the steepest slope and it is the most influential variable.

The component and component-plus-residual plot of x_9, Figure 7A.15, is unexpected. It shows that the importance of this variable may well depend on a single "far-out" observation, point 33. Had our eyesight been better, we would have seen in the component-effects table, Figure 7A.5, that the component effect of this observation was three times that of any other observation. The purpose of the plots is to make the effect of each observation easier to see.

The plot of x_{10}, Figure 7A.16, does not look unusual.

In Pass 2 we add a squared term to x_3 to determine if it is needed and append an indicator variable to observation 33 to see if b_9 is dependent on this single observation.

The LINWOOD program (with a 1 in column 61 of the control card) allows an indicator-variable-observations card to be read containing the observation number and the assigned variable number into which a 1 is placed before transformations.

Although it is not really needed in this problem, we mention here a simple transformation for avoiding high correlation between x and x^2. We use $(x - \bar{x})$ and $(x - d)^2$ as independent variables, where

$$d = \frac{\sum_{j=1}^{N} x_j^2 (x_j - \bar{x})}{2 \sum_{j=1}^{N} (x_j - \bar{x})^2}.$$

This formula is found by requiring the covariation between $(x_j - \bar{x})$ and $(x_j - d)^2$ to be zero. Thus $\sum_{j=1}^{N} (x_j - \bar{x})(x_j - d)^2 = 0$. Solving directly for d

gives the expression shown above. We thank O. Dykstra for pointing this out to us. For the present case, $\bar{x} = 0.54$ and $d = 0.59$, so we use $(x_3 - 0.54)$ as our linear term, and $(x_3 - 0.59)^2$ as our quadratic term.

A summary of the fit of the full equation, Pass 2, is given in Figure 7A.17. It shows no unacceptable $R(I)$ squared values and a slight reduction in the residual mean square. The cumulative frequency plot of the residuals, not shown, is comparable to Figure 7A.3 of Pass 1.

7.5   Search of All 2^K Possible Subset Equations

From the C_p search of all 2^{12} potential equations, we see in Figure 7A.18 that variables 1, 2, 3, 4, 5, 6, 8, 10 and $x_3{}^2$ are in all four of the equations with the smallest C_p's. Variable 7 is omitted in the two smallest C_p equations. Either x_9 can be deleted with the inclusion of an indicator variable for observation *33* (deleting the influence of observation *33* on the rest of the equation), or x_9 can be retained with observation *33* present. The inclusion of x_9 depends on the validity of observation *33*. A decision could perhaps be made if we could review the conditions under which point *33* was taken, or if we could repeat its conditions in the plant. Such a high value of x_9 seems unlikely.

This example shows one reason for questioning the practice of calculating and displaying only the equation at each value of p that has the minimum value of C_p. As in this case, a candidate equation with an inconsequentially higher C_p value would have been missed. Often an equation with a slightly higher C_p value will contain a reasonable combination of variables while the minimum C_p equation does not.

7.6 $t_{K,i}$-directed search

Although a $t_{K,i}$-directed search (described in Chapter 6) is not essential with $K = 12$, we observe that it would have provided us with the same set of candidate equations as the full search, but with less effort. Table 7.2 shows the variables rearranged in the order of their decreasing t-values in the full equation. The ellipsis "Cumulative C_p" means that the equation used to obtain each C_p-value includes all preceding variables as well as the new one added on each line.

In this case (but not always) the turning point in the column of C_p-values tells us correctly which variables should be in the "basic set." A search is then carried out on all combinations of the remaining variables. In less clear-cut cases, we have found that we must include in our basic set one *less* variable or, more rarely, two less, To be conservative, the computer search subroutine has been programmed to include in the basic set two fewer variables than the minimum cumulative C_p requires.

TABLE 7.2
CUMULATIVE p- AND C_p-VALUES

Variables	$t_{K,i}$	Cumulative p	C_p
6	4.6	2	
8	4.0	3	
2	3.4	4	
5	3.3	5	
3	3.2	6	29
x_3^2	2.2	7	23
1	1.8	8	18
4	1.8	9	17
10	1.7	10	12.4
7	1.2	11	12.4
point 33	0.6	12	11.0*
9	0.2	13	13.0

* Minimum value.

In this example, with variables 1, 2, 3, x_3^2, 4, 5, 6, 8, and 10 in the "basic set" of variables (two less than the minimum C_p set), the search of all combinations of the remaining variables (7, point 33, 9) leads to the same lower C_p candidates as did the search of all 12 variables (4096 equations). Thus with the $t_{K,i}$-directed search the C_p-values of only 16 equations had to be calculated.

Pass 3 is the fit of the candidate equation in which x_7 and x_9 are omitted and x_3^2 and an indicator variable for point 33 are included. A summary is given in Figures 7A.20 and 7A.21. The fit is somewhat better than in Pass 1 ($F = 25$ vs. 21; $s^2 = 0.103$ vs. 0.110; $R_y^2 = 0.909$ vs. 0.896). We choose to eliminate variable x_9 because its apparent effect is due to a single observation, point 33, and, as observed earlier, such a high value for x_9 seems unlikely.

7.7   Test for an Outlier

Benken and Draper have examined in considerable detail the influence of the location of the x_i in factor space upon the variance of the residuals. While they write, "In many situations, little is lost by failing to take into account the differences in variances," they suggest that each residual be studentized by dividing by the square root of its variance.

The estimate of the variance of the jth residual is calculated as follows:

$$v_j = \frac{1}{W} + \left[\sum_{i=1}^{K} \sum_{i'=1}^{K} \frac{(x_{ij} - \bar{x}_i)(x_{i'j} - \bar{x}_{i'})(c_{ii'})}{q_i q_{i'}} \right]$$

where $W = \sum_{j=1}^{N} w_j$,   the sum of the weights, w_j, of N observations

$$\bar{x}_i = \frac{\sum_{j=1}^{N} x_{ij} w_j}{W},$$ the weighted mean of each variable

$c_{ii'}$,   the ii' term in the weighted inverse matrix, $(XWX)^{-1}$

$$q_i = \left[\sum_{j=1}^{N} (x_{ij} - \bar{x}_i)^2 w_j \right]^{1/2},$$ the root weighted sum of squares of each variable when $b_0 =$ calculated value

or $$q_i = \left[\sum_{j=1}^{N} (x_{ij})^2 w_j \right]^{1/2},$$ when $b_0 = 0$

$$\text{variance (fitted } Y_j) = v_j \, \text{RMS}$$

$$\text{variance (residual of fitted } Y_j) = \left(\frac{1}{w_j - v_j} \right) \text{RMS}$$

$$\text{variance (residual of new observation } j) = \left(\frac{1}{w_j + v_j} \right) \text{RMS.}$$

In weighted least squares, $w_j =$ weight given, otherwise $w_j = 1$ and $W = N$. *Note*: The variance of the residual of a *new* observation is used in the cross-verification portion of LINWOOD where a second set of data is fitted to the coefficients obtained using the first set of data.

The studentized residual is then calculated as follows:

$$\text{studentized residual} = \frac{\text{residual}}{(\text{variance (residual)})^{1/2}}.$$

In cases where there is *one* outlier, Lund has constructed critical value tables of studentized residuals at alpha equal to 0.10, 0.05, and 0.01 (confidence levels of 90%, 95%, and 99%, respectively) using p (number of parameters) and N (number of observations). These tables are reproduced in Appendix 7D with permission of the author and journal.

In the ten-variable example, observation *1* in Pass 3, Figure 7A.20, has a studentized residual value of 2.0. At the 90% level with $p = 11$ and $N = 36$, the critical value is 2.8, so point *1* cannot be considered an outlier.

7.8   Interior Analysis III.   Weighted-Squared-Standardized-Distance to Look for Far-Out Points of Influential Variables

The residuals, d_j, are interior statistics, but they concern Y and do not provide any insight into the conditions, x, under which the data were taken. We need some way of distinguishing data taken under extreme conditions from the rest. This capability will be useful for spotting points that are especially influential, points that are in error, and perhaps points that indicate curvature.

It is obvious that we should not combine x_i of different dimensions and units directly, in order to find extreme points in factor space. Minimal scaling will require dimensionless transforms of the x-coordinates. The simplest suggestion might be to put

$$X_{ij} = \frac{x_{ij} - \bar{x}_i}{s_i}.$$

where $\bar{x}_i$ is the average of the x_{ij}, and s_i is the root-mean-square of x_i.

We could then compute a "standardized squared distance from mean" for each point, $D_j{}^2$:

$$D_j{}^2 = \sum_{i=1}^{K} X_{ij}{}^2$$

The principal disadvantage of this measure of distance from the X centroid is its equal weighting of coordinates that may differ widely in their importance. Indeed, if several factors have no effect whatever on the response, their contribution to this $D_j{}^2$ may mislead us into believing a point to be outermost when it is not extreme in any important respect.

Wishing to gauge each x_i-coordinate by its apparent effect in moving responses from $\bar{y}$, we might use

$$X_{ij} = b_i(x_{ij} - \bar{x}_i).$$

Since this has the dimensions of y, the corresponding squared distance will need some normalizing factor to make it unit-free. We suggest $s_y{}^2$.

Our "weighted squared standardized distance" (WSSD) for point j is now:

$$\text{WSSD}_j = \sum_{i=1}^{K} \left[\frac{b_i(x_{ij} - \bar{x}_i)}{s_y} \right]^2 = \frac{1}{s_y{}^2} \sum_{i=1}^{K} C_{ij}{}^2$$

This quantity, although it has the disadvantage of its uncertainty (due to the presence of the b_i), does weight the x-coordinates as fairly as we can at each stage.

The WSSD can be used to spot points which may be controlling the global statistics, including the b_i and the R_i^2. To take the simplest pathological case, suppose that $K = 2$, and that *two* points have much larger WSSD's than all the rest, These two points will control the b_i and their large correlation. None of the standard global statistics will detect this, nor will the residuals, which will be *small* for the two outlying points.

If, again, we have about $(K + 1)$ remote points, they may well control all b_i $(i = 1, 2, \ldots, K)$ with small correlations. Looking at the usual statistics will not reveal anything wrong. Indeed, there may not be anything wrong, but if a small minority of points is controlling all the estimated b_i, the data analyst will surely want to know this. It may happen that the data taken on "bad" days, under extreme conditions, do not provide the best equation for predicting more nearly normal operation.

Once the outlying points are spotted, it is easy to carry through an additional pass, to see whether the interior data are telling a story compatible with that told by the outer points.

To return to the data under study, points *19* and *20* are clearly quite far out. In difficult cases we would want to see all K components, C_{ij}, which added together give $(Y_j - \overline{Y})$ and whose squares, summed and divided by s_y^2, give WSSD$_j$. These are all given in Figure 7A.5.

The present situation, however, is simple enough so that inspection suffices.

We note the location of each component value with respect to its maximum and minimum values. Points *19* and *20* have the largest negative components in x_3, x_6, and x_{10} and the largest positive components in x_2, x_5, x_7, and x_8. Thus they are far out in component-space. Their calculated WSSD values (given for Pass 1 in Figure 7A.2 and for Pass 3 in Figure 7A.20) are both large.

To test whether points *1*, *19*, or *20* respond differently from the remainder of the data or just extend the range of the data, in Pass 4 we add an indicator variable (via the indicator-variable-observations card) for each of the three points. A summary of the fit is given in Figure 7A.22. A C_p seach and C_p plot, Figures 7A.23 and 7A.24, indicate that the three observations have a different response level from the others, and that x_3^2 is not needed (apparently *19* and *20* caused the curvature). Without the influence of point *1* on the other variables, x_4 can be deleted and x_7 should be included as an influential variable. Again x_9 can be deleted if point *33* is excluded (with its own indicator variable). A summary of the fit of this candidate equation, Pass 5, is given in Figure 7A.25.

The hypothesis that the additional coefficients in Pass 5 do not noticeably improve the fit of Pass 3 is tested by an F ratio, the mean square due to the extra variables divided by the residual mean square of the equation containing the additional coefficients.

Pass	Coefficients	Degrees of Freedom	Residual Sum of Squares	Residual Mean Square
3	p = 11	25	0.2664	
5	$p + q$ = 13	23	0.1870	0.0081
	q = 2	2	0.0794	0.0397

$$F(2, 23) = \frac{0.0397}{0.0081} = 4.9.$$

From the F-tables, $F(0.975, 2, 23)$ is 4.3. Thus, if we use the 97.5% level as our criterion, the calculated F-ratio value is significant. The data are therefore incompatible with the hypothesis that the additional coefficients do not improve the fit.

Pass 5 provides the following statistics on the coefficients of the indicator variables for points *1*, *19*, and *20*:

Point	Coefficient	Standard Error	t-Value
1	−0.30	0.10	3.2
19	−0.33	0.16	2.0
20	−0.40	0.16	2.6

Points *19* and *20* represent the first two weeks of operation after a three-week shutdown (or at least omission of data). Point *1* may well have suffered from the same start-up problems. Since the three coefficients have similar magnitudes, in Pass 6 we test the hypothesis that a *single* start-up indicator variable can replace the three indicator variables for points *1*, *19*, and *20*. To accomplish this we assign all three observations to the same start-up indicator variable (via the indicator-variable-observations card). Also to test if the start-up conditions require a different coefficient for x_3, we include the interaction term, (start-up)(x_3), in addition to all of the variables used in Pass 2. A summary of Pass 6 is given in Figure 7A.27. This is the fit of the "full" equation. We will consider the results of the C_p search for subset candidate equations in the following section.

7.9 Interior Analysis IV. Variance Ratio to Look for Far-Out Points of Uninfluential Variables

In defining studentized residuals, Section 7.7, both the variance of the fitted Y and the variance of the residual were defined. In the Table of Functions

Related to the Variance of Fitted Y (requested by a 1, 2, or 3 in column 51 of the LINWOOD control card), these functions are given for each observation as well as the ratio of the variance of the fitted Y to the variance of the residual, as suggested by R. D. Cook.

In unweighted least squares,

$$\text{Ratio of } \frac{V(Y)}{V(R)} = \frac{(v)\,\text{RMS}}{(1-v)\,\text{RMS}} = \frac{v}{1-v} \qquad \text{for a fitted observation}$$

$$= \frac{(v)\,\text{RMS}}{(1+v)\,\text{RMS}} = \frac{v}{1+v} \qquad \text{for a new observation.}$$

(See Section 7.7 for definition of v.)

Since the variance of the residual is directly dependent on the variance of the fitted Y, the ratio for a fitted observation emphasizes the value of v as it

v	$V(Y)/V(R)$
0.50	1
0.67	2
0.75	3
0.80	4
0.89	8
0.91	10
0.95	20
0.975	40
0.987	80
0.990	100
1.000	∞

approaches 1. The ratio measures on a more visible scale the relative sensitivity of the estimate of the coefficients to potential far-out observations.

Unlike WSSD which takes into account the influence of each variable, the variance ratio does not. Thus the ratio is useful to spot far-out observations including those far-out in uninfluential variables. These observations can then be checked by deletion or by adding an indicator variable to see if they are controlling and hence making an influential variable appear uninfluential.

As an *exercise* to demonstrate the potential usefulness of the variance ratio, we add an additional (fictitious) observation to the ten-variable data set. Except for x_6, the mean value is used for each variable. For x_6, a value of 16 is used, ten times its mean value. The dependent variable is given a value of 120, the same as that of observation 2. Using a form of equation similar to Pass 1, this single new observation reduces the relative influence of the most influential variable, x_6, to 0.0 and its t-value to 0.0. Even though the WSSD value of this observation is 0, the variance ratio has an extremely high value of 305. The component-plus-residual plot confirms that the new observation is far-out and controlling. As we know from Pass 1, with this observation deleted x_6 is again the most influential of all the variables.

In the functions related to the variance of fitted Y table, Figure 7A.6, we see a hint that observation *33* in Pass 1 is a far-out observation. It has a variance ratio of 6 while the next largest ratio is 1.6.

In Pass 6, Figure 7A.28, we see in the functions table that point *1* has a variance ratio of 66. This is a trivial finding because we are dealing with only three start-up points. Point *1* is far-out from points *19* and *20* and controlling. Without point *1*, the start-up slope of x_3 between points *19* and *20* would have been considerably different than the slope with point *1* present. Had there been several more start-up points, this might have been a valuable signal.

The results of the C_p search and the C_p plot of Pass 6, Figures 7A.29 and 7A.30, indicate that the start-up observations continue to have a different response level (different from b_0), but the (start-up)(x_3) interaction term is not needed. The slope of x_3 is apparently the same during start-up (points *1*, *19*, and *20*) as in normal operations. Again x_3^2, x_4, and x_9 can be deleted and x_7 included.

The fit of this candidate equation is shown as Pass 7 in Figures 7A.31 and 7A.32. Again we test the hypothesis that the additional coefficients do not improve the fit.

Pass	Coefficients	Degrees of Freedom	Residual Sum of Squares	Residual Mean Square
7	p = 11	25	0.1908	
5	$p + q$ = 13	23	0.1870	0.0081
	q = 2	2	0.0038	0.0019

$$F(2, 23) = \frac{0.0019}{0.0081} = 0.2.$$

The F-ratio value is clearly not significant. The data are therefore considered to be compatible with the hypothesis that the additional coefficients do not improve the fit. The one start-up indicator variable effectively replaces the three indicator variables for points *1*, *19*, and *20*.

7.10 Interior Analysis V. Error Estimation from Near Neighbors

Up to this point we have continually depended on some measure of lack of fit for our estimate of random error. We have in fact always assumed that some "full" equation gave us an *over*fit, so that the RMS estimated a variance measuring only random error. But how can we gain some assurance that this is the case? We have just been working on a problem in which a new term, $x_3{}^2$, turned up, which reduced our estimate by about 20%. It is disturbing to reflect that one or more large terms may be lying hidden in the residual sum of squares. The classical requirement for safe inference—a good estimate of random error from randomized replicates—can hardly ever be met by historical data. Must we then admit that no safe inferences can be drawn from any such data? We do not think so.

Some data, although containing no exact replicates, will have some "near replicates," that is, pairs of observations taken far apart in time but under nearly the same x-conditions. It is of little importance if two points differ widely in their levels of an independent variable that has negligible influence. But it is important, if points differ widely in their values for important x_i—those which make a large difference in Y.

As a measure of (squared) distance in "component-effect space" between any two points j and j', we will use:

$$\text{WSSD}_{jj'} = \sum_{i=1}^{K} \left[\frac{b_i(x_{ij} - x_{ij'})}{s_y} \right]^2 = \frac{1}{s_y{}^2} \sum_i (C_{ij} - C_{ij'})^2.$$

The purpose of the divisor $s_y{}^2$ is not to "studentize" the statistic $\text{WSSD}_{jj'}$ (since the s^2 is often an upward-biased estimate of the true variance), but rather to render the distances dimensionless and small. The numerator terms are all correlated random variables and are also somewhat biased when the equation is, as we now suspect, incomplete. It will not be advisable to flood the $\text{WSSD}_{jj'}$ with many uninfluential terms, since these bring with them more random variability and so tend to inflate the sum. For these two partly compensating reasons, we will not set a critical value, say 1, for $\text{WSSD}_{jj'}$, in the hope that lesser values might locate pairs that would yield an estimate of σ^2 of bounded bias.

We will look for pairs of points with smallest $\text{WSSD}_{jj'}$ in order to correct their (presumably small) y-differences, $(y_j - y_{j'})$, by their corresponding equation differences, $(Y_j - Y_{j'})$, to get a single local corrected estimate of random error. Now

$$(y_j - y_{j'}) - (Y_j - Y_{j'}) = (y_j - Y_j) - (y_{j'} - Y_{j'})$$
$$= d_j - d_{j'},$$

where d_j is obviously the residual for point j. For ease in computation we will use the absolute values of these differences in residuals as if they were *ranges* of pairs of observations:

$$|d_j - d_{j'}| \equiv \Delta d_{jj'} \equiv \Delta d.$$

TABLE 7.3
DETAILS OF NEAR-NEIGHBOR CALCULATIONS
(10 Variables, Pass 7)

| Obser-vation | Ordered Fitted Y | Residual | Delta Residuals and the Weighted Standardized Squared Distances of Near Neighbors | | | | | | | |
| | | | Adjacent | | 1 Apart | | 2 Apart | | 3 Apart | |
			Delta	WSSD	Delta	WSSD	Delta	WSSD	Delta	WSSD
20	1.44	−0.04	0.08	51.48	0.01	40.80	0.07	5.96	0.13	39.15
4	1.51	0.03	0.09	1.96(10)	0.01	75.16	0.06	8.34	0.02	34.86
9	1.62	−0.05	0.08	62.01	0.14	4.72	0.07	34.38	0.01	8.71
19	1.69	0.03	0.06	60.27	0.01	73.17	0.07	73.08	0.08	152.75
16	1.80	0.09	0.07	23.05	0.13	1.80(9)	0.02	27.54	0.04	5.42
1	1.80	0.02	0.06	17.64	0.09	27.91	0.11	16.89	0.17	15.48
14	1.87	−0.04	0.15	17.49	0.17	1.37(5)	0.11	1.73(7)	0.06	2.59
38	1.90	0.11	0.02	14.09	0.25	13.06	0.20	14.25	0.01	6.67
13	1.96	0.13	0.28	1.01(1)	0.23	1.96(11)	0.04	12.28	0.16	24.74
12	2.00	−0.15	0.05	3.17	0.23	9.77	0.12	22.83	0.17	7.30
10	2.02	−0.10	0.19	10.03	0.07	19.68	0.12	5.15	0.02	19.21
11	2.03	0.09	0.12	3.45	0.08	5.99	0.18	4.47	0.01	3.76
7	2.04	−0.03	0.05	9.98	0.06	3.47	0.13	5.83	0.06	18.24
2	2.06	0.02	0.10	9.69	0.08	1.50(6)	0.02	6.55	0.13	4.99
36	2.09	−0.08	0.18	8.08	0.12	19.94	0.03	5.91	0.05	17.73
5	2.10	0.10	0.07	5.42	0.22	2.64	0.13	9.29	0.20	2.30(15)
15	2.11	0.04	0.15	9.18	0.07	4.65	0.14	13.14	0.04	9.71
21	2.12	−0.11	0.09	10.07	0.01	2.75	0.19	12.65	0.21	13.34
17	2.13	−0.03	0.07	13.14	0.11	5.86	0.03	24.42	0.07	37.32
6	2.13	−0.10	0.18	8.39	0.10	16.32	0.00	21.46	0.10	8.41
23	2.20	0.08	0.08	31.84	0.18	44.82	0.08	10.43	0.01	6.33
37	2.20	0.00	0.10	6.23	0.00	19.62	0.08	17.20	0.09	16.13
27	2.21	−0.10	0.10	29.87	0.19	26.43	0.01	31.14	0.13	3.96
33	2.21	0.00	0.08	10.17	0.09	7.74	0.03	18.61	0.04	8.36
35	2.22	0.08	0.17	3.39	0.06	12.80	0.13	8.94	0.01	16.27
34	2.23	−0.09	0.12	17.85	0.05	7.09	0.17	22.17	0.06	16.03
29	2.24	−0.03	0.07	22.06	0.05	2.04(13)	0.06	45.00	0.01	3.35
32	2.32	−0.04	0.12	20.84	0.02	15.44	0.06	21.46	0.07	46.39
31	2.34	0.08	0.11	47.10	0.06	2.86	0.05	23.72	0.09	4.35
22	2.34	−0.03	0.05	37.55	0.05	43.87	0.02	31.90	0.10	40.90
24	2.41	0.02	0.01	15.00	0.03	1.97(12)	0.05	1.36(4)	0.12	4.95
3	2.44	0.03	0.04	11.10	0.04	9.91	0.13	9.41	0.04	12.12
30	2.50	−0.01	0.08	1.35(3)	0.09	2.99	0.07	1.79(8)		
25	2.52	0.07	0.17	2.25(14)	0.01	1.21(2)				
28	2.53	−0.10	0.17	5.65						
26	2.60	0.06								

Since the expected value of the range for pairs of independent observations from a normal distribution is 1.128σ, we will average n of our Δd's and multiply by 0.886 ($= 1/1.128$) to get a running estimate, s_n, of σ.

Since the number of distances for N points is $\binom{N}{2}$, which increases roughly as N^2, a *double* sorting is proposed to first screen out points that cannot be neighbors, that is, those with widely separated Y-values. Points with nearly the same Y-values *may* be neighbors, or they may be on or near single contours of constant Y but far apart in some x_i.

Having ordered the data by increasing Y, we now compute the squared distances for the $(N - 1)$ pairs with adjacent Y-values. We then compute the same statistic for pairs separated by one, by two, and by three intervening Y-values. It is for these $(4N - 10)$ points, then, that we compute WSSD.

We now print the $\Delta_f d$ ordered by their increasing WSSD and cumulate these by the formula:

$$s_n = 0.886 \left(\frac{N}{N - p} \right)^{\frac{1}{2}} \frac{\sum\limits_{f=1}^{n} \Delta_f d}{n}$$

where n is a new index going from 1 for the closest pairs to $(4N - 10)$ for the pair with the largest WSSD in our set and where the square root of $(N/(N - p))$ roughly accounts for the degrees of freedom relative to the number of observations. The details of this double sorting are shown in Table 7.3, where the 15 closest pairs are indicated in parentheses, adjacent to their WSSD.

The computer printout, Figure 7A.33, shows the successive estimates of the standard deviation, s_n, at the left. The third column gives the WSSD between the two points identified in the fourth and fifth columns. We see s_n stabilizing at about 0.11 which agrees with s_n found from the best-fitting equation. We conclude that there is no further evidence of lack of fit.

7.11 Cross Verification

One of the most rigorous tests of a fitted equation is cross verification with a second sample taken at another time. The LINWOOD program allows the user to fit the first sample of data, and to retain the fitted coefficients. A second sample of data is then fitted to an equation with the same variables. For comparison, the data are also fitted using the retained coefficients. The component and component-plus-residual plots use the components obtained from the first equation's coefficients and the residuals from the second sample of data. We can then see which observations fit well and which do not. Reexamining the conditions under which the data with the larger residuals were taken may lead to the recognition of variables that were not changed over a sufficient range in the first sample of data.

The tests given below to judge compatibility of two sets of data were proposed by G. C. Chow. The first set of data contains N observations; the second, L observations. Let us consider:

A Sum of squares of $N + L$ residuals from the fit of $N + L$ observations, with $N + L - p$ degrees of freedom.

B Sum of squares of N residuals from the fit of N observations, with $N - p$ degrees of freedom.

C Sum of squares of L residuals from the fit of L observations, with $L - p$ degrees of freedom.

Here $p = K$ variables for $b_0 = 0$ or

$$p = K + 1 \text{ variables for } b_0 = \text{calculated value.}$$

For $L > p$, the ratio of $(A - B - C)/p$ to $(B + C)/(N + L - 2p)$ will be distributed as $F(p, N + L - 2p)$ under the hypothesis that both groups of observations belong to the same fitted equation. For $L \leqslant p$, the ratio of $(A - B)/L$ to $B/(N - p)$ will be distributed as $F(L, N - p)$ under the hypothesis that both groups of observations belong to the same fitted equation.

In the ten-variable example, we would be delighted if additional data could be obtained to check whether or not an observation of x_9 as high as observation *33* was possible and could be duplicated. Observations in the region of points *19* and *20* not associated with start-up would also be most helpful. Unfortunately the process has since been changed and such information cannot now be obtained.

To demonstrate the compatibility test, however, we can fit as in Pass 8, the N observations taken under "normal" operation, and cross verify as in Pass 9 with the $N + L$ points taken under both "normal" and "start-up" conditions. Observation *33* is omitted from both sets of data. The results are given in Figures 7A.34 through 7A.39.

The RSS of A (Pass 9) is 0.31906 and $N + L$, 32 + 3, is *35*. The RSS of B (Pass 8) is 0.18704 and N is 32. The number of parameters, p, in both A and B is 9. The ratio of $(A - B)/L$ to $B/(N - p) = (0.31906 - 0.18704)/3$ to $0.18704/(32 - 9) = 5.4$. From the F tables $F(\alpha, L, N - p) = F(0.01, 3, 23) = 4.76$. Thus the hypothesis that the start-up and normal operation observations belong to the same fitted equation is rejected.

It can be seen in the cumulative frequency plot, Figure 7A.40, that the responses of the three start-up observations are dramatically different from those taken under normal operation.

In the component and component-plus-residual plots of Pass 9, Figures 7A.41 through 7A.48, we are able to see (1) the relative slopes of the component effects of each variable under normal operations (from Pass 8), (2) the scatter of the normal operations residuals about the component lines, and (3) the location and lack of fit of the start-up observations. It is easy to see why these observations were so influential.

7.12 Other Examples of Error Estimation from Near Neighbors

Six-variable example of Chapter 6

Since variables A, B, D, F (x_1, x_2, x_4, x_6) have clear effects, whereas C and E (x_3 and x_5) have small ones (relative influence = 0.03 for each), it makes sense to drop the latter two variables and to look for near neighbors in the resulting four-dimensional influence space.

As in the preceding section, the data are first ordered by their Y-values (see Figure 7B.1 at *right*), and the WSSD are compated for all pairs within four steps of each other. The second ordering starts with the smallest of these $(4N - 10)$ distances. The corresponding cumulative standard deviation, s_n, is listed in the first column of this figure. Again we see that the s_n-values go as low as 0.10, which is quite close to the RRMS of 0.11, and then increase steadily as WSSD becomes greater than 1.

Our conclusion is that the four-variable equation fits well. There is no evidence to the contrary.

Stack loss example of Chapter 5

Inspection of the ordering by fitted Y (the second and third right-hand columns in Figure 7C.2) shows that points *15* to *19* are in a single clump. Points *10* to *14* are in another clump, and *5* to *8* in another. In the third to fifth left-hand columns we see that many of the exact replicates (WSSD = 0.0) were taken on successive days: points *7* and *8*, *10* and *11*, *15* and *16*, and *17* and *18*. Our s_n, then, is reflecting the "autocorrelation" of operations on successive days under unchanged conditions. These pairs all meet our near-neighbor requirement but they are obviously too close together in time to be considered independent.

The only way to look at the actual effects of real changes in operating conditions is to average these adjacent points and to fit the resulting six pieces of data! Figures 7C.3–7C.7 summarize this little exercise. The soundest conclusion is that only x_1 affects y, with coefficient 0.97 (standard error 0.07, based on 4 degrees of freedom), and that the system standard deviation was about 1.5. The thinness of the evidence needs no emphasis, but it should be pointed out that this is the fault of the data. Even though the original table appeared to give 21 three-factor data points, we find that a linear equation in one factor amply fits the *six* different stable conditions listed.

The conclusions that we reported at the end of Chapter 5 are somewhat modified:

1. After removal of the four transitional points we have but *six* really different conditions, with points *5–8*, *10–14*, and *15–19* judged to be local nests with their own error, s_w, of 1.0 (11 degrees of freedom).

2. The six points remaining are well fitted by:

$$Y = -41.2 + 0.97x_1.$$

3. The residual root mean square, s, from this equation is 1.5 (with, alas, 4 degrees of freedom).

We have overdone this study, hoping that readers will understand our purposes and that they will regard only charitably our final reduction of the problem almost to absurdity.

7.13 Additional Examples of Using Component and Component-Plus-Residual Plots

These examples are taken with permission of *Technometrics* from F. S. Wood's paper, Volume 15, Number 4.

Example of the use of plots in choosing the form of equation

In the User's Manual we give an example using the plot options in the LINWOOD program to choose an acceptable form of equation. Data are generated from a known quadratic equation with two independent variables. Both residual plots and component and component-plus-residual plots are used to show the need for the squared term.

Distribution of observations over each independent variable

The distribution of the observations over each x_i has an important influence on the values of the coefficients. Observations far from the center of the data are more influential than those near the center. The component plots of the x_i are useful to observe the distribution of data. If the distribution is very skew, it is important to judge whether or not the responses of the extreme observations are consistent with the remainder of the data.

The use of indicator variables allows us to see whether or not this is so. Each extreme observation is given an indicator variable of 1 to indicate its presence. We will call them the "indicated observations." All other observations will have a value of 0 for the indicator variable.

Let us consider adding the indicator variable x_2 to equation

(7.1) $$Y' = b_0' + b_1'x_1$$

to give

(7.2) $$Y = b_0 + b_1x_1 + b_2x_2 + b_3x_1x_2.$$

The coefficient b_2 represents the offset of the indicated observations from the value of b_0. The coefficient b_3 indicates how much the slope for the indicated observations differs from b_1, the slope for the remainder of the data.

A C_p search is then made to determine if there are subsets of equations that will fit the data as well as equation (7.2). If either or both x_2 and x_1x_2 are needed (influential), the responses for the indicated observations are not compatible with the remainder of the data. They may be found to represent conditions which are unusual or which are taken under some specific set of

TABLE 7.4
DATA FOR FOUR-VARIABLE EXAMPLE

OBSV.	X1	X2	X3	X4	Y
1	55.33	1.72	54.	1.66219	92.19
2	59.13	1.20	53.	1.58399	92.74
3	57.39	1.42	55.	1.61731	91.88
4	56.43	1.78	55.	1.66228	92.80
5	55.98	1.58	54.	1.63195	92.56
6	56.16	2.12	56.	1.68034	92.61
7	54.85	1.17	54.	1.58206	92.33
8	52.83	1.50	58.	1.54998	92.22
9	54.52	0.87	57.	1.55230	91.96
10	54.12	0.88	57.	1.57818	92.17
11	51.72	0.0	56.	1.60401	92.75
12	51.29	0.0	58.	1.59594	92.89
13	53.22	1.31	58.	1.54814	92.79
14	54.76	1.67	58.	1.63134	92.55
15	53.34	1.81	59.	1.60228	92.42
16	54.84	2.87	60.	1.54949	92.43
17	54.03	1.19	60.	1.57841	92.77
18	51.44	0.42	59.	1.61183	92.60
19	53.54	1.35	59.	1.51081	92.30
20	57.88	1.28	62.	1.56443	92.30
21	60.93	1.22	62.	1.53995	92.48
22	59.59	1.13	61.	1.56549	91.61
23	61.42	1.49	62.	1.41330	91.30
24	56.60	2.10	62.	1.54777	91.37
25	59.94	2.29	61.	1.65523	91.25
26	58.30	3.11	62.	1.29994	90.76
27	58.25	3.10	63.	1.19975	90.90
28	55.53	2.88	64.	1.20817	90.43
29	59.79	1.48	62.	1.30621	90.83
30	57.51	0.87	60.	1.29842	92.18
31	62.82	0.88	59.	1.42483	91.73
32	62.57	0.42	59.	1.45056	91.10
33	60.23	0.12	59.	1.54357	91.74
34	65.08	0.10	60.	1.68940	91.46
35	65.58	0.05	59.	1.74695	91.44
36	65.64	0.05	60.	1.74915	91.56
37	65.28	0.42	60.	1.78053	91.90
38	65.03	0.65	59.	1.78104	91.61
39	67.84	0.49	54.	1.72387	92.09
40	73.74	0.0	54.	1.73496	90.64
41	72.66	0.0	55.	1.71566	91.09
42	71.31	3.44	55.	1.60325	90.51
43	72.30	4.02	55.	1.66783	90.24
44	68.81	6.88	55.	1.69836	91.01
45	66.61	2.31	52.	1.77967	91.90
46	63.66	2.99	52.	1.81271	91.92
47	63.85	0.24	50.	1.81485	92.16
48	67.25	0.0	53.	1.72526	91.36
49	67.19	0.0	52.	1.86782	92.16
50	62.34	0.0	48.	2.00677	92.68
51	62.98	0.0	47.	1.95366	92.88
52	69.89	0.0	55.	1.89387	92.59
53	73.13	0.0	57.	1.81651	91.35
54	65.09	1.01	57.	1.45939	90.29
55	64.71	0.61	55.	1.38934	90.71
56	64.05	1.64	57.	1.33945	90.41
57	63.97	2.80	60.	1.42094	90.43
58	70.48	4.64	60.	1.57680	89.87
59	71.11	3.56	60.	1.41229	89.98
60	69.05	2.51	60.	1.54605	90.00
61	71.99	1.28	55.	1.55182	89.66
62	72.03	1.28	56.	1.67390	90.08
63	69.90	2.19	56.	1.67265	90.67
64	72.16	0.51	56.	1.55242	90.59
65	70.97	0.09	55.	1.45728	91.06
66	70.55	0.05	52.	1.26174	90.69
67	69.73	0.05	54.	1.28802	91.11
68	69.93	0.05	55.	1.36399	90.32
69	70.60	0.0	55.	1.42210	90.36
70	75.54	0.0	55.	1.67219	90.57
71	49.14	0.0	40.	2.17140	94.17
72	49.10	0.0	42.	2.31909	94.39
73	44.66	4.99	42.	2.14314	93.42
74	44.64	3.73	44.	2.08081	94.65
75	4.23	10.76	41.	2.17070	97.61
76	5.53	7.99	40.	1.99418	97.08
77	17.11	5.06	47.	1.61437	95.12
78	67.60	1.84	55.	1.64758	91.86
79	64.81	2.24	54.	1.69552	91.61
80	63.13	1.60	52.	1.65118	92.17
81	63.48	3.46	52.	1.48216	91.56
82	62.25	3.56	50.	1.49734	92.16

conditions. Hence the indicator variables should be retained to represent those conditions. If they were not taken under abnormal conditions, either the indicated observations should be omitted as outliers or terms should be included in the equation to allow for curvature. If, on the other hand, these variables are not found to be influential, the original equation is appropriate over the entire range of observations.

Two examples are given.

Four-variable example—distribution of independent variables

Table 7.4 shows data obtained in a process variable study of a refinery unit involving four independent variables. The first three represent feed compositions while the fourth is the log of a combination of process conditions. The dependent variable, y, is the octane number of the product produced. The results of fitting an equation with these four variables are given in Figures 7D.1–7D.10. In the plot of the fitted Y versus the residuals, Figure 7D.3, we note that most of the fitted values are in the refinery unit's design range of 90–94 octanes. However a few values ranged as high as 97 octane. Economically, in large refineries where an increase of one octane number in the final product is worth several million dollars per year, this is a very significant increase and well worth an investigation into the conditions that produced these abnormally high numbers. It was found after a considerable search that all of the samples giving octane numbers above 95 were taken when a specific refinery feed preparation unit was shut down (observations 75–77). The following values were obtained excluding and including observations 75–77:

| | Excluding 75–77 | | 75–77 | All Data, Including 75–77 | |
	Values	Range	Values	Range	Increase in Range (%)
y	89.6–94.6	5	95.1–97.6	8	60
x_1	45–75	30	4–17	71	135
x_2	0–7	7	5–11	11	55
x_3	40–64	24	40–47	24	0
x_4	1.2–2.3	1.1	1.6–2.2	1.1	0

Figures 7D.4 and 7D.5 are component and component-plus-residual plots for x_1 and x_2. Again the $b_i(x_i - \bar{x}_i)$ component is denoted by a C at each observation. A $+$ symbol denotes the component-plus-residual at each observation. A $*$ symbol is used to denote both when they occur in the same position. All are plotted versus each x_i.

It is clear that observations *75–77* greatly extended the range of these variables.

In Pass 2 an indicator variable is used to determine if the responses of observations *75–77* are consistent with the remainder of the data. Figure 7D.8 gives a summary of this fit. The equation used is

$$Y = b_0 + b_1 x_1 + b_2 x_2 + b_3 x_3 + b_4 x_4 + b_5 x_5 + b_6 x_1 x_5 + b_7 x_2 x_5$$

where x_5 is the indicator variable, $x_5 = 1$ for observations *75–77*, $x_5 = 0$ otherwise.

Figure 7D.9 is the printout of the C_p search and Figure 7D.10 is the C_p versus p plot. None of the equations containing any combination of variables x_5, $x_1 x_5$, or $x_2 x_5$ fits the data any better than the equation without them. Therefore, observations *75–77* merely extend the effective range of y, x_1, and x_2.

Since the feed preparation unit that was temporarily shut down for repairs was an important factor in the production of gasoline, facilities were added to make the overall feed stock composition closer to what it had been when the unit was shut down. This expenditure has had a high rate of return.

Eleven-variable example—inner and outer observations

Table 7.5 is a printout of the data from a marketing price-volume study. The observation number is the day of the year in 1970. 124 observations were taken over a period of 6 months. Holiday data were omitted as being non-representative of normal price-volume relationships. Figure 7E.1, Pass 1, gives the statistics for the fit of equation

$$Y = b_0 + b_1 x_1 + \cdots + b_5 x_5 + b_6 x_6 + \cdots + b_9 x_9 + b_{10} x_{10} + b_{11} x_{11}$$

where x_1 through x_5 are indicator variables for months 2 through 6; x_6 through x_9 are indicator variables for days 2 through 5 of each week; x_{10} is the price per gallon of gasoline; and x_{11} is the differential price to competition. Coefficients b_1 through b_5 indicate how much sales differed in months 2 through 6 from month 1. Likewise coefficients b_6 through b_9 indicate how much sales differed in days 2 through 5 from day 1 of each week.

Figure 7E.2 is a component and component-plus-residual plot of x_{10}. There appears to be an artificial barrier with the number of observations piling up between 37 to 38 cents per gallon. On investigation it was found that a special sales situation had occurred at this upper price level.

Figure 7E.3 is a component and component-plus-residual plot of x_{11}. There are a few outer observations that extend the range of x_{11} over threefold, from (0.7 to 3.8) to (-1.5 to 8.5).

In order to examine the effects of both the special sales situation and the outer observations, indicator variables x_{12} and x_{13} are used. x_{12} is given the

TABLE 7.5
Data for Eleven-Variable Example

OBSV	PRICE	PRICE DIFF	VOLUME	OBSV	PRICE	PRICE DIFF	VOLUME	OBSV	PRICE	PRICE DIFF	VOLUME
48	36.48	1.94	2.7412	109	37.58	3.80	2.6551	169	28.16	-1.47	2.8745
50	34.84	1.63	2.8028	111	35.42	2.11	2.7731	170	27.74	1.27	2.5289
52	33.77	1.41	2.8C65	113	34.53	1.53	2.7917	171	27.69	1.33	2.8579
53	33.62	1.31	2.6911	114	34.32	1.54	2.8482	172	27.71	1.38	2.7896
55	32.87	1.09	2.7589	115	34.21	1.58	2.8136	176	37.78	0.98	2.7372
57	32.31	C.98	2.8096	116	34.23	1.65	2.7226	177	37.75	0.72	2.8621
58	31.93	1.13	2.8639	118	33.91	1.52	2.7589	178	37.71	0.74	2.7945
59	31.73	1.17	2.8549	120	33.60	1.58	2.7853	179	37.71	0.74	2.7C76
6C	31.66	1.15	2.69C2	121	33.32	1.53	2.8698	181	37.71	1.20	2.8222
62	31.21	1.03	2.8122	122	33.24	1.63	2.8388	183	37.71	1.72	2.8549
64	30.68	1.25	2.8041	125	33.17	1.58	2.7459	184	37.69	1.80	2.9101
65	30.23	1.23	2.84C7	127	32.76	1.65	2.7875	186	37.69	1.81	2.7528
66	29.87	1.01	2.8293	128	32.32	1.47	2.8028	188	37.26	1.60	2.8293
67	25.84	1.02	2.6875	129	32.00	1.71	2:8871	190	36.60	1.55	2.8109
69	25.46	1.1C	2.84C0	130	31.95	1.61	2.8169	191	36.04	1.43	2.8871
71	29.21	1.34	2.7875	132	31.76	1.89	2.7760	192	35.72	1.76	2.8325
72	28.82	1.25	2.8814	134	31.31	1.62	2.8062	193	35.74	1.84	2.7497
73	28.74	1.37	2.847C	135	31.00	1.55	2.8287	195	34.79	1.84	2.7952
74	28.68	1.32	2.7251	136	3C.94	1.53	2.8982	197	33.43	1.21	2.8267
76	33.38	6.50	2.7474	137	3C.91	1.53	2.8663	198	33.11	1.16	2.9201
78	33.78	5.37	2.7CC7	139	32.75	3.62	2.7945	199	33.09	1.25	2.85C0
79	33.83	3.34	2.7868	141	37.54	8.50	2.7364	200	33.C2	1.22	2.7612
80	33.87	2.39	2.7723	142	34.68	5.72	2.7973	202	32.67	1.46	2.8344
81	33.87	2.02	2.6609	143	31.04	2.33	2.8331	204	31.90	1.29	2.8351
83	33.86	0.81	2.6474	144	3C.67	2.03	2.7738	205	31.76	1.48	2.8802
85	32.49	0.81	2.8149	146	3C.79	1.39	2.8306	206	31.61	1.40	2.8082
86	32.32	0.84	2.8579	148	29.54	1.37	2.8722	207	31.57	1.37	2.7664
87	32.27	0.78	2.8382	149	25.49	1.36	2.9117	209	31.11	1.38	2.8122
92	37.71	0.79	2.7574	151	29.49	1.37	2.7308	211	3C.58	1.98	2.8274
93	37.71	0.77	2.8222	153	29.37	1.33	2.8293	212	3C.95	2.C5	2.8791
94	37.76	0.79	2.7853	155	29.11	1.22	2.85C0	213	29.38	1.74	2.8555
95	37.76	1.53	2.7193	156	29.02	1.27	2.9015	214	29.28	1.67	2.7474
97	37.78	1.76	2.7825	157	29.0C	1.31	2.6470	216	28.98	1.28	2.8293
99	37.75	1.94	2.7627	158	28.97	1.32	2.7551	218	28.50	0.77	2.8195
100	37.75	2.03	2.85C6	160	28.57	1.20	2.8312	219	28.52	1.79	2.8987
101	37.75	1.94	2.8CC7	162	28.19	1.05	2.8274	220	28.39	1.66	2.8075
102	37.67	2.50	2.6875	163	28.00	1.05	2.5079	221	28.35	1.60	2.7566
104	37.75	2.74	2.7789	164	27.93	1.04	2.8457	223	28.02	1.5C	2.8162
106	37.75	3.19	2.7582	165	27.89	1.00	2.7664	225	27.71	1.48	2.7938
107	37.67	3.68	2.8109	167	27.79	-0.10	2.8525	226	27.48	1.46	2.8865
108	37.61	3.68	2.8035					227	27.32	1.31	2.8401
								228	27.31	1.31	2.7782

value of 1 for each day the special sales situation occurred, 0 otherwise; x_{13} is given the value of 1 for each of the outer 6 observations of x_{11}, 0 otherwise. The fitting equation of Pass 2 is

$$Y = b_0 + b_1 x_1 + \cdots + b_9 x_9 + b_{10} x_{10} + b_{11} x_{11}$$
$$+ b_{12} x_{10} x_{12} + b_{13} x_{11} x_{12} + b_{14} x_{11} x_{13}.$$

Coefficients b_{12} and b_{13} measure how much the slopes from fitting the observations taken under special sales conditions differ from coefficients b_{10} and b_{11} for the remainder of the data. Coefficient b_{14} in turn measures how much the slope from fitting the outer x_{11} observations differ from coefficient b_{11}. Figure 7E.4 gives the statistics of this fit.

A C_p search is made of all the variables to determine not only the influence of $x_{10}x_{12}$, $x_{11}x_{12}$, and $x_{11}x_{13}$, but also the importance of the various month and day indicator variables. Had we wished to keep all of the month and day variables, as well as x_{10} and x_{11}, in the basic equation (in order to search only for the influence of the three new variables), a fractional C_p search could have been made (via a 1 in column 50 of the control card and with the numbers 12, 13, and 14 on a fractional C_p search card). From an abbreviated printout of the searches, Figure 7E.5, and the C_p versus p plot, Figure 7E.6, we see that variables $x_{10}x_{12}$ and $x_{11}x_{13}$ are not influential. However, $x_{11}x_{12}$ is influential and is included in the basic set of variables in the C_p search. Thus the effect of the price differential to competition is different during the special sales situations. The equations progressively deleting $x_{10}x_{12}$, $x_{11}x_{13}$, and x_7 (day 3) fall on the slope 2, constant residual-sum-of-squares line. From this we conclude that, with the remaining variables present, x_7 is not influential. (The level of day 3 sales is not different from day 1 sales.) The three equations containing the remaining variables, deleting x_1 (month 2), deleting x_5 (month 6), and deleting both x_1 and x_5, give the lower C_p values.

Although deleting x_1 gives the equation with the smallest C_p value, there is seldom one "best" equation. The selection of an equation should depend on several considerations: the stability of the coefficients when other variables are deleted, the overall fit of the residuals, and the purpose of fitting the equation.

Separate fits of each of these three candidate equations reveal that the remaining coefficients are relatively stable as x_1 and x_5 are deleted (see Passes 3, 4, and 5, Table 7.6). Nothing unusual is observed in the plots of the residuals. Thus the choice depends on the purpose of fitting the equation. The C_p statistic is a measure of total squared error (bias plus random). As all three equations have essentially the same C_p value, retaining x_1 and x_5 may reduce bias but at the expense of increasing the variance of prediction. If the equation, is needed for interpolation in the region of factor space spanned by the data

TABLE 7.6

STATISTICS FOR CANDIDATE EQUATIONS: ELEVEN-VARIABLE EXAMPLE

TERM	PASS 3			PASS 4			PASS 5		
	COEF.	S.E.COEF.	T	COEF.	S.E.COEF.	T	COEF.	S.E.COEF.	T
CONSTANT	2.993	-	-	2.983	-	-	3.006	-	-
x1	0.009	0.007	1.3	-	-	-	-	-	-
x2	0.038	0.006	6.2	0.035	0.006	6.1	0.032	0.006	5.9
x3	0.024	0.007	3.6	0.021	0.006	3.4	0.017	0.006	3.0
x4	0.049	0.006	7.8	0.045	0.005	8.2	0.044	0.005	8.0
x5	0.014	0.008	1.7	0.012	0.008	1.6	-	-	-
x6	-0.072	0.005	14.0	-0.072	0.005	13.9	-0.072	0.005	13.8
x7	-	-	-	-	-	-	-	-	-
x8	0.066	0.005	12.9	0.067	0.005	12.5	0.067	0.005	12.8
x9	0.022	0.005	4.2	0.022	0.005	4.2	0.022	0.005	4.2
x10	-0.0058	0.0009	6.5	-0.0054	0.0008	6.5	-0.0060	0.0007	8.4
x11	-0.015	0.002	6.6	-0.015	0.002	7.0	-0.015	0.002	7.0
x11x12	0.007	0.003	2.9	0.008	0.003	3.2	0.009	0.003	3.5
NO. OF OBSERVATIONS	124			124			124		
NO. OF IND. VARIABLES	11			10			9		
RESIDUAL DEG. OF FREEDOM	112			113			114		
F-VALUE	72			79			86		
RESIDUAL ROOT MEAN SQR.	0.02100			0.02105			0.02119		
RESIDUAL MEAN SQUARE	0.00044			0.00044			0.00045		
RESIDUAL SUM OF SQUARES	0.04937			0.05008			0.05117		
MULT. CORREL. COEF.SQRD	0.8762			0.8744			0.8717		
P	12			11			10		
CP	9.2			8.8			9.2		

it may pay to delete these variables and accept some bias in order to get a lower average error of prediction.

Since the purpose of fitting the equation in this case is to summarize the data with minimal bias, we select the equation which retains both x_1 and x_5, Pass 3. The results of fitting this equation are given in Figures 7E.7 to 7E.11. The residual root mean square error of the fit in logs (0.02100) corresponds to slightly less than 5% of the volume sold per day. Figure 7E.11 is the resulting component and component-plus-residual plot of x_{11}. The spread of the residuals about the component line is much smaller than in Figure 7E.3.

From this study of the component and component-plus-residual plots we have found (1) that the effect of the differential price to competition changes depending on whether or not special sales conditions exist and (2) that the outer data are similar in their response and extend the effective range of the differential price to competition over threefold, from 3 to 10 cents per gallon. This is useful marketing information.

7.14 Other Conditions Identified by Component and Component-Plus-Residual Plots

Nested data are suggested by the appearance of a number of observations at each of several levels of an independent variable. The sequence of the observations will sometimes tell whether or not the observations are nested. Hard-to-change variables are often held constant while easy-to-change variables are moved. In Chapter 8 we deal with a set of data in which several observations on one variable are made on different samples of crude oil. The characteristics of the crudes in turn are measured by three other variables. The recognition of nesting and the appropriate analysis are discussed in detail.

If the observations are made at two or more discrete levels, the plots may signal that the variable is not continuous and cannot be either interpolated or extrapolated (such as men and women, and sweet and sour foods). Occasionally such grouping may indicate that additional effort is needed to obtain information under intermediate conditions. For example, observations were obtained in the Men-In-Space program both inside and outside the capsule. After viewing the component and component-plus-residual plots, valuable information was obtained by conducting additional experiments in the doorway of the capsule.

7.15 Summary

1. A component-effects table showing the effect of each variable on each observation is presented. Its columns list the variables in order of their decreasing relative influences. Its rows are ordered by decreasing values of the effect of the most influential variable. This table makes explicit the many cases of combined effects and of compensating effects present in most sets of historical data.

2. Component and component-plus-residual plots versus each x_i allow us to see both the distribution of the observations over the range of each independent variable, and the influence of each observation on each component of the equation.

3. In cases where there is *one* suspected outlier, critical value tables for studentized residuals are given that allow us to test whether or not the suspected observation is compatible with the equation being fitted.

4. A measure of the *relevant* squared distance of data points from their centroid, WSSD$_j$, is introduced. It allows for the apparent differences in the influences of each x_i on Y. This measure is useful in looking for controlling far-out points of influential variables.

5. The ratio of the variance of fitted Y to the variance of the residual is used to spot controlling far-out observations of uninfluential variables. These observations may be erroneously causing influential variables to appear uninfluential.

6. A measure of "distance between points in influence space," WSSD$_{jj'}$, is used first to find near replicates and then to find less biased estimates of the error standard deviation. These estimates are used to uncover evidence of lack of fit.

7. Cross verification provides a rigorous test of the "final" fitted equation with a second sample of data taken at a different time. Component and component-plus-residual plots of these data are instructive. The component

is retained from the equation fitted with the first set of data, while the residuals are calculated from the second set of data. Observations that do not fit the earlier equation will deserve further study.

8. In the ten-variable example, using component and component-plus-residual plots we see (1) that the slope of x_4 is increased while that of x_7 is decreased by observation 1, (2) that x_3^2 may be needed, and (3) that the slope of x_9 may well depend on a single far-out observation, point 33.

9. All 2^{12} C_p-values are computed. Only those equations with C_p-values less than the full equation are plotted. With point 1 present, x_7 can be deleted. With an indicator variable for far-out point 33, x_9 can be deleted. x_3^2 is needed.

10. The use of the $t_{K,i}$ from the full equation to spot likely subsets of good equations is demonstrated in detail.

11. Using the critical value tables for studentized residuals, observation 1 is not found to be an outlier, so it is retained.

12. The weighted standardized squared distance (WSSD) is used to spot two points (19 and 20), far-out in influence space. They are observed to have been taken after a three-week shutdown of the unit (or at least omission of data). An indicator variable is used for each start-up point, 1, 19, and 20, to determine whether or not its individual response is the same as, or different from, the remainder of the data. Without the three points, the x_3^2 and x_4 terms are no longer needed, but x_7 is.

13. To test the hypothesis that points 1, 19, and 20 all suffered from similar start-up problems, a single indicator variable (entered alone and as a cross product term with x_3) is used to determine if the three points respond differently in either level or slope of x_3 from those taken under normal operating conditions. A C_p search suggests that their effects are the same. A separate start-up coefficient for x_3 is not needed.

14. A single fictitious observation, far-out in x_6 space, is used to transform this variable from one having the greatest relative influence to one with none. The ratio of the variance of fitted Y to the variance of the residual indicates that this is a far-out observation. A component and component-plus-residual plot confirms the fact and suggests that the fictitious observation may be controlling.

15. When we estimate the standard deviation of y from near neighbors, we find no evidence of lack of fit of our final equation.

16. Cross verification is used to show (1) the fit of the data taken under normal operating conditions and (2) the fit of the data taken under both normal and start-up conditions using the equation from the first set of data. Component and component-plus-residual plots show the location of the start-up observations and the magnitudes of their residuals compared to observations taken under normal conditions.

17. Acceptably close neighbors are found in the six-variable data of

Chapter 6. They give the same standard deviation as the fitting equation, which is therefore judged to fit well.

18. The stack loss data of Chapter 5 are next studied. A smaller standard deviation is found from near neighbors, but this is due to adjacency in time of the neighboring points. We conclude that there are really only six different stable conditions in these data, which a single variable fits admirably.

19. Component and component-plus-residual plots versus each independent variable have provided insight into (1) the appropriate form of the equation, (2) the distribution of observations over the range of each independent variable, and (3) the influence of each observation on each component of the equation. With the aid of indicator variables and the C_p search technique, observations far out in x can be tested to determine if their responses are similar to those of the remainder of the data. This has often resulted either in a new form of the equation or in a new appreciation of the limitations and strengths of the fitted equation.

20. A four-variable example using these plots shows that three observations greatly extend the information about two variables and provide a basis for improving both the quality and profitability of the product under study.

21. Component and component-plus-residual plots in an eleven-variable example indicate that an unrecognized interaction might occur during a special sales situation. A C_p search confirms the suspicion and an improved insight into the price-volume relationships results. Far-out observations are found to have responses similar to inner observations, justifying the use of the equation over the extended range.

22. The above examples show that *global* techniques need to be augmented by *interior analysis*: to check the fit and the influence of each observation on each component of the equation, to ascertain the effect of observations far out in influence space, and to estimate error from near neighbors in order to determine the lack of fit of the equation. Without such criteria, the analyst does not know whether he has a poor or a good equation. He does not have a reliable basis for deciding when he should stop fitting.

23. For clarity, each aspect of *interior analysis* has been considered individually. As the techniques become more familiar, they can be applied simultaneously. For example, we could have gone directly from Pass 1 to Pass 4 in the ten-variable example without the intermediate steps. Others may have looked at Pass 1 and gone another route. The order does not seem to be important as long as all possible combinations are considered as in Pass 4 or Pass 6. We then have the advantage of selecting from *all* of the possible candidate equations and hence we avoid seeing only selected branches of the tree, the hazard of stepwise procedures.

APPENDIX 7A

COMPUTER PRINTOUTS OF TEN-VARIABLE EXAMPLE (page 150)

Pass	Figures	Pass	Figures
1	7A.1 –7A.16	6	7A.27–7A.30
2	7A.17–7A.19	7	7A.31–7A.33
3	7A.20–7A.21	8	7A.34–7A.36
4	7A.22–7A.24	9	7A.37–7A.48
5	7A.25–7A.26		

APPENDIX 7B

COMPUTER PRINTOUT OF SIX-VARIABLE EXAMPLE (page 199)

Pass 2   Figure 7B.1

APPENDIX 7C

COMPUTER PRINTOUTS OF STACK LOSS EXAMPLE (page 200)

Pass	Figures
13	7C.1 and 7C.2
28	7C.3–7C.6
30	7C.7

APPENDIX 7D

COMPUTER PRINTOUTS OF FOUR-VARIABLE OCTANE EXAMPLE (page 207)

Pass	Figures
1	7D.1–7D.7
2	7D.8–7D.10

APPENDIX 7E

COMPUTER PRINTOUTS OF ELEVEN-VARIABLE EXAMPLE (page 218)

Pass	Figures
1	7E.1–7E.3
2	7E.4–7E.6
3	7E.7–7E.12

APPENDIX 7F

CRITICAL VALUES FOR STUDENTIZED RESIDUAL TO TEST FOR SINGLE OUTLIER (page 232)

10VARIABLES   PASS 1

DATA INPUT   10 INDEPENDENT VARIABLES   1 DEPENDENT VARIABLE(S)

OBSV.	SEC.	1-11-21	2-12-22	3-13-23	4-14-24	5-15-25	6-16-26	7-17-27	8-18-28	9-19-29	10-20-30
1	1	7.890	79.000	0.570	2.500	0.190	1.422	91.700	5.000	26.000	4.100
	0	66.000									
2	0	8.680	83.000	0.500	1.520	0.190	1.642	87.800	4.490	7.000	2.700
		120.000									
3	0	5.610	82.000	0.680	1.690	0.220	2.310	87.000	8.460	26.000	1.600
	1	293.000									
4	0	14.750	78.000	0.510	1.210	0.180	1.150	87.000	1.770	6.000	4.300
		35.000									
5	0	7.260	86.000	0.510	1.450	0.200	1.661	86.400	5.530	13.000	2.600
	1	160.000									
6	0	6.620	86.000	0.490	1.220	0.200	1.759	87.600	3.610	3.000	0.100
	0	106.000									
7	0	6.250	86.000	0.500	1.200	0.230	1.892	88.900	4.160	6.000	3.700
	0	104.000									
9	0	14.550	77.000	0.550	2.910	0.170	1.031	90.200	2.860	3.300	3.500
	1	37.000									
10	0	9.440	74.000	0.500	1.840	0.200	1.348	86.800	5.130	10.000	5.300
	1	84.000									
11	0	6.540	84.000	0.530	1.430	0.220	1.638	91.800	3.490	10.030	3.900
		132.000									
12	1	8.390	75.000	0.630	1.450	0.190	1.305	92.200	3.450	11.030	4.430
	0	71.000									
13	0	9.680	72.000	0.600	1.590	0.190	1.335	90.200	3.410	12.000	4.100
		123.000									
14	0	10.090	76.000	0.620	1.470	0.160	1.180	90.500	4.140	8.030	6.300
	1	67.000									
15	0	8.600	82.000	0.540	1.640	0.200	1.370	89.900	7.800	26.000	6.600
	0	141.000									
16	0	11.480	77.000	0.650	1.930	0.180	1.031	91.400	4.650	28.000	6.600
	1	77.000									
17	0	8.840	73.000	0.590	1.580	0.200	1.347	91.400	5.030	13.000	11.900
	1	125.000									
19	0	12.060	72.000	0.780	1.970	0.090	0.641	83.200	9.310	4.000	0.200
	0	52.000									
20	1	12.210	65.000	0.840	1.310	0.120	0.618	85.300	6.690	35.000	3.600
	0	25.000									
21	1	6.720	80.300	0.510	2.870	0.220	1.731	87.500	5.120	10.000	5.870
	1	102.000									
22	0	8.650	79.000	0.540	1.460	0.160	1.381	89.800	7.970	75.300	3.800
	0	206.000									
23	0	8.010	76.000	0.510	1.920	0.170	1.466	88.800	3.650	19.000	5.400
	0	190.000									
24	0	5.960	70.000	0.480	1.650	0.230	1.594	90.200	4.170	16.330	7.600
	1	270.000									
25	0	5.860	71.000	0.430	1.960	0.220	2.131	89.100	5.490	6.300	6.030
	0	393.000									
26	0	5.630	71.000	0.430	1.510	0.220	2.204	98.800	4.290	30.000	10.400
	1	458.000									
27	1	6.190	71.000	0.440	1.620	0.260	2.101	90.100	1.530	17.330	13.700
		129.000									

Figure 7A.1

150

```
28  0    5.900   72.000  0.420  2.050  0.240  87.600  7.660  12.000   9.400
         6.570                  2.063
29  0  268.000   74.000  0.440  1.040  0.240  88.500  2.560   9.000  17.900
         5.840                  1.909
30  0  188.000   73.000  0.420  1.780  0.220  87.200  5.480  33.000  10.700
         5.840                  2.017
31  0  310.000   69.000  0.420  1.900  0.230  86.400  2.060  22.300  12.900
         8.230                  2.011
32  1  260.000   72.000  0.460  2.020  0.170  89.400  1.830  50.230  12.500
         8.120                  1.585
33  1  190.000   82.000  0.470  1.760  0.190  90.100  0.650 185.000   6.500
         9.250                  1.706
34  1  164.000   83.000  0.500  1.310  0.190  94.900  3.120  28.000  11.700
         8.310                  1.674
35  1  138.000   82.000  0.470  1.340  0.200  90.300  5.130   4.300   9.600
         7.800                  1.667
36  0  202.000   85.000  0.630  1.860  0.220  92.200  3.160  24.000   8.500
         6.580                  1.962
37  0  102.000   77.000  0.620  1.510  0.240  95.000  1.130  15.000  16.600
         8.970                  1.959
38  1  160.000   76.000  0.620  1.920  0.240  95.500  2.130  22.000   8.800
       101.000                  1.592
```

DATA TRANSFORMATIONS

POSITION	CODE	OPERATION
1		NONE
2		NONE
3		NONE
4		NONE
5		NONE
6		NONE
7		NONE
8		NONE
9		NONE
10		NONE
11	2	COMMON LOG

CONSTANT	LOCATION	OMIT	VARIABLE	NAME
	1	0	1	X1
	2	0	2	X2
	3	0	3	X3
	4	0	4	X4
	5	0	5	X5
	6	0	6	X6
	7	0	7	X7
	8	0	8	X8
	9	0	9	X9
	10	0	10	X10
	11	0	11	LOG Y

SUMS OF VARIABLES
```
2.97370D+02  2.77200D+03  1.94000D+01  6.13900D+01  7.23000D+00  5.78770D+01  3.22490D+03  1.56310D+02  8.24000D+02  2.47300D+02
7.59426D+01
```

MEANS OF VARIABLES
```
8.26028D+00  7.70000D+00  5.38889D-01  1.70528D+00  2.00833D-01  1.60769D+00  8.95806D+01  4.34194D+00  2.28880D+01  6.86940D+00
2.10952D+00
```

ROOT MEAN SQUARES OF VARIABLES
```
2.37164D+00  5.26986D+00  9.87863D-02  4.18306D-01  3.39222D-02  4.13473D-01  2.58853D+00  2.10839D+00  3.13350D+01  4.35384D+00
2.85905D-01
```

SIMPLE CORRELATION COEFFICIENTS, R(I,I FROM F)
```
 1   1.000
 2  -0.041   1.000
 3   0.511  -0.003   1.000
 4   0.118  -0.157   0.003   1.000
 5  -0.708   0.064  -0.550  -0.067   1.000
 6  -0.868   0.095  -0.645  -0.089   0.841   1.000
 7  -0.069   0.241  -0.018  -0.028   0.378   0.134   1.000
 8  -0.000   0.006   0.334   0.080  -0.165  -0.198  -0.476   1.000
 9  -0.044   0.091  -0.078   0.019  -0.135   0.035   0.074  -0.198   1.000
10  -0.355  -0.297  -0.437  -0.093   0.539   0.450  -0.184  -0.458   0.346   1.000
11  -0.812  -0.056  -0.630  -0.102   0.560   0.814   0.040   0.399   0.160   0.434   1.000
```

Figure 7A.1 *(continued)*

LINEAR LEAST-SQUARES CURVE FITTING PROGRAM

1 OVARIABLES PASS 1   DEP VAR 1: LOG Y   MIN Y = 1.398D+00   MAX Y = 2.661D+00   RANGE Y = 1.263D+00

ALL 10 VARIABLES ARE PRESENT.
NO OBSERVATIONS ARE OMITTED.
CHECK COMPONENT-PLUS-RESIDUAL PLOTS.
  SHOULD X3 BE SQUARED?
  IS X9 DEPENDENT ON A SINGLE OBSERVATION?

IND.VAR(I)	NAME	COEF.B(I)	S.E. COEF.	T-VALUE	R(I)SQRD	MIN X(I)	MAX X(I)	RANGE X(I)	RFL.INF.X(I)
0	X1	9.61476D-01	1.75D-02	1.6	0.7971	5.610D+00	1.475D+01	9.140D+00	0.20
1	X2	-2.81065D-02	4.50D-03	2.4	0.3830	6.900D+01	8.800D+01	1.900D+01	0.16
2	X3	-1.05210D-02	2.78D-01	3.6	0.5404	4.200D-01	8.400D-01	4.200D-01	0.33
3	X4	-9.94835D-01	4.65D-02	1.2	0.0988	1.040D+00	2.910D+00	1.870D+00	0.08
4	X5	-5.46460D-02	1.35D-02	2.9	0.8346	9.000D-02	2.600D-01	1.700D-01	0.53
5	X6	-3.95060D+00	1.34D-01	4.1	0.8862	6.180D-01	2.310D+00	1.690D+00	0.53
6	X7	5.44901D-01	1.08D-02	2.6	0.5523	9.320D+01	9.550D+01	2.300D+01	0.27
7	X8	2.76181D-02	1.25D-02	3.8	0.5005	6.500D-01	6.310D+00	8.660D+00	0.33
8	X9	4.80904D-02	6.85D-04	1.3	0.2451	1.000D+00	1.850D+02	1.820D+02	0.13
9	X10	7.57204D-03	6.5CD-03	1.2	0.5740	1.000D-01	1.790D+01	1.790D+01	0.11

NO. OF OBSERVATIONS   36
NO. OF IND. VARIABLES   10
RESIDUAL DEGREES OF FREEDOM   25
F-VALUE   21.7
RESIDUAL ROOT MEAN SQUARE   0.11029816
RESIDUAL MEAN SQUARE   0.01216569
RESIDUAL SUM OF SQUARES   0.30414222
TOTAL SUM OF SQUARES   2.94185656
MULT. CORREL. COEF. SQUARED   .8966

-----ORDERED BY COMPUTER INFLU-----

IDENT.	OBSV.	WSS DISTANCE	OBS. Y	FITTED Y	RESIDUAL
7 14	2.	1.	1.820	2.038	-0.218
7 21	2.	1.	2.079	2.055	0.025
7 28	3.	19.	2.467	2.386	0.081
8 4	4.	10.	1.544	1.575	-0.031
8 11	5.	1.	2.204	2.097	0.107
8 18	6.	2.	2.025	2.125	-0.099
8 25	7.	5.	2.017	2.085	-0.068
9 8	8.	13.	1.568	1.567	0.001
9 15	9.	3.	1.924	1.995	-0.071
9 22	10.	1.	2.121	2.033	0.088
9 29	11.	4.	1.851	1.930	-0.079
10 6	12.	3.	1.490	1.506	-0.016
10 13	13.	14.	1.827	1.810	0.016
10 20	14.	1.	2.149	2.097	0.052
11 10	15.	11.	1.886	1.732	0.154
11 17	16.	1.	2.397	2.069	0.029
12 1	17.	52.	1.716	1.710	0.006
12 8	18.	20.	1.398	1.546	-0.148
12 15	19.	44.	2.003	2.037	-0.029
12 22	20.	2.	2.314	2.314	-0.016
12 29	21.	6.	2.279	2.279	0.155
	22.	6.	2.431	2.416	0.015
	23.	10.	2.591	2.527	0.064
	24.	11.	2.661	2.626	0.025
	25.	12.	2.111	2.280	-0.169
	26.	11.	2.428	2.537	-0.105
	27.	7.	2.274	2.299	-0.015
	28.	7.	2.451	2.596	-0.018
	29.	7.	2.415	2.324	-0.051
	30.	1.	2.279	2.279	0.000
	31.	32.	2.245	2.295	-0.146
	32.	3.	2.103	2.240	0.066
	33.	1.	2.105	2.082	-0.074
	34.	1.	2.009	2.196	-0.008
	35.	1.	2.004	2.204	0.066
	36.	6.	2.004	1.517	0.000

-----ORDERED BY RESIDUALS-----

OBSV.	OBS. Y	FITTED Y	ORDERED RESID.	STUDN.RESID.	SEQ
13	2.050	1.908	0.184	1.8	1
23	2.279	2.120	0.159	1.8	2
16	1.866	1.732	0.154	1.5	3
5	2.204	2.097	0.107	1.1	4
31	2.415	2.324	0.091	0.9	5
11	2.121	2.033	0.088	1.0	6
38	2.405	2.317	0.088	0.9	7
25	2.467	2.386	0.081	1.2	8
9	2.240	2.167	0.066	0.7	9
35	2.591	2.527	0.064	0.7	10
15	2.097	2.052	0.035	0.4	11
26	2.661	2.626	0.029	0.3	12
17	2.079	2.055	0.025	0.2	13
6	2.073	2.055	0.019	0.2	14
33	2.431	2.416	0.015	0.2	15
24	1.710	1.710	0.001	0.1	16
19	1.716	1.710	0.006	0.1	17
32	1.568	1.567	0.001	0.1	18
9	2.279	2.279	0.001	0.0	19
22	2.279	2.279	0.000	0.0	20
29	2.314	2.330	-0.015	-0.2	21
30	2.509	2.527	-0.016	-0.2	22
21	2.314	2.330	-0.016	-0.2	23
2	2.037	2.038	-0.028	-0.3	24
7	1.567	1.567	-0.031	-0.4	25
20	2.017	2.085	-0.069	-0.7	26
10	2.405	2.537	-0.071	-0.7	27
8	2.017	2.085	-0.071	-0.7	28
36	1.099	1.930	-0.074	-0.8	30
21	2.062	2.062	-0.079	-0.7	29
16	1.951	1.930	-0.079	-0.8	30
23	2.009	2.025	-0.079	-0.8	31
28	2.428	2.537	-0.109	-1.2	32
34	2.140	2.205	-0.146	-1.7	33
20	1.358	1.546	-0.148	-1.9	34
27	2.111	2.280	-0.169	-1.7	35
36	1.820	2.038	-0.218	-2.3	36

Figure 7A.2

152

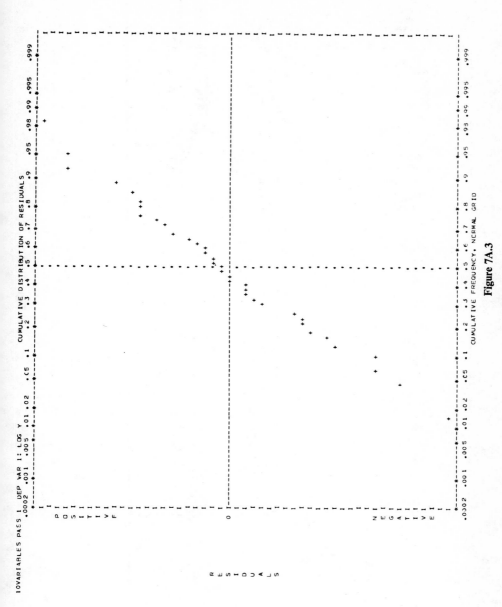

Figure 7A.3

153

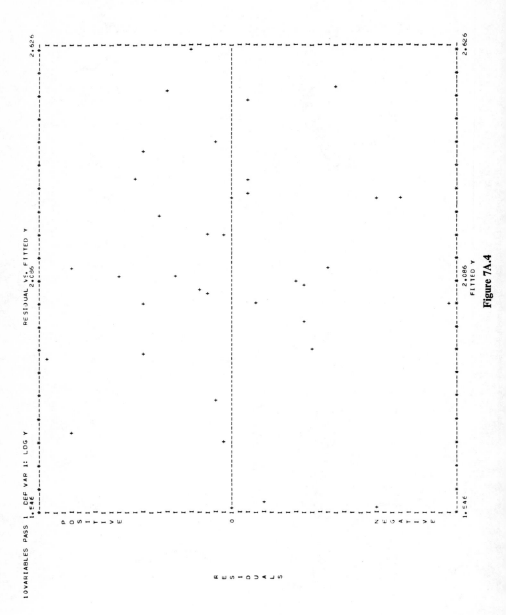

Figure 7A.4

154

10VARIABLES PASS 1   DEP VAR 1: LOG Y

COMPONENT EFFECT OF EACH VARIABLE ON EACH OBSERVATION (IN UNITS OF Y)
(VARIABLES ORDERED BY THEIR RELATIVE INFLUENCE --- OBSERVATIONS ORDERED BY INFLUENCE OF MOST INFLUENTIAL VARIABLE)

SEC. OBSV.   VARIABLES

SEC.	OBSV.	6 X6	5 X5	3 X3	8 X8	7 X7	1 X1	2 X2	9 X9	10 X10	4 X4
1	3	0.38	-0.08	-0.14	0.20	-0.07	0.07	-0.05	0.00	-0.04	0.00
2	26	0.32	-0.08	0.11	-0.00	-0.02	0.07	0.07	0.01	-0.03	0.01
3	25	0.29	-0.12	0.11	0.06	-0.01	0.07	0.07	0.01	-0.01	-0.01
4	27	0.27	-0.23	0.10	-0.14	0.02	0.06	0.07	-0.01	0.03	0.00
5	28	0.25	-0.16	0.12	0.16	-0.06	0.07	0.05	-0.01	0.02	-0.02
6	30	0.22	-0.08	0.12	0.05	-0.07	0.07	0.04	0.01	0.03	-0.00
7	31	0.22	-0.12	0.12	-0.11	-0.09	0.07	0.09	-0.00	0.05	-0.01
8	24	0.21	-0.12	0.06	-0.01	0.02	0.06	0.09	0.01	0.01	0.00
9	36	0.19	-0.08	-0.09	-0.06	0.07	0.01	-0.09	0.00	0.01	-0.01
10	37	0.19	-0.16	-0.08	-0.14	0.15	0.05	0.0	-0.01	0.07	0.01
11	29	0.16	-0.16	0.10	-0.05	-0.03	0.05	0.03	-0.01	0.01	0.04
12	7	0.15	-0.12	0.04	-0.01	-0.02	0.06	-0.12	-0.01	0.07	0.03
13	6	0.10	0.00	0.05	-0.04	-0.06	0.05	-0.05	-0.02	-0.02	0.03
14	21	0.07	-0.08	0.03	0.04	0.01	0.04	-0.03	-0.01	-0.01	-0.06
15	33	0.05	0.04	0.07	-0.18	-0.05	0.00	-0.05	0.14	-0.00	-0.00
16	34	0.04	0.04	0.04	-0.06	0.15	-0.03	-0.05	-0.02	0.04	0.02
17	35	0.03	0.00	0.07	0.04	0.02	-0.00	-0.05	-0.02	0.02	0.02
18	5	0.03	0.00	0.03	0.06	-0.03	-0.03	-0.10	0.01	-0.03	0.01
19	2	0.02	0.04	0.04	0.01	-0.05	-0.01	-0.07	-0.01	-0.03	0.01
20	11	0.02	-0.08	0.01	-0.04	0.06	0.05	-0.08	-0.01	-0.02	0.02
21	38	-0.01	-0.16	0.08	-0.11	0.16	-0.02	0.01	-0.00	0.01	-0.01
22	32	-0.01	0.12	0.08	-0.12	-0.01	0.00	0.05	0.02	0.04	-0.02
23	23	-0.08	0.12	0.03	-0.03	-0.02	0.01	0.01	0.00	-0.01	-0.01
24	1	-0.10	0.04	-0.03	0.03	0.06	0.01	-0.02	0.05	-0.02	-0.04
25	22	-0.12	0.16	-0.00	0.17	0.01	-0.01	-0.02	0.05	-0.02	0.01
26	15	-0.13	0.00	-0.00	0.17	0.01	-0.01	-0.05	0.00	0.02	0.00
27	10	-0.14	0.00	0.04	0.04	-0.02	-0.03	0.03	-0.01	-0.00	-0.01
28	17	-0.14	0.00	-0.06	0.03	-0.05	-0.02	0.04	-0.01	0.04	0.01
29	13	-0.15	0.04	-0.06	-0.04	0.02	-0.04	0.05	-0.01	-0.02	-0.02
30	12	-0.16	0.04	-0.09	-0.04	0.07	-0.00	0.02	-0.01	-0.02	-0.01
31	14	-0.23	0.04	-0.08	-0.01	-0.03	-0.05	-0.01	-0.01	-0.00	-0.04
32	4	-0.25	0.08	0.03	-0.12	-0.07	-0.18	-0.01	-0.02	-0.03	0.03
33	9	-0.31	0.12	-0.01	-0.07	0.02	-0.18	0.0	-0.01	-0.02	-0.07
34	16	-0.31	0.08	-0.11	0.01	0.05	-0.09	0.0	0.00	-0.03	-0.01
35	19	-0.53	0.44	-0.24	0.24	-0.18	-0.11	0.05	-0.02	-0.05	-0.01
36	20	-0.54	0.32	-0.30	0.11	-0.12	-0.11	0.09	0.01	-0.05	0.02

Figure 7A.5

155

IOVARIABLES PASS 1   DEP VAR I: LOG Y

FUNCTIONS RELATED TO THE VARIANCE OF FITTED Y.   OBSERVATIONS ORDERED BY COMPUTER INPUT.

IDENT.	OBSV.	RESIDUAL	STUDENTIZED RESIDUAL	V(Y)/RMS	V(R)/RMS	V(Y)/V(R)
7 14	1	-0.218	-2.3	0.2324	0.7676	0.3
7 21	2	0.025	0.2	0.1375	0.8625	0.2
7 28	3	0.081	1.2	0.6228	0.3772	1.7
8 4	4	-0.031	-0.4	0.5975	0.4025	1.5
8 11	5	0.107	1.1	0.1645	0.8355	0.2
8 18	6	-0.099	-1.1	0.3004	0.6996	0.4
8 25	7	-0.068	-0.7	0.2633	0.7367	0.4
9 8	9	0.001	0.0	0.5386	0.4614	1.2
9 15	10	-0.071	-0.7	0.1299	0.8701	0.1
9 22	11	0.088	0.9	0.2190	0.7810	0.3
9 29	12	-0.079	-0.8	0.2130	0.7870	0.3
10 6	13	0.184	1.8	0.1454	0.8546	0.2
10 13	14	0.016	0.2	0.1056	0.8944	0.1
10 20	15	0.052	0.6	0.2938	0.7062	0.4
11 3	16	0.154	1.5	0.1522	0.8478	0.2
11 10	17	0.029	0.3	0.1668	0.8332	0.2
12 1	19	0.006	0.1	0.4931	0.5069	1.0
12 8	20	-0.148	-1.9	0.4836	0.5164	0.9
12 15	21	-0.028	-0.3	0.4334	0.5666	0.8
12 22	22	-0.016	-0.2	0.3532	0.6468	0.5
12 29	23	0.159	1.6	0.1881	0.8119	0.2
1 5	24	0.015	0.2	0.1796	0.8204	0.2
1 12	25	0.064	0.7	0.2530	0.7470	0.3
1 19	26	0.035	0.4	0.2109	0.7891	0.3
1 26	27	-0.169	-1.7	0.2071	0.7929	0.3
2 2	28	-0.109	-1.2	0.3014	0.6986	0.4
2 9	29	-0.015	-0.2	0.4040	0.5960	0.7
2 16	30	-0.018	-0.2	0.1444	0.8556	0.2
2 23	31	0.091	1.0	0.2843	0.7157	0.4
3 1	32	0.000	0.0	0.3415	0.6585	0.5
3 8	33	0.019	0.5	0.8586	0.1414	6.1
3 15	34	-0.146	-1.7	0.4130	0.5870	0.7
3 22	35	0.066	0.7	0.1811	0.8189	0.2
4 12	36	-0.074	-0.8	0.3093	0.6907	0.4
4 21	37	0.008	0.1	0.3760	0.6240	0.6
4 28	38	0.088	1.0	0.3018	0.6982	0.4

Figure 7A.6

156

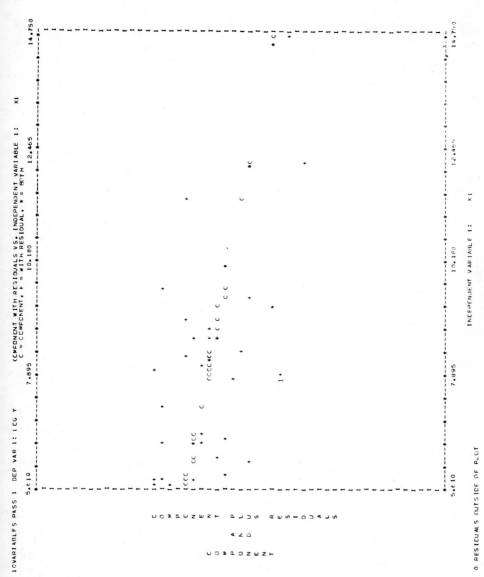

Figure 7A.7

157

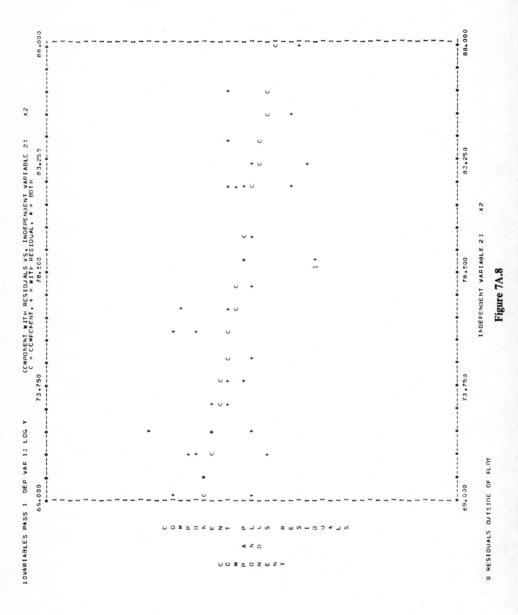

Figure 7A.8

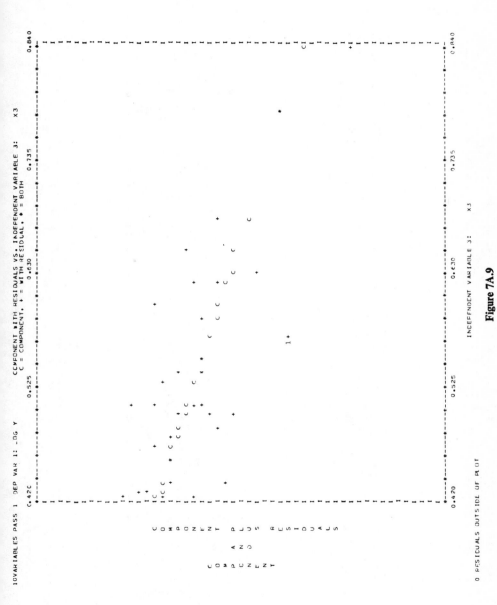

Figure 7A.9

159

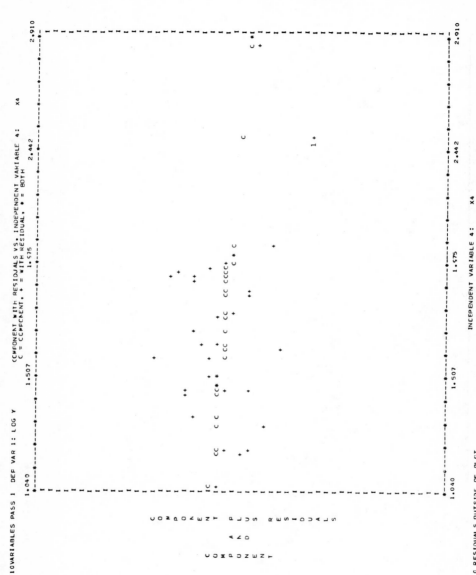

Figure 7A.10

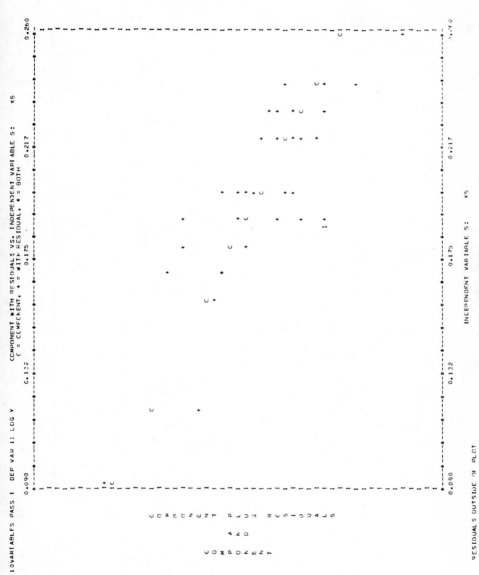

Figure 7A.11

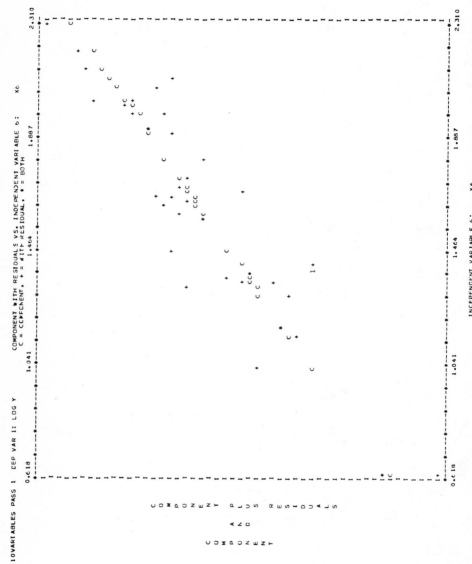

Figure 7A.12

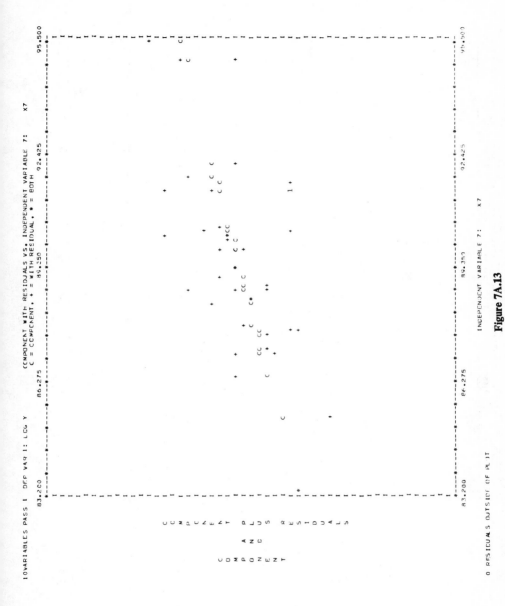

Figure 7A.13

163

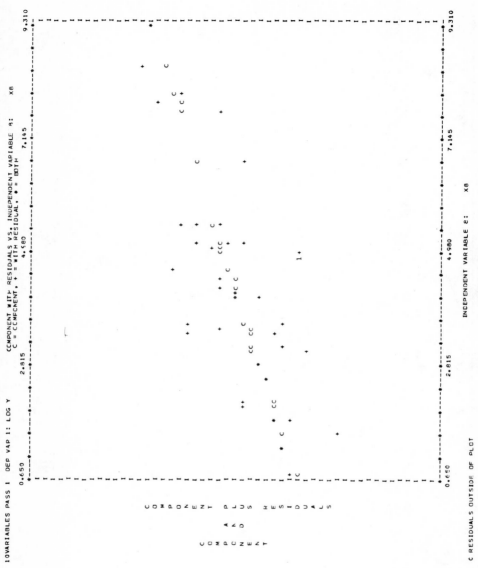

Figure 7A.14

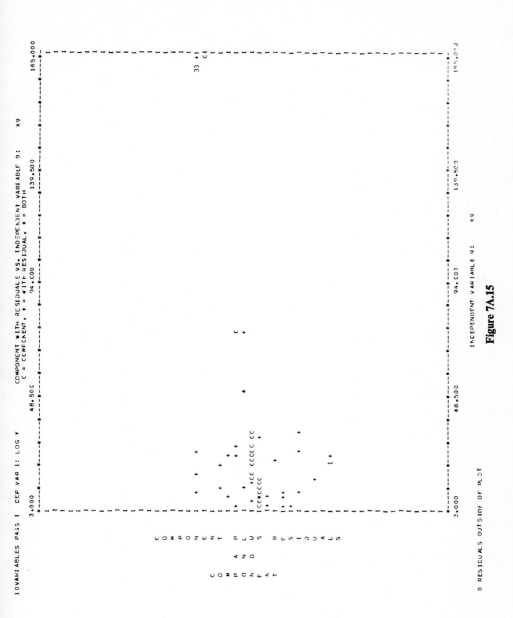

Figure 7A.15

165

Figure 7A.16

166

IOVARIABLES PASS 2   DEP VAR 1: LOG Y   MIN Y = 1.398D+00   MAX Y = 2.661D+00   RANGE Y = 1.263D+00

FULL EQUATION FOR CP SEARCH -
10 VARIABLES PLUS (X2 - 0.55)SQRD ARE PRESENT.
INDICATOR VARIABLE FOR OBSERVATION 33, EFFECT ON X9?

INDVAR(I)	NAME	COEF.B(I)	S.E. COEF.	T-VALUE	R((I)SQRD	MIN X(I)	MAX X(I)	RANGE X(I)	REL.INF.X(I)
		2.14804D+00							
1	X1	-3.015000-02	1.64D-02	1.8	0.7977	5.6100+00	1.4750+01	9.1400+00	0.22
2	X2	-1.64290D-02	4.80D-03	3.4	0.5350	6.9000+01	8.8000+01	1.9900+01	0.25
3	X3 - 54	-8.59049D-01	2.690-01	3.2	0.5656	-1.2000-01	3.0000-01	4.2000-01	0.29
4	X4	-8.66750D-02	4.56D-02	1.8	0.1547	1.0400+00	2.9100+00	1.8700+00	0.12
5	X5	-4.51590D+00	1.35D+00	3.3	0.8533	9.0000-02	2.6000-01	1.7000-01	0.61
6	X6	5.89747D-01	1.29D-01	4.6	0.8916	6.1800-01	2.3100-01	1.6920+00	0.79
7	X7	1.47320D-02	1.27D-02	4.4	0.7147	8.3200+01	9.5500+01	1.2300+01	0.14
8	X8	5.11039D-02	1.48D-03	4.0	0.5691	6.5000-01	9.3100+00	8.6600+00	0.35
9	X9	3.25627D-04	1.480-03	0.2	0.8575	3.0000+00	1.8200+02	1.8200+01	0.05
10	X10	1.11179D-02	6.40D-03	1.7	0.6043	1.0000-01	1.7900+01	1.7800+01	0.16
11	X2SQRD	-5.01049D+00	2.24D+00	2.2	3.6564	6.7710-36	6.2500-02	6.2500-02	0.25
12	PT 33	1.7CE2D-01	2.76D-01	0.6	0.8550	0.0	1.0000+00	1.0000+00	0.13

NO. OF OBSERVATIONS   36
NO. OF IND. VARIABLES   12
RESIDUAL DEGREES OF FREEDOM   23

F-VALUE   20.9
RESIDUAL ROOT MEAN SQUARE   0.10375111
RESIDUAL MEAN SQUARE   0.01076425
RESIDUAL SUM OF SQUARES   0.24757874
TOTAL SUM OF SQUARES   2.9416566E
MULT. CORREL. COEF. SQUARED   .5168

		---ORDERED BY COMPUTED INPUT---					---ORDERED BY RESIDUALS---			
IDENT.	OBSV. W55	DISTANCE	OBSV.	FITTED Y	RESIDUAL		OBSV.	OBS. Y	FITTED	CRDERED RESID.	STUD.RESID.	SEQ
7	14	2.	1	1.820	2.025	-0.206	16	1.886	1.732	0.155	1.6	1
7	21	2.	2	2.079	2.061	0.078	38	2.004	1.878	0.126	1.5	2
7	28	2.	3	2.467	2.445	0.022	11	2.121	1.997	0.123	1.4	3
8	4	13.	4	1.544	1.607	-0.063	23	2.279	2.156	0.123	1.3	4
8	11	3.	5	2.204	2.089	0.115	5	2.204	2.089	0.115	1.2	5
8	18	5.	6	2.025	2.132	-0.107	13	2.050	1.976	0.114	1.3	6
8	25	1E.	7	2.017	2.055	-0.038	25	2.591	2.494	0.097	1.1	7
9	8	3.	8	1.568	1.530	0.015	31	2.415	2.321	0.094	1.1	8
9	15	1E.	9	1.924	1.919	0.005	26	2.305	2.212	0.093	1.1	9
9	22	3.	10	2.121	1.597	0.123	35	2.661	2.617	0.044	0.5	10
9	29	5.	11	1.651	1.561	-0.105	15	2.149	2.108	0.041	0.5	11
10	6	7.	12	1.976	1.976	0.034	3	2.467	2.445	0.022	0.4	12
10	13	14.	13	1.626	1.660	-0.034	9	1.568	1.550	0.019	0.3	13
10	20	6.	14	2.108	2.108	0.041	2	2.079	2.061	0.018	0.2	14
11	3	14.	15	1.886	1.732	0.155	30	2.491	2.479	0.012	0.1	15
11	10	4.	16	2.097	1.732	-0.017	32	2.279	2.275	0.004	C.1	16
12	1	6E.	17	1.715	1.733	-0.012	33	2.215	2.215	0.0	0.0	17
12	8	20	18	1.358	1.410	-0.012	20	1.398	1.410	-0.012	-0.3	18
12	15	3.	19	2.009	2.041	-0.022	24	2.431	2.447	-0.015	-0.2	19
12	22	5.	20	2.314	2.349	-0.035	37	2.204	2.219	-0.015	-0.2	20
12	29	13.	21	2.179	2.156	0.123	19	1.716	1.733	-0.017	-0.3	21
1	5	8.	22	2.311	2.447	0.097	21	2.009	2.041	-0.029	-0.4	22
1	12	15.	23	2.591	2.494	0.097	14	1.860	2.041	-0.032	-0.4	23
1	19	19.	24	2.661	2.437	-0.217	22	2.314	2.349	-0.035	-0.4	24
1	26	19.	25	2.111	2.505	0.005	7	2.017	2.055	-0.038	-0.4	25
2	2	14.	26	2.428	2.334	-0.126	17	2.017	2.055	-0.038	-0.6	26
2	9	10.	27	2.274	2.374	-0.077	7	2.017	2.057	-0.048	-0.5	27
2	16	9.	28	2.491	2.491	0.012	36	2.009	2.068	-0.060	-0.7	28
2	23	13.	29	2.413	2.371	0.004	4	1.544	1.607	-0.063	-1.0	29
3	1	5.	30	2.479	2.275	0.155	28	2.505	2.428	-0.077	-0.9	30
3	8	8.	31	2.215	2.215	0.0	34	2.140	2.221	-0.081	-1.1	31
3	15	33.	32	2.140	2.215	-0.081	10	1.924	2.019	-0.095	-1.0	32
3	22	14.	33	2.305	2.212	0.093	6	2.025	2.132	-0.107	-1.2	33
4	12	8.	34	2.409	2.068	-0.068	12	1.851	1.961	-0.109	-1.2	34
4	21	12.	35	2.294	2.212	0.082	27	2.111	2.237	-0.126	-1.4	35
4	28	6.	36	2.004	1.879	0.126	1	1.820	2.025	-0.206	-2.3	36

Figure 7A.17

167

CP VALUES FOR THE SELECTION OF VARIABLES

10 VARIABLES PASS 2

NUMBER OF OBSERVATIONS 36
NUMBER OF VARIABLES IN FULL EQUATION 12
NUMBER OF VARIABLES IN BASIC EQUATION 0
REMAINDER OF VARIABLES TO BE CONSIDERED 12

P	CP	VARIABLES IN EQUATION											
11	10.5	X1	X2	X3	X4	X5	X6		X8	X9	X10	X3SQRD	
11	10.7	X1	X2	X3	X4	X5	X6		X8		X10	X3SQRD	PT33
12	11.0	X1	X2	X3	X4	X5	X6	X7	X8		X10	X3SQRD	PT33
12	11.4	X1	X2	X3	X4	X5	X6	X7	X8	X9	X10	X3SQRD	
10	11.5	X1	X2	X3	X4	X5	X6		X8	X9	X10	X3SQRD	
10	11.9	X1	X2	X3	X4	X5	X6		X8		X10	X3SQRD	PT33
9	12.8	X1	X2	X3		X5	X6		X8		X10	X3SQRD	
13	13.0	X1	X2	X3	X4	X5	X6	X7	X8	X9	X10	X3SQRD	PT33

Figure 7A.18

168

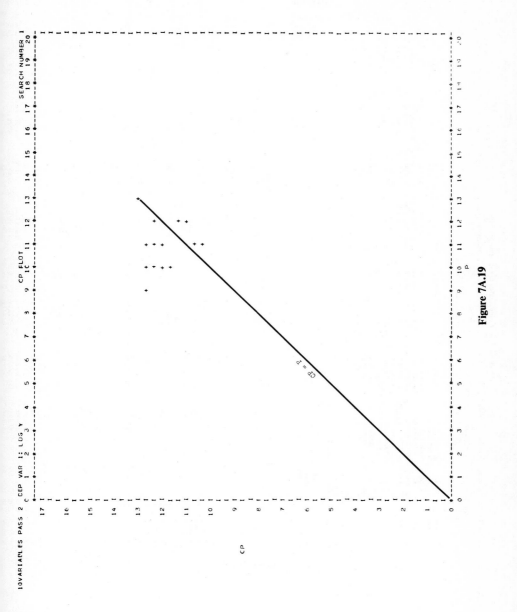

Figure 7A.19

169

LINEAR LEAST-SQUARES CURVE FITTING PROGRAM

10 VARIABLES PASS 3   DEP VAR 1: LOG Y   MIN Y = 1.398D+00   MAX Y = 2.661D+00   RANGE Y = 1.2C3D+00

CANDIDATE EQUATION -
VARIABLES X7 AND X9 DELETED. (X3 - 0.59) SQRD PRESENT.
INDICATOR VARIABLE FCR OBSERVATION 33.

IND.VAR(I)	NAME	COEF.B(I)	S.E. COEF.	T-VALUE	R(I)SQRD	MIN X(I)	MAX X(I)	RANGE X(I)	REL.INF.X(I)
0		-3.46603D+00							
1	X1	-3.26221D-02	1.620D-02	2.0	0.7940	5.610D+00	1.475D+01	9.140D+00	0.24
2	X2	-1.62201D-02	4.800D-03	3.3	0.5240	6.900D+01	8.600D+01	1.900D+01	0.24
3	X3-54	-7.24505D-01	2.450D-01	3.0	0.4815	-1.2C0D-01	3.0000-01	4.200D-01	0.24
4	X4	-8.13865D-02	4.540D-02	1.8	0.1545	1.040D+00	2.910D+00	1.870D+00	0.12
5	X5	-4.33803D+00	1.170D+00	3.7	0.8057	9.0C0D-02	2.600D-01	1.700D-01	0.58
6	X6	5.73246D-01	1.230D-01	4.6	0.8831	6.120D+00	2.310D+00	1.692D+00	0.77
7	X8	4.75376D-02	1.190D-02	4.0	0.5130	6.500D-01	9.310D+00	8.660D+00	0.33
8	X10	1.49858D-02	5.640D-03	2.7	0.4958	1.0C0D-01	1.790D+01	1.780D+01	0.21
9	X35Q3	-6.65C850D+00	1.870D+00	3.5	0.5112	6.771D-36	6.250D-02	6.250D-02	0.33
10	PT 33	2.34708D-01	1.210D-01	1.9	0.2564	0.0	1.000D+00	1.000D+00	0.19

NO. OF OBSERVATIONS   36
NO. OF IND. VARIABLES   10
RESIDUAL DEGREES CF FREECCM   25
F-VALUE   25.1
RESIDUAL ROOT MEAN SQUARE   0.10322869
RESIDUAL MEAN SQUARE   0.01065581
RESIDUAL SUM OF SQUARES   0.2EE35920
TOTAL SUM OF SQUARES   2.9415E866
MULT. CORREL. COEF. SCUARED   .9054

---ORDERED BY COMPUTER INPUT--- / ---ORDERED BY RESIDUALS---

IDENT.	OBSV.	WSS DISTANCE	OBS. Y	FITTED Y	RESIDUAL	OBSV.	OBS. Y	FITTED Y	ORDERED RESID.	STUD.RESID.	RESID. SEQ
7 14	1	3.	1.820	2.003	-0.184	38	2.004	1.838	0.166	1.8	1
7 21	2	2.	2.079	2.074	C.CC5	16	1.88	1.725	0.162	1.7	2
7 28	3	12.	2.467	2.472	-0.005	11	2.181	1.981	0.139	1.5	3
8 11	4	3.	1.544	1.640	-0.095	25	2.591	2.462	0.129	1.4	4
8 18	5	3.	2.204	2.098	0.107	23	2.279	2.166	0.113	1.2	5
8 25	6	8.	2.025	2.143	-0.117	24	2.204	2.098	0.107	1.1	6
9 8	7	18.	2.017	2.069	-0.052	18	2.090	1.985	0.105	1.1	7
9 15	8	3.	1.568	1.541	C.C27	13	2.3C5	2.202	0.104	1.1	8
9 22	9	3.	1.924	2.024	-0.100	35	2.661	2.601	0.060	0.7	9
9 29	10	4.	2.121	1.981	C.139	26	2.415	2.357	0.058	0.6	10
10 13	11	4.	1.851	1.949	-0.097	31	2.149	2.107	0.042	0.5	11
10 20	12	7.	1.825	1.877	-0.051	15	1.568	1.385	0.027	0.3	12
11 3	13	5.	1.88	1.725	0.162	9	1.368	1.385	0.013	0.1	13
11 10	14	4.	1.716	1.767	-0.066	20	2.491	2.479	0.012	0.1	14
12 1	15	6.	1.398	1.385	0.013	30	2.279	2.272	0.007	0.1	15
12 8	16	22.	2.009	2.078	-0.069	12	2.079	2.074	0.005	0.1	16
12 15	17	8.	2.314	2.314	0.013	33	2.215	2.215	0.0	0.0	17
12 22	18	8.	2.279	2.166	0.112	22	2.314	2.314	0.0	-0.0	18
12 29	19	6.	2.431	2.436	-0.004	24	2.431	2.436	-0.004	-0.0	19
1 5	20	13.	2.591	2.462	0.126	3	2.467	2.472	-0.005	-0.1	20
1 12	21	15.	2.661	2.601	0.060	37	2.2C4	2.223	-0.019	-0.2	21
1 19	22	18.	2.111	2.227	-0.117	34	2.140	2.160	-0.020	-0.2	22
1 26	23	9.	2.423	2.495	-0.067	14	1.826	1.877	-0.051	-0.5	23
2 2	24	9.	2.274	2.347	-0.073	19	1.716	1.767	-0.051	-0.5	24
2 9	25	10.	2.491	2.476	C.C12	7	2.017	2.069	-0.052	-0.6	25
2 16	26	12.	2.415	2.357	0.058	36	2.0C9	2.069	-0.060	-0.7	26
2 23	27	5.	2.279	2.272	0.307	17	2.017	2.163	-0.063	-0.7	27
3 1	28	9.	2.215	2.215	0.0	28	2.426	2.455	-0.067	-0.8	28
3 8	29	12.	2.160	2.160	-0.020	21	2.0C9	2.078	-0.069	-0.8	29
3 15	30	2.	2.305	2.202	0.1C4	4	1.544	1.640	-0.073	-0.8	30
3 22	31	7.	2.039	2.069	-0.060	32	1.851	1.949	-0.095	-0.8	31
4 21	32	1.	2.204	2.223	-0.019	10	2.024	2.111	-0.095	-1.0	32
4 28	33	5.	2.004		-0.164	6	2.025	2.143	-0.100	-1.0	33
	34					27	2.111	2.227	-0.117	-1.1	34
	35					5	2.025	2.143	-0.117	-1.3	35

Figure 7A.20

170

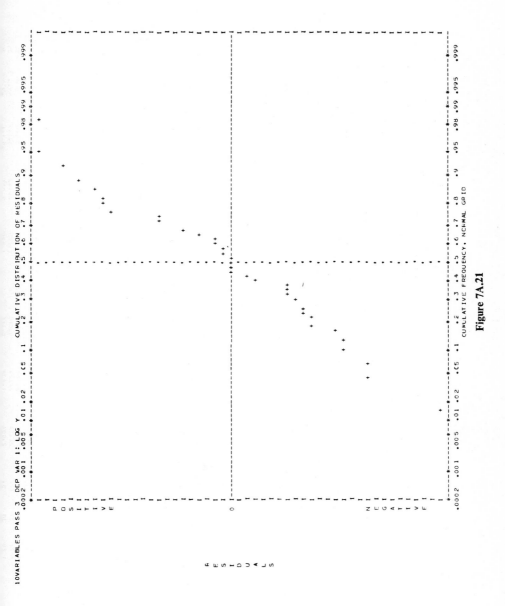

Figure 7A.21

LINEAR LEAST-SQUARES CURVE FITTING PROGRAM

FULL EQUATION FCR CP SEARCH –
10 VARIABLES PLUS (X2 – 0.5515QRD ARE PRESENT.
INDICATOR VARIABLE FCR CRSERVATION 33, EFFECT ON X9?
INDICATOR VARIABLES FOR STARTING OBSERVATICNS 1, 19 AND 20, WHAT IS
THE EFFECT CN X3SQRD AND OTHER VARIABLES?

IND.VAR(I)	NAME	COEF.B(I)	S.E. COFF.	T-VALUE	R(I)SQRD	MIN X(I)	MAX X(I)	RANGE X(I)	REL.INF.X(I)
0		1.65943D+00							
1	X1	-4.56320D-02	1.63C-02	2.8	0.8288	5.6100+00	1.4750+01	9.1400+00	0.33
2	X2	-1.35622D-02	4.69D-03	2.8	0.6138	6.9000+01	8.8000+01	1.9003+01	0.20
3	X3–-54	-4.30056D-01	4.70C-01	0.9	0.8813	-1.2C0D-01	3.0000+01	4.2000-01	0.14
4	X4	-4.CC716D-02	4.480-02	0.9	0.2716	1.0400+00	2.9100+00	1.8700+00	0.06
5	X5	-5.06825D+00	1.440+00	3.5	0.8925	2.6000-01	2.6000-01	1.7000-01	0.68
6	X6	-4.70570C-01	1.40C-01	3.4	0.9234	6.1800-01	2.3100+0C	1.6920+00	0.63
7	X7	2.127C7D-02	1.440+00	1.8	0.7329	8.3200+01	9.5500+01	1.2300+01	0.21
8	X8	4.77261C-02	1.220-02	3.9	0.6122	6.5C00-01	9.3100+00	8.6600+00	0.33
9	X9	1.112C9D-04	1.580-03	0.1	0.8952	3.0000+00	8.5000+02	1.8200+02	0.02
10	X10	8.65183D-03	6.360-03	1.4	0.6663	1.0000-01	1.7900+01	1.7800+01	0.12
11	X3SQRD	1.12789D+03	5.680+00	0.2	0.9556	6.2500-02	6.2500-02	6.2500-02	0.06
12	PT 33	-1.57806D-01	2.280-01	0.5	0.8990	0.0	1.0000+00	1.0000+00	0.12
13	PT 1	-2.97219D-01	1.090-01	2.5	0.9106	0.0	1.0000+0C	1.0000+00	0.21
14	PT 19	-3.77014D-01	3.210-01	1.2	0.9106	0.0	1.0000+00	1.0000+00	0.20
15	PT 20	-5.22953D-01	4.650-01	1.1	0.9574	0.0	1.0000+00	1.0000+00	0.41

NO. CF CBSERVATICNS   36
NO. OF IND. VARIABLES   15
RESIDUAL DEGREES OF FREEDOM   20.6
F-VALUE   20.6
RESIDUAL ROOT MEAN SQUARE   0.C54CC425
RESIDUAL MEAN SQUARE   0.008946133
RESIDUAL SUM CF SQUARES   0.17522657
TOTAL SUM OF SQUARES   2.54156C56
MULT. CORREL. COFF. SCUARED   .9391

IDENT.	OBSV.	WSS DISTANCE	OBS. Y	FITTED Y		RESIDUAL	OBSV.	OBS. Y	FITTED Y		ORDERED RESID.	STUD.RESID.	SEQ
7	14	10.	1.820	1.820		0.020	13	2.050	1.975		0.114	1.4	1
7	21	2.	2.079	2.059		0.011	38	2.004	1.894		0.111	1.6	2
7	28	21.	2.467	2.456		0.011	5	2.224	2.100		0.104	1.2	3
8	11	16.	1.544	1.542		0.002	11	2.121	2.023		0.098	1.2	4
8	18	3.	2.204	2.100		0.104	35	2.279	2.187		0.092	1.1	5
8	25	7.	2.025	2.139		-0.114	35	2.305	2.221		0.084	1.1	6
9	8	82.	2.017	2.046		-0.029	31	2.415	2.337		0.078	1.0	7
9	15	10.	1.568	1.582		-0.012	25	2.551	2.514		0.077	1.0	8
9	22	3.	1.924	1.924		-0.005	16	1.886	1.810		0.076	1.0	9
9	29	4.	2.121	2.123		-0.005	26	2.601	2.605		0.056	0.7	10
10	6	13.	1.851	2.021		-0.170	15	2.149	2.108		0.041	0.5	11
10	13	4.	2.090	1.575		0.114	24	2.431	2.410		0.022	0.3	12
10	20	6.	2.149	1.807		-0.06C	3	2.279	2.059		0.020	0.2	13
11	10	12.	1.886	2.108		0.041	2	2.467	2.456		0.011	0.2	14
11	17	3.	2.097	2.131		-0.024	29	2.274	2.266		0.008	0.1	15
12	9	87.	1.716	1.716		0.0	37	2.204	2.197		0.008	0.1	16
12	16	3.	1.398	1.298		0.0	1	1.544	1.542		0.002	0.0	17
12	15	47.	2.009	2.070		-0.061	19	1.716	1.716		0.0	0.0	18
12	23	10.	2.314	2.348		-0.035	20	1.398	1.398		0.0	0.0	19
12	29	25.	2.491	2.515		-0.017	33	2.491	2.491		0.0	0.0	20
1	25	5.	2.431	2.410		0.024	30	2.563	2.613		-0.012	-0.1	21
2	12	12.	2.591	2.514		0.077	9	1.568	1.582		-0.013	-0.2	22
2	26	13.	2.661	2.605		0.056	17	2.017	2.046		-0.029	-0.4	23
1	19	20.	2.111	2.217		-0.107	32	2.279	2.309		-0.030	-0.4	24
2	28	15.	2.428	2.522		-0.054	17	2.057	2.131		-0.034	-0.4	25
2	2	29.	2.274	2.266		0.008	23	2.314	2.348		-0.035	-0.7	26
2	16	10.	2.491	2.503		-0.012	14	1.826	1.887		-0.060	-0.7	27
2	23	12.	2.415	2.415		0.078	21	2.3C9	2.070		-0.061	-1.0	28
3	1	5.	2.279	2.3C9		-0.030	36	2.009	2.076		-0.067	-0.9	29
3	15	4.	2.140	2.215		-0.083	34	2.140	2.223		-0.083	-1.2	30
3	22	1.	2.039	2.221		0.024	28	2.428	2.522		-0.094	-1.3	31
4	30	6.	2.076	2.076		0.0	1	1.444	1.524		-0.095	-1.1	32
4	27	13.	2.204	2.197		0.008	27	2.217	2.217		-0.107	-1.4	33
4	21	8.	2.004	1.894		0.111	12	1.851	2.021		-0.170	-2.3	36

Figure 7A.22

CP VALUES FOR THE SELECTION OF VARIABLES

10 VARIABLES PASS 4

```
NUMBER OF OBSERVATIONS                   36
NUMBER CF VARIABLES IN FULL EQUATION     15
NUMBER OF VARIABLES IN BASIC EQUATION     7
REMAINDER OF VARIABLES TO BE CONSIDERED   8
```

```
BASIC SET OF VARIABLES:     X1      X2      X3      X5      X6    X8    X10
VARIABLES TO BE SEARCHED:   X4      X7      X9    X3SQRD  PT33   PT1   PT19
                            PT20
```

VARIABLES IN EQUATION: BASIC SET PLUS

```
P   CP
12  10.4   X7   X5      X3SQRD           PT1
12  10.5   X7           X3SQRD  PT33     PT1
* 13 10.9  X7                   PT33     PT1   PT19   PT20
11  11.2        X9      X3SQRD           PT1
13  11.2   X7   X5                       PT1   PT19   PT20
14  12.0 X4 X7                   PT33    PT1   PT19   PT20
16  16.0 X4 X7  X9      X3SQRD   PT33    PT1   PT19   PT20
```

* HAS LOWEST RMS VALUE

Figure 7A.23

173

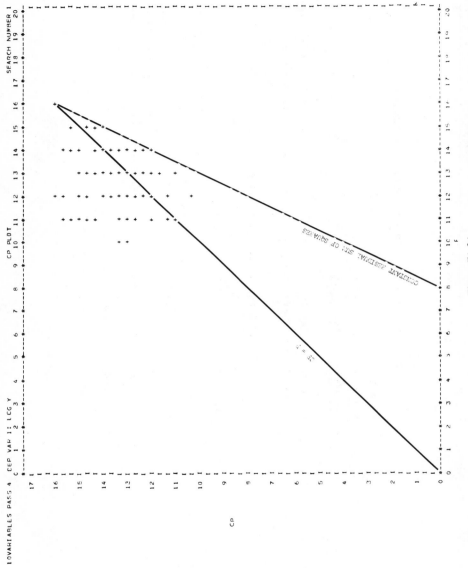

Figure 7A.24

LINEAR LEAST-SQUARES CURVE FITTING PROGRAM

10VARIABLES PASS 5   DEP VAR 1: LOG Y   MIN Y = 1.39RD+00   MAX Y = 2.661D+00   RANGE Y = 1.263D+00

CANCIEATE EQUATION –
VARIABLES X4 AND X9 DELETED, X35CPD ACT PRESENT.
INDICATOR VARIABLE FOR OBSERVATICN 23.
INDICATCR VARIABLES FOR STARTING OBSERVATIONS 1, 19 AND 20.

INC.VAR(I)	NAME	COEF.+B(I)	S.E. COEF.	T-VALUE	R(I)SCRC	MIN X(I)	MAX X(I)	RANGE X(I)	REL.INF.X(I)
0		1.94380D+00							
1	X1	-4.76127C-02	1.47D-02	3.2	0.8077	5.610D+00	1.4750+C1	9.140D+00	0.34
2	X2	-1.30071D-02	3.72D-03	3.5	0.3954	6.900D+01	8.6800+01	1.9000+01	0.20
3	X3	-5.08374D-01	3.02D-01	1.7	0.7384	4.22CD-01	8.4000-01	4.2000-01	0.17
4	X5	-5.12385D+00	1.09D+00	4.7	0.8311	9.0C0D-02	2.6900-01	1.7000-01	0.70
5	X6	4.2033BD-01	1.10D-01	4.4	0.886A	6.1800D-01	2.3100+00	1.6920+00	0.64
6	X7	2.07099C-02	1.03C-02	2.0	0.6747	4.3200+01	5.5500+01	1.2300+01	0.20
7	X9	4.71924D-02	1.07D-02	4.4	0.5432	6.5000-01	5.3100+00	8.6600+00	0.32
8	X10	9.78458C-03	5.12D-C3	1.8	0.5972	1.0CCD+00	1.7900+01	1.7800+01	0.14
9	PT 33	1.66331D-01	1.04C-01	1.6	0.2296	0.0	1.000D+C0	1.000D+00	0.13
10	PT 1	-3.04C220-01	5.67D-02	3.2	0.1053	0.0	1.000D+00	1.000D+00	0.24
11	PT 19	-3.26370C-01	1.6CD-01	2.0	0.3747	0.0	1.000D+00	1.000D+00	0.26
12	PT 20	-4.030C5D-01	1.5CD-01	2.6	0.6545	0.0	1.000D+C0	1.000D+00	0.32

NC. OF OBSERVATICNS   36
NO. OF IND. VARIABLES   12
RESIDUAL DECREES OF FREEDOM   23
F-VALUE   28.2
RESIDUAL ROOT MEAN SQUARE   0.05C17647
RESIDUAL MEAN SQUARE   0.0CE1221C
RESIDUAL SUM OF SCLARES   0.18703558
TOTAL SUM OF SQUARES   2.54156E56
MULT. COEFEL. CCEF. SQUARED   .9364

IDENT.	OBSV.	-ORDERED BY COMPUTER INPUT-- WSS DISTANCE OBS., Y	FITTED Y	RESIDUAL		OBSV.	-ORDERED BY RESIDUALS- OBS., Y	FITTED Y	CRDERED RFSID.	STUD.RESID.	SEQ
7 14	13.		1.820	2.050	0.0	13	2.050	1.967	0.123	1.5	1
7 21	2.	2.079	2.057	0.022	38	2.304	1.898	0.106	1.5	2
7 28	3.	2.467	2.450	0.C17	5	2.264	2.099	0.105	1.3	3
8 4	21.	1.544	1.514	0.030	11	2.121	2.021	0.100	1.3	4
8 11	5.	2.204	2.099	0.1C6	35	2.305	2.213	0.093	1.1	5
8 18	4.	2.025	2.126	-0.1C1	16	1.86E	1.804	0.082	1.0	6
8 25	9.	2.017	2.C38	-0.021	23	2.27%	2.202	0.077	1.0	7
9 8	10.	1.568	1.621	-0.053	25	2.551	2.517	0.074	1.0	8
9 15	2.	1.924	2.021	-0.097	31	2.415	2.341	0.074	1.0	9
9 22	7.	2.121	2.021	0.100	26	2.6C1	2.595	0.066	0.8	10
9 29	4.	1.851	1.967	-0.115	15	2.144	2.148	0.040	0.5	11
10 6	4.	2.090	2.004	0.C86	4	1.544	1.514	0.030	0.5	12
10 13	7.	1.824	1.673	-0.C47	29	2.274	2.246	0.028	0.4	13
10 20	15.	2.109	2.109	0.040	2	2.079	2.057	0.022	0.3	14
11 3	16.	1.886	1.6C4	0.082	24	2.431	2.421	0.021	0.3	15
11 10	17.	2.097	2.130	-0.C34	9	2.4E7	2.450	0.017	0.4	16
12 1	55.	1.716	1.716	0.0	37	2.204	2.203	0.001	0.0	17
12 8	80.	1.398	1.398	0.0	1	1.820	1.820	0.0	0.0	18
12 15	3.	2.333	2.337	-0.117	19	1.716	1.716	0.0	0.0	19
12 22	42H	2.314	2.337	-2.023	20	1.398	1.398	0.0	0.0	20
12 29	4.	2.279	2.202	0.C77	33	2.215	2.215	0.0	0.0	21
1 5	1C.	2.431	2.410	0.021	30	2.491	2.500	-0.C09	-0.1	22
1 12	1C.	2.591	2.517	0.C74	7	2.017	2.038	-0.021	-0.3	23
1 19	15.	2.661	2.595	0.C66	22	2.314	2.337	-C.023	-0.3	24
1 26	7%.	2.111	2.210	-0.100	17	2.130	2.130	-0.034	-0.4	25
2 2	17.	2.442H	2.527	-C.C55	32	2.279	2.324	-0.045	-0.6	26
2 9	11.	2.274	2.246	0.C28	14	1.8C8	1.873	-0.047	-0.7	27
2 16	30.	2.451	2.500	-0.C09	9	1.568	1.621	-0.053	-0.7	28
2 23	13.	2.415	2.341	0.C74	34	2.140	2.216	-0.076	-1.2	29
3 1	4.	2.279	2.324	-C.045	36	2.009	2.090	-0.082	-1.1	30
3 8	E.	2.215	2.215	C.0	10	1.924	2.021	-0.097	-1.2	31
3 15	34.	2.140	2.216	-C.076	28	2.428	2.527	-0.099	-1.3	32
4 5	E.	2.140	2.213	0.093	27	2.111	2.210	-0.100	-1.3	33
4 12	7.	2.305	2.209	-0.082	6	2.111	2.126	-0.101	-1.3	34
4 22	15.	2.204	2.204	0.001	21	2.209	2.126	-0.117	-1.4	35
4 21	17.	2.204	2.203	0.001	21	1.851	2.0CE	-0.155	-2.0	36
4 28	9.	2.034	1.698	C.1CE	12					

Figure 7A.25

175

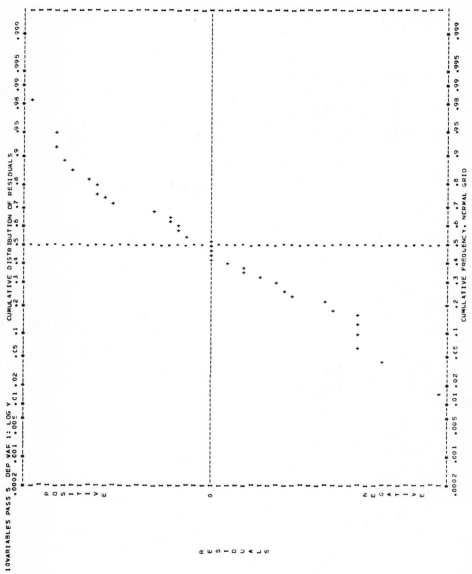

Figure 7A.26

176

LINEAR LEAST-SQUARES CURVE FITTING PROGRAM

10VARIABLES PASS 6 DEP VAR 1: LOG Y   MIN Y = 1.3580D+00   MAX Y = 2.661D+00   RANGE Y = 1.263D+00

FULL EQUATION FOR CP SEARCH -
10 VARIABLES PLUS (X3 - 0.591)SGED ARE PRESENT.
INDICATOR VARIABLE FOR OBSERVATION 33, EFFECT ON X9?
INDICATOR VARIABLE FOR STARTING OBSERVATIONS 1, 19 AND 20. WHAT IS THE
EFFECT ON X35ORD AND OTHER VARIABLES?
DO THE STARTING OBSERVATIONS HAVE A DIFFERENT X3 SLOPE?

IND.VAR(I)	NAME	CCEF+B(I)	S+E. COEF.	T-VALUE	R((I)SORD	MIN X(I)	MAX X(I)	RANGE X(I)	REL.INF+X(I)
0		1.68E543+00							
1	X1	-4.519010-02	1.600D-02	2.8	0.8284	5.610+00	1.4750+01	9.1400+00	0.33
2	X2	-1.42234D-02	4.640D-03	3.1	0.5860	6.9000+01	8.8000+01	1.9000+01	0.21
3	X3-54	-4.61890-01	4.470-01	1.1	0.8733	-1.200D-01	3.0000-01	4.2000-01	0.16
4	X4	-3.63932D-02	4.390-02	0.9	0.8678	1.4730+00	2.9100+00	1.8700+00	0.06
5	X5	-5.363150+00	1.300+00	4.1	0.8729	2.6600-01	1.0000-01	1.6920+00	0.72
6	X6	4.981740-01	1.270+00	3.9	0.9106	6.1800-01	2.3100+00	1.6920+00	0.67
7	X7	2.016890-02	1.160-02	1.7	0.7243	8.3200+01	5.5550+01	1.2300+01	0.07
8	X8	4.567290-02	1.140-02	4.4	0.5727	6.5000-01	9.3100+00	8.6600+00	0.34
9	X9	-2.441700-04	1.400-03	0.2	0.8712	3.0CD+00	1.8500+02	1.8200+02	0.04
10	X35ORD	5.56690-03	5.760-03	1.7	0.6065	1.0000-01	1.7900+01	1.7800+01	0.14
11	PT 33	-1.82374D-01	5.010+00	0.0	0.9449	6.2500-02	6.2500-02	6.2500-02	0.01
12	PT3	-2.24790D-01	2.54C-01	0.9	0.8616	0.0	1.0000+00	1.0000+00	0.18
13	START	-2.477850-01	1.290-01	1.9	0.8122	0.0	1.0000+00	1.0000+00	0.20
14	ST#X3	-5.46902D-01	1.580+00	0.3	0.9753	0.0	3.0000-01	3.0000-01	0.13

NO. OF OBSERVATIONS   36
NC. OF IND. VARIABLES   14
RESIDUAL DEGREES OF FREEDOM   21
F-VALUE   22.8
RESIDUAL ROOT MEAN SQUARE   0.0C201C75
RESIDUAL MEAN SQUARE   0.CCEE5100
RESIDUAL SUM OF SQUARES   0.1E1E70S5
TOTAL SUM OF SQUARES   2.94156556
MULT. CORREL. COEFF. SQUARED   .9382

IDENT.	OBSV.	---ORDERED BY COMPUTED INPUT--- OBS. Y	FITTED Y	RESIDUAL	OBSV.	OBS. Y	---ORDERED BY RESIDUALS--- FITTED Y	RESIDUAL	---ORDERED BY RESIDUALS--- ORDERED RESID.	STUD. RESID.	SEQ
7 14	1.	2.020	1.8720	-0.100C	38	2.204	1.880	0.125	0.125	1.7	1
7 21	2.	2.079	2.C64	0.015	13	2.C50	1.890	0.110	0.110	1.4	2
7 28	3.	2.467	2.459	0.008	11	2.121	2.015	0.105	0.103	1.3	3
8 4	4.	1.544	1.542	0.0C2	5	2.204	2.101	0.103	0.103	1.2	4
8 11	5.	2.204	2.101	0.103	16	1.886	1.798	0.089	0.089	1.1	5
8 18	6.	2.025	2.139	-0.114	31	2.415	2.331	0.084	0.084	1.1	6
8 25	7.	2.017	2.041	-C.024	23	2.279	2.196	0.083	0.083	1.0	7
9 15	8.	1.568	1.587	-0.018	35	2.305	2.226	0.079	0.079	1.0	8
9 29	9.	1.924	2.023	-0.098	25	2.591	2.515	0.076	0.076	0.9	9
9 6	10.	2.121	2.016	0.105	26	2.661	2.605	0.056	0.056	0.7	10
10 13	11.	1.851	1.880	-0.165	15	2.449	2.111	0.039	0.039	0.5	11
10 20	12.	2.090	1.817	-0.010	19	2.716	1.689	0.027	0.027	0.5	12
10 23	13.	2.149	2.111	0.039	24	2.416	2.4C4	0.015	0.015	0.2	13
11 3	14.	1.896	1.738	-0.085	2	2.079	2.459	0.008	0.008	0.2	14
11 10	15.	2.097	2.143	-0.046	29	2.467	2.459	0.007	0.007	0.1	15
12 8	16.	1.716	1.649	0.027	37	1.544	1.542	0.002	0.008	0.0	16
12 15	17.	1.398	1.419	-0.021	33	2.204	2.202	0.002	0.002	0.0	17
12 22	18.	2.009	2.078	-0.069	30	2.215	2.215	0.0	0.0	0.0	18
12 29	19.	2.314	2.344	-0.021	1	2.451	2.497	-0.005	-0.005	-0.1	19
1 24	20.	2.279	2.196	0.083	9	1.820	1.826	-0.006	-0.006	-0.5	20
1 31	21.	2.431	2.416	0.015	20	1.568	1.537	-0.018	-0.018	-0.3	21
1 19	22.	2.591	2.515	0.076	32	1.358	1.419	-0.021	-0.021	-0.3	22
1 26	23.	2.611	2.615	-0.004	22	2.017	2.041	-0.024	-0.024	-0.3	23
2 2	24.	2.428	2.522	-0.094	32	2.308	2.308	-0.029	-0.029	-0.4	24
2 9	25.	2.274	2.267	0.007	17	2.314	2.344	-0.031	-0.031	-0.6	25
2 16	26.	2.491	2.497	-0.005	22	2.097	2.143	-0.046	-0.046	-0.6	26
2 23	27.	2.415	2.331	0.084	14	1.826	1.887	-0.061	-0.061	-0.7	27
3 1	28.	2.279	2.208	0.0	36	2.009	2.075	-0.066	-0.066	-0.7	28
3 8	29.	2.215	2.215	-0.029	21	2.009	2.078	-0.069	-0.069	-1.1	29
3 15	30.	2.279	2.222	-0.082	34	2.140	2.222	-0.082	-0.082	-1.1	30
3 22	31.	2.395	2.226	-0.079	27	2.111	2.204	-0.094	-0.094	-1.3	31
4 12	32.	2.009	2.075	-0.066	28	2.428	2.522	-0.094	-0.094	-1.3	32
4 11	33.	2.264	2.292	0.0C2	10	1.924	2.023	-0.098	-0.098	-1.3	33
4 28	34.	2.393	1.880	0.125	12	2.025	2.139	-0.114	-0.114	-1.5	34
	35.					1.651	2.016	-0.165	-0.165	-2.2	35
	36.										36

Figure 7A.27

10 VARIABLES   PASS 6   DEP VAR 1: LOG Y

FUNCTIONS RELATED TO THE VARIANCE OF FITTED Y.   OBSERVATIONS ORDERED BY COMPUTER INPUT.

IDENT.	OBSV.	RESIDUAL	STUDENTIZED RESIDUAL	V(Y)/RMS	V(R)/RMS	V(Y)/V(R)
7 14	1	-0.006	-0.5	0.9850	0.0150	65.7
7 21	2	0.015	0.2	0.1395	0.8605	0.2
7 28	3	0.008	0.2	0.7474	0.2526	3.0
8 4	4	0.002	0.0	0.7002	0.2998	2.3
8 11	5	0.103	1.2	0.1691	0.8309	0.2
8 18	6	-0.114	-1.5	0.3041	0.6959	0.4
8 25	7	-0.024	-0.3	0.2901	0.7099	0.4
9 8	8	-0.018	-0.3	0.5727	0.4273	1.3
9 15	9	-0.098	-1.1	0.1487	0.8513	0.2
9 22	10	0.105	1.3	0.2539	0.7461	0.3
9 29	11	-0.165	-2.2	0.3579	0.6421	0.6
10 6	12	0.110	1.4	0.2338	0.7662	0.3
10 13	13	-0.061	-0.7	0.2070	0.7930	0.3
10 20	14	0.039	0.5	0.3171	0.6829	0.5
11 3	15	0.089	1.2	0.4099	0.5901	0.7
11 10	16	-0.046	-0.6	0.2929	0.7071	0.4
12 1	17	0.027	0.5	0.6962	0.3038	2.3
12 8	18	-0.021	-0.5	0.8162	0.1838	4.4
12 15	19	-0.069	-1.1	0.5320	0.4680	1.1
12 22	20	-0.031	-0.6	0.6884	0.3116	2.2
12 29	21	0.083	1.0	0.2367	0.7633	0.3
1 5	22	0.015	0.2	0.3203	0.6797	0.5
1 12	23	0.076	1.0	0.3527	0.6473	0.5
1 19	24	0.056	0.7	0.2150	0.7850	0.3
1 26	25	-0.094	-1.2	0.2833	0.7167	0.4
2 2	26	-0.094	-1.3	0.3754	0.6246	0.6
2 9	27	0.007	0.1	0.4365	0.5635	0.8
2 16	28	-0.005	-0.1	0.2055	0.7945	0.3
2 23	29	0.084	1.1	0.3329	0.6671	0.5
3 1	30	-0.029	-0.4	0.4645	0.5355	0.9
3 8	31	0.0	0.0	1.0000	0.0000	*****
3 15	32	-0.082	-1.3	0.5036	0.4964	1.0
3 22	33	0.079	1.0	0.2967	0.7033	0.4
4 12	34	-0.066	-0.9	0.3265	0.6735	0.5
4 21	35	0.002	0.0	0.4317	0.5683	0.8
4 28	36	0.125	1.7	0.3569	0.6431	0.6

Figure 7A.28

CF VALUES FOR THE SELECTION OF VARIABLES

10VARIABLES PASS 6

NUMBER OF OBSERVATIONS   36
NUMBER OF VARIABLES IN FULL EQUATION   14
NUMBER OF VARIABLES IN BASIC EQUATION   8
REMAINDER OF VARIABLES TO BE CONSIDERED   6

BASIC SET OF VARIABLES: X1   X2   X5   X6   X7   X8   X10
                          START
                                        X4   X9   X3SQRD PT33   ST*X3

VARIABLES TO BE SEARCHED: X3

P CP   VARIABLES IN EQUATION: BASIC SET PLUS

* 11 8.1   X3   PT33
   10 8.4   X3
   11 8.7   X3   X9
   12 9.7   X3   X4   FT33
   12 9.8   X3   PT33   ST*X3
   12 9.9   X3   X3SQRD PT33
   15 15.0 X3   X4   X9   X3SQRD PT33   ST*X3

* HAS LOWEST RMS VALUE

Figure 7A.29

179

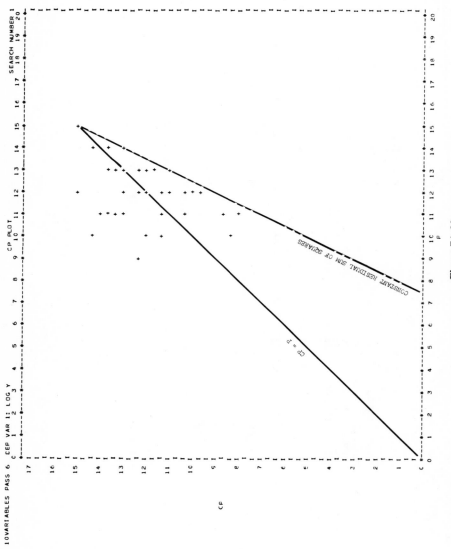

Figure 7A.30

MIN Y = 1.398D+00   MAX Y = 2.661D+00   RANGE Y = 1.263D+00

CANDIDATE EQUATION -
VARIABLES X4 AND X9 OMITTED, X3SQRD NOT PRESENT.
INDICATOR VARIABLE FOR OBSERVATION 33,
INDICATOR VARIABLE FOR STARTING OBSERVATIONS 1, 19 AND 20.

IND.VAR(I)	NAME	COEF.B(I)	S.E. COEF.	T-VALUE	R(I)SQRD	MIN X(I)	MAX X(I)	RANGE X(I)	REL.INF.X(I)
0		1.7618D+00							
1	X1	-4.70184D-02	1.420D-02	3.3	0.8068	5.610D+00	1.475D+01	9.140D+00	0.34
2	X2	-1.25550D-02	3.540D-03	3.6	0.3762	6.900D+01	8.800D+01	1.900D+01	0.19
3	X3	-5.87199D-01	2.370D-01	2.5	0.6037	4.200D-01	8.400D-01	4.200D-01	0.20
4	X5	-5.21053D+00	1.02D+00	5.1	0.8178	9.000D-02	2.600D-01	1.700D-01	0.70
5	X6	4.66037D-01	1.06D-01	4.6	0.8861	1.692D+00	2.310D+00	1.692D+00	0.65
6	X7	2.27222D-02	8.490D-03	2.7	0.5484	8.320D+01	9.550D+01	1.230D+01	0.22
7	X8	4.81849D-02	1.03D-02	4.7	0.5338	6.500D+01	9.310D+01	8.660D+00	0.33
8	X10	0.58288D-03	5.120D-03	1.5	0.5619	1.000D-01	1.790D+01	1.780D+01	0.14
9	PT 33	1.64273D-01	1.01D-01	1.6	0.2227	0.0	1.000D+00	1.000D+00	0.13
10	START	-3.22920D-01	7.500D-02	4.1	0.5557	0.0	1.000D+00	1.000D+00	0.26

NO. OF OBSERVATIONS	36
NO. OF IND. VARIABLES	10
RESIDUAL DEGREES OF FREEDOM	25
F-VALUE	36.0
RESIDUAL ROOT MEAN SQUARE	0.06733372
RESIDUAL MEAN SQUARE	0.00202417
RESIDUAL SUM OF SQUARES	0.19085418
TOTAL SUM OF SQUARES	2.54155556
MULT. COPREL. COEF. SQUARED	.9351

ORDERED BY COMPUTER INPUT

IDENT.	OBSV.	WSS DISTANCE	OBS. Y	FITTED Y	RESIDUAL
7	14	14.	1.820	1.802	0.018
7	21	2.	2.079	2.059	0.020
7	28	3.	2.467	2.441	0.026
8	4	23.	1.544	1.510	0.034
8	11	3.	2.204	2.102	0.102
8	18	4.	2.125	2.127	-0.102
8	25	7.	2.017	2.042	-0.025
9	8	26.	1.568	1.621	-0.052
9	15	3.	1.924	2.021	-0.096
9	22	4.	2.121	2.026	0.095
9	29	5.	1.851	2.000	-0.149
10	13	4.	1.824	1.959	0.131
10	20	6.	2.143	1.665	-0.039
11	3	6.	1.886	1.797	0.089
11	10	7.	2.097	2.126	-0.029
12	17	3.	1.716	1.689	0.027
12	8	102.	1.398	1.443	-0.045
12	15	20.	2.323	2.341	-0.027
12	22	78.	2.314	2.200	-0.027
12	29	12.	2.279	2.411	0.020
13	23	6.	2.431	2.523	0.069
12	15	10.	2.591	2.598	-0.062
11	25	12.	2.561	2.126	-0.101
1	26	20.	2.111	2.212	0.029
2	23	16.	2.428	2.323	-0.011
2	29	13.	2.274	1.621	-0.076
3	16	10.	2.491	2.090	-0.044
2	23	15.	2.415	2.140	0.0
3	1	7.	2.279	2.111	-0.050
3	15	4.	2.143	2.025	0.085
2	22	35.	2.305	2.428	0.003
4	12	8.	2.090	2.309	0.106
4	21	17.	2.004		
4	23	10.	2.004	1.878	

ORDERED BY RESIDUALS

OBS. Y	FITTED Y	OBSV.	RESIDUAL	ORDERED RESID.	STUD.RESID.	SEQ
2.050	1.959	13	0.131	0.131	1.6	1
2.004	1.898	38	0.106	0.106	1.4	2
2.204	2.102	5	0.102	0.102	1.3	3
2.121	2.026	11	0.095	0.095	1.2	4
1.886	1.797	16	0.089	0.089	1.1	5
2.305	2.221	35	0.085	0.085	1.1	6
2.415	2.336	31	0.079	0.079	1.1	7
2.279	2.200	23	0.079	0.079	1.0	8
2.591	2.523	25	0.069	0.069	0.9	9
2.661	2.598	26	0.062	0.062	0.8	10
2.149	2.113	15	0.037	0.037	0.5	11
1.544	1.510	4	0.034	0.034	0.6	12
2.274	2.245	29	0.029	0.029	0.4	13
1.716	1.689	19	0.027	0.027	0.4	14
2.467	2.441	3	0.026	0.026	0.5	15
2.431	2.411	24	0.020	0.020	0.3	16
2.079	2.059	21	0.020	0.020	0.2	17
1.820	1.802	1	0.018	0.018	0.4	18
2.215	2.201	37	0.0	0.003	0.0	19
2.215	2.215	33	0.0	0.0	0.0	20
2.491	2.502	30	-0.011	-0.011	-0.1	21
2.017	2.042	7	-0.025	-0.025	-0.3	22
2.314	2.341	22	-0.027	-0.027	-0.4	23
2.057	2.126	17	-0.062	-0.029	-0.4	24
1.826	1.865	14	-0.101	-0.039	-0.5	25
2.279	2.323	32	-0.104	-0.044	-0.6	26
1.358	1.443	20	-0.029	-0.045	-0.7	27
1.568	1.621	9	-0.011	-0.052	-0.8	28
2.009	2.090	36	-0.076	-0.081	-1.1	29
2.140	2.230	34	-0.044	-0.096	-1.2	30
1.924	2.021	10	0.0	-0.101	-1.3	31
2.111	2.212	27	-0.050	-0.102	-1.3	32
2.025	2.127	6	0.085	-0.104	-1.4	33
2.428	2.532	28	0.061	-0.115	-1.4	34
2.309	2.124	21	0.003		-1.4	35
1.851	2.000	12	0.105	-0.149	-2.0	36

Figure 7A.31

181

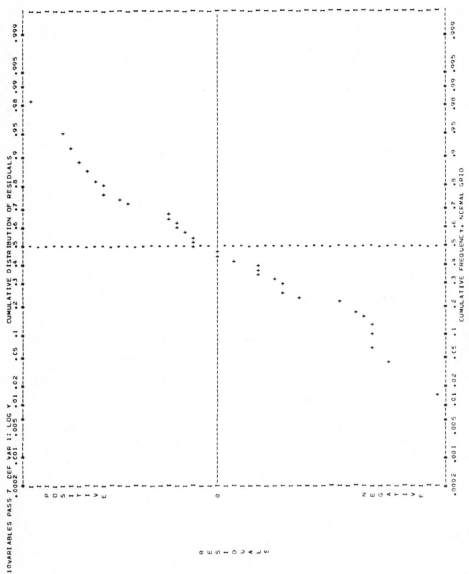

Figure 7A.32

LINEAR LEAST-SQUARES CURVE FITTING PROGRAM

10 VARIABLES   PASS 7   DEP VAR 1: LOG Y   RESIDUAL ROOT MEAN SQUARE OF FITTED EQUATION: 0.09

STANDARD DEVIATION ESTIMATED FROM RESIDUALS OF NEIGHBORING OBSERVATIONS (OBSERVATIONS 1 TO 4 APART IN FITTED Y ORDER).

NO.	CUMULATIVE STD DEV	WSSD	ORDERED BY WSSD OBSV.	OBSV.	DEL RESIDUALS	WSSD	ORDERED BY FITTED Y DEL RESIDUALS	FITTED Y	OBSV.	SEQ.
1	0.30	1.01	13	12	0.28	51.48	0.08	1.44	20	1
2	0.15	1.21	25	26	0.01	1.96	0.09	1.51	4	2
3	0.13	1.35	30	25	0.08	62.01	0.08	1.62	9	3
4	0.11	1.36	24	25	0.05	60.27	0.06	1.69	19	4
5	0.12	1.37	14	13	0.17	23.05	0.07	1.80	16	5
6	0.12	1.50	2	5	0.08	17.64	0.06	1.80	1	6
7	0.12	1.73	14	12	0.11	17.49	0.15	1.87	14	7
8	0.11	1.79	30	26	0.07	14.09	0.02	1.90	38	8
9	0.11	1.80	16	14	0.13	1.01	0.28	1.96	13	9
10	0.11	1.96	4	9	0.09	3.17	0.05	2.00	12	10
11	0.12	1.96	13	10	0.23	10.03	0.19	2.02	10	11
12	0.12	1.97	24	30	0.03	3.45	0.12	2.03	11	12
13	0.11	2.04	29	31	0.05	9.98	0.05	2.04	7	13
14	0.12	2.25	25	28	0.17	5.69	0.10	2.06	2	14
15	0.12	2.30	5	6	0.20	8.08	0.18	2.09	36	15
16	0.12	2.55	14	10	0.06	5.42	0.07	2.10	5	16
17	0.13	2.64	5	21	0.22	5.18	0.15	2.11	15	17
18	0.12	2.75	21	6	0.01	10.07	0.09	2.12	21	18
19	0.12	2.86	31	24	0.06	13.14	0.07	2.13	17	19
20	0.12	2.99	30	28	0.09	8.39	0.18	2.13	6	20
21	0.11	3.17	12	10	0.05	31.84	0.08	2.20	23	21
22	0.11	3.25	29	24	0.01	6.23	0.10	2.20	37	22
23	0.11	3.39	35	34	0.17	29.87	0.10	2.21	27	23
24	0.11	3.45	11	7	0.12	10.17	0.08	2.21	33	24
25	0.11	3.47	7	36	0.06	2.39	0.17	2.22	35	25
26	0.11	3.76	11	5	0.01	17.85	0.12	2.23	34	26
27	0.11	3.96	27	29	0.13	22.05	0.07	2.24	29	27
28	0.11	4.35	31	30	0.09	20.84	0.12	2.32	32	28
29	0.11	4.47	11	36	0.18	47.10	0.11	2.34	31	29
30	0.11	4.65	15	17	0.07	37.55	0.05	2.34	22	30
31	0.11	4.72	9	16	0.14	15.00	0.01	2.41	24	31
32	0.11	4.55	24	28	0.12	11.10	0.04	2.44	3	32
33	0.11	4.99	2	21	0.12	1.35	0.08	2.50	30	33
34	0.11	5.15	10	2	0.12	2.25	0.17	2.52	25	34
35	0.11	5.42	5	15	0.07	5.65	0.17	2.53	28	35
36	0.11	5.42	16	13	0.04	40.80	0.01	2.60	26	36

Figure 7A.33

10 VARIABLES PASS 8   DEP VAR 1: LCG Y   MIN Y = 1.5640+00   MAX Y = 2.6610+00   RANGE Y = 1.1170+00

EQUATION FOR "NORMAL" OPERATION –
VARIABLES X4 ANC X9 CMITTED.
STARTING OBSERVATIONS 1, 19, 20 AND CRSERVATION 33 DELETED.
SAVE COEFFICIENTS TO CHECK FIT CF "START-UP" OBSERVATIONS IN
CROSS VERIFICATION. PASS 9.

IND.VAR(I)	NAME	COEF.B(I)	S.E. COEF.	T-VALUE	R(I)SQRD	MIN X(I)	MAX X(I)	RANGE X(I)	REL.INF.X(I)
0		1.94388D+00							
1	X1	-4.47C7C-02	1.47C-02	3.2	0.7706	5.6100+C0	1.4750+C1	9.1400+00	0.39
2	X2	-1.30071D-02	3.72D-03	3.5	0.3121	6.9000+01	8.8000+01	1.9000+01	0.22
3	X3	-5.C837AC-01	3.020-01	1.7	0.5010	4.2C0D-01	6.8000-01	2.6000-01	0.12
4	X5	-5.1838SD+00	1.C9D+00	4.7	0.5565	1.600D-01	1.0000-01	1.0000-01	0.46
5	X6	4.80338D-01	1.10D-01	4.4	0.6262	1.031D+C0	2.310D+00	1.2790+00	0.55
6	X7	2.07099D-02	1.03D-02	2.0	0.5479	8.C400+C1	5.5500+01	9.1000+00	0.17
7	X8	4.71924D-02	1.C7D-02	4.4	0.3582	1.330D+00	8.4600+00	7.1300+00	0.30
8	X10	9.78458C-03	5.52D-03	1.8	0.5260	1.0C0D-01	1.7900+01	1.7800+01	0.16

NO. OF OBSERVATIONS	32
NC. OF IND. VARIABLES	8
RESIDUAL DEGREES CF FREEDOM	23
F-VALUE	26.9
RESIDUAL ROOT MEAN SQUARE	0.09017647
RESIDUAL MEAN SQUARE	0.CC813216
RESIDUAL SUM OF SQUARES	0.1870395E
TOTAL SUM CF SQUARES	2.13321345
MULT. CORREL. COEF. SCLARED	.9123

*** COEFFICIENTS SAVEC AS REQUESTED ***

IDENT.	OBSV.	WSS DISTANCE	OBS. Y	FITTED Y	RESIDUAL	FITTED Y	OBSV.	CES. Y	FITTED Y	CRDERED RESID.	STUD.RESID.	SEQ
7	21	2. 24.	2.079	2.057	0.022	2.050	13	2.C79	1.967	0.123	1.5	1
2	28	2. 25.	2.467	2.450	0.017	2.004	38	2.467	1.898	0.106	1.5	2
8	4	3.	1.544	1.514	0.C30	2.0C4	5	1.544	2.099	0.106	1.3	3
8	11	3.	2.204	2.099	0.106	2.121	11	2.204	2.021	0.100	1.3	4
8	18	7.	2.025	2.126	-0.101	2.305	35	2.017	2.213	0.093	1.3	5
8	25	29.	2.038	2.038	-0.021	1.886	16	1.804	1.804	0.082	1.1	6
9	2	3.	2.017	2.021	-0.053	2.279	23	2.202	2.202	0.077	1.0	7
9	8	29.	1.568	1.621	-0.C57	2.591	25	2.591	2.517	0.074	1.0	8
9	15	3.	1.924	2.021	-0.100	2.415	31	2.341	2.341	0.074	1.0	9
9	22	3.	2.121	2.005	0.077	2.661	26	2.595	2.595	0.066	1.0	10
9	29	6.	1.651	1.567	0.100	2.149	15	2.109	2.109	0.040	0.8	11
10	5	6.	2.090	1.873	0.123	1.544	4	1.514	1.514	0.030	0.5	12
10	12	7.	2.109	2.109	-0.047	2.075	29	2.246	2.246	0.028	0.4	13
10	19	18.	1.936	1.804	0.C40	2.274	2	2.057	2.057	0.022	0.3	14
11	3	4.	2.057	2.130	-0.034	2.431	24	2.410	2.410	0.021	0.3	15
11	10	2.	2.126	2.126	-0.C82	2.467	37	2.450	2.450	0.017	0.4	16
11	17	14.	2.337	2.337	-0.C17	2.204	30	2.203	2.203	-0.009	0.0	17
12	15	2.	2.410	2.410	-0.023	2.017	7	2.500	2.500	-0.021	-0.1	18
12	22	14.	2.517	2.517	0.077	2.314	22	2.038	2.038	-0.023	-0.3	19
12	29	11.	2.591	2.591	0.074	2.097	17	2.337	2.130	-0.034	-0.3	20
1	5	11.	2.661	2.595	0.021	2.279	32	2.130	2.130	-0.045	-0.4	21
1	12	6.	2.431	2.517	0.074	1.826	14	2.324	1.873	-0.047	-0.6	22
1	19	14.	2.661	2.595	0.C66	2.140	9	1.873	1.621	-0.053	-0.7	23
1	26	8.	2.527	2.527	-0.099	2.009	34	1.621	2.216	-0.076	-1.2	24
2	2	14.	2.246	2.246	-0.009	1.924	36	2.216	2.090	-0.082	-1.1	25
2	9	8.	2.500	2.491	0.077	2.428	10	2.090	2.021	-0.097	-1.2	26
2	16	30.	2.491	2.500	-0.045	2.111	28	2.021	2.527	-0.059	-1.3	27
2	23	10.	2.415	2.341	-0.076	1.924	27	2.527	2.210	-0.100	-1.3	28
3	1	7.	2.279	2.324	0.001	2.428	6	2.111	2.126	-0.097	-1.3	29
3	15	5.	2.305	2.090	0.001	2.111	21	2.126	2.126	-0.100	-1.3	30
4	21	11.	2.204	2.203	-0.117	2.025	1	2.126	2.126	-0.117	-1.4	31
4	28	7.	2.004	1.898	0.106	1.851	12	2.009	2.006	-0.155	-2.0	32

Figure 7A.34

184

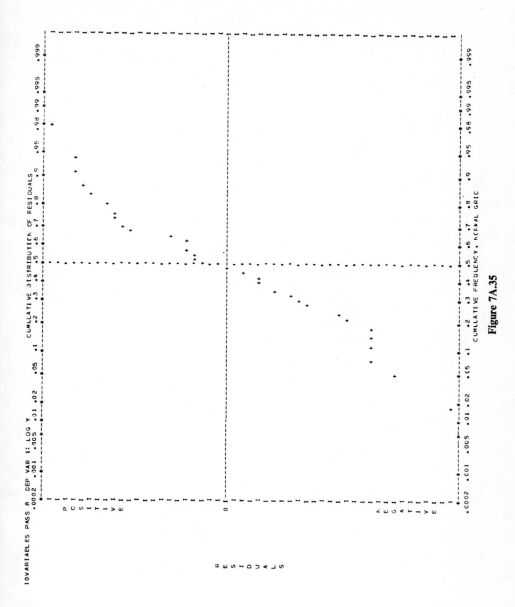

Figure 7A.35

185

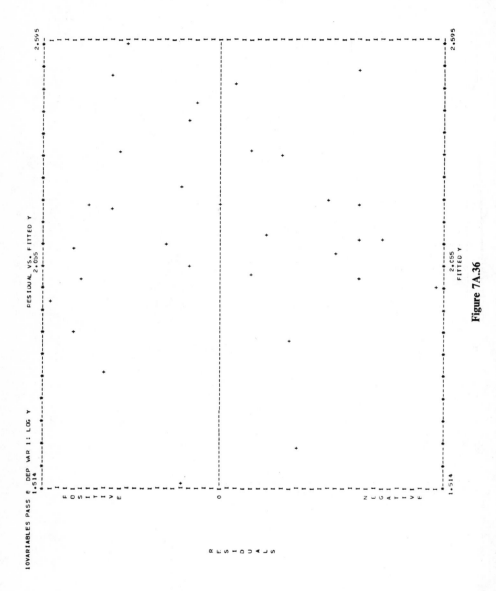

Figure 7A.36

LINEAR LEAST-SQUARES CURVE FITTING PROGRAM

10 VARIABLES   PASS 9   DEP VAR 1: LOG Y   MIN Y = 1.398D+00   MAX Y = 2.661D+00   RANGE Y = 1.263D+00

CROSS VERIFICATION OF EQUATION FOR "NORMAL" OPERATION WITH DATA
FROM BOTH "START-UP" AND "NORMAL" OPERATION -
OBSERVATION 33 DELETED.

IND.VAR(I)	NAME	CCEF.B(I)	S.E. COEF.	T-VALUE	R(I)SQRD	MIN X(I)	MAX X(I)	RANGE X(I)	REL.INF.X(I)
0		8.31219D-01							
1	X1	-3.13906D-02	1.73D-02	1.8	0.7916	5.610D+00	1.475D+01	9.140D+00	0.23
2	X2	-1.01624D-03	4.43D-03	2.3	0.3386	6.900D+01	8.800D+01	1.900D+01	0.15
3	X3	-5.55569D-01	2.77D-01	3.5	0.5248	4.200D-01	8.400D-01	4.200D-01	0.32
4	X5	-4.72152D+00	1.26D+00	3.4	0.8086	9.000D-02	2.600D-01	1.700D-01	0.58
5	X6	5.53252C-01	1.33D-01	4.2	0.8831	6.180D-01	2.310D+00	1.692D+00	0.74
6	X7	2.83100D-02	1.06D-02	2.7	0.5358	8.320D+01	9.550D+01	1.230D+01	0.28
7	X8	4.83206C-02	1.30D-02	3.7	0.4876	1.330D+00	9.310D+0C	7.980D+00	0.31
8	X10	5.53317D-03	6.50D-03	1.5	0.5618	1.000D-01	1.790D+01	1.790D+01	0.13

NC. OF OBSERVATIONS   35
NO. OF IND. VARIABLES   8
RESIDUAL DEGREES OF FREEDOM   26
F-VALUE   26.6
RESIDUAL ROOT MEAN SQUARE   0.11C77707
RESIDUAL MEAN SQUARE   0.01227156
RESIDUAL SUM OF SQUARES   0.31906052
TOTAL SUM OF SQUARES   2.92015465
MULT. CORREL. COEF. SQUARED   .8911

IND.VAR(I)	NAME	COEF.B(I)	MIN X(I)	MAX X(I)	RANGE X(I)	REL.INF.X(I)

*** COEFFICIENTS FROM PREVIOUS FIT ***

IND.VAR(I)	NAME	COEF.B(I)	MIN X(I)	MAX X(I)	RANGE X(I)	REL.INF.X(I)
0		1.94388D+00				
1	X1	-4.76127D-02	5.610D+00	1.475D+01	9.140D+00	0.34
2	X2	-1.30071D-02	6.900D+01	8.800D+01	1.900D+01	0.20
3	X3	-5.08374C-01	4.200D-01	8.400D-01	4.200D-01	0.17
4	X5	-5.18365D+00	9.000D-02	2.600D-01	1.700D-01	0.70
5	X6	4.80338C-01	6.180D-01	2.310D+00	1.692D+00	0.64
6	X7	2.07659D-02	8.320D+01	9.550D+01	1.230D+01	0.20
7	X8	4.71534D-02	1.330D+00	9.310D+00	7.980D+00	0.30
8	X10	9.78458D-03	1.000D-01	1.790D+01	1.790D+01	0.14

FIT OF SECOND SAMPLE USING COEFFICIENTS FROM FIRST SAMPLE OF DATA

NO. OF OBSERVATIONS   35
NO. OF PARAMETERS   9
RESIDUAL DEGREES CF FFEEDOM   23
RESIDUAL ROOT MEAN SQUARE   0.15446545
RESIDUAL MEAN SQUARE   0.02385956
RESIDUAL SUM OF SQUARES   0.54877024
TOTAL SUM OF SQUARES   2.53C15465
MULT. CORREL. COEF. SQUARED   .75C6

Figure 7A.37

187

LINEAR LEAST-SQUARES CURVE FITTING PROGRAM

10VARIABLES PASS 9 DEP VAR 1: LOG Y

FUNCTIONS RELATED TO THE VARIANCE OF FITTED Y. OBSERVATIONS ORDERED BY COMPUTER INPUT.

IDENT.	OBSV.	RESIDUAL	STUDENTIZED RESIDUAL	V(Y)/RMS	V(R)/RMS	V(Y)/V(R)
7 14	1	-0.305	-3.2	0.1497	1.1497	0.1
7 21	2	0.022	0.2	0.1294	1.1294	0.1
7 28	3	0.017	0.1	0.7466	1.7466	0.4
8 4	4	0.030	0.3	0.5492	1.5492	0.4
8 11	5	0.106	1.1	0.1689	1.1689	0.1
8 18	6	-0.101	-1.0	0.2669	1.2669	0.2
8 25	7	-0.021	-0.2	0.2845	1.2845	0.2
9 8	9	-0.053	-0.5	0.3641	1.3641	0.3
9 15	10	-0.097	-1.0	0.1359	1.1359	0.1
9 22	11	0.100	1.0	0.2670	1.2670	0.2
9 29	12	-0.155	-1.5	0.2502	1.2502	0.2
10 6	13	0.123	1.3	0.1889	1.1889	0.2
10 13	14	-0.047	-0.5	0.1555	1.1555	0.1
10 20	15	0.040	0.4	0.2906	1.2906	0.2
11 3	16	0.082	0.8	0.2027	1.2027	0.2
11 10	17	-0.034	-0.3	0.2157	1.2157	0.2
12 1	19	-0.326	-2.0	2.1616	3.1616	0.7
12 8	20	-0.403	-2.6	1.9767	2.9767	0.7
12 15	21	-0.117	-1.2	0.1432	1.1432	0.1
12 22	22	-0.023	-0.2	0.3061	1.3061	0.2
12 29	23	0.077	0.8	0.2139	1.2139	0.2
1 5	24	0.021	0.2	0.1810	1.1810	0.2
1 12	25	0.074	0.7	0.2780	1.2780	0.2
1 19	26	0.066	0.7	0.2026	1.2026	0.2
1 26	27	-0.100	-1.0	0.2729	1.2729	0.2
2 2	28	-0.099	-1.0	0.3251	1.3251	0.2
2 9	29	-0.028	0.3	0.3778	1.3778	0.3
2 16	30	-0.009	-0.1	0.1414	1.1414	0.1
2 23	31	0.074	0.7	0.2920	1.2920	0.2
3 1	32	-0.045	-0.4	0.3811	1.3811	0.3
3 15	34	-0.076	-0.7	0.4846	1.4846	0.3
3 22	35	0.093	0.9	0.1689	1.1689	0.1
4 12	36	-0.082	-0.8	0.2902	1.2902	0.2
4 21	37	0.001	0.0	0.3724	1.3724	0.3
4 28	38	0.106	1.0	0.3527	1.3527	0.3

Figure 7A.38

LINEAR LEAST-SQUARES CURVE FITTING PROGRAM

10VARIABLES PASS 9   DEP VAR 1: LOG Y

	ORDERED BY DISTANCE		COMPUTER	INPUT				ORDERED BY FITTED Y	ORDERED BY RESIDUALS		
IDENT.	OBSV.	WSS DISTANCE	OBS. Y	FITTED Y	RESIDUAL	OBSV.	OBS. Y	FITTED Y	ORDERED RESID.	STUD. RESID.	SEQ
7	14	1.	1.820	2.124	-0.305	13	2.090	1.967	0.123	1.3	1
7	21	1.	2.079	2.057	0.022	38	2.004	1.898	0.106	1.0	2
7	28	7.	2.467	2.450	0.017	5	2.204	2.099	0.106	1.1	3
8	4	9.	1.544	1.514	0.030	11	2.121	2.021	0.100	1.0	4
8	11	1.	2.204	2.099	0.106	35	2.305	2.213	0.093	0.9	5
8	18	1.	2.025	2.126	-0.101	16	1.886	1.804	0.082	0.8	6
8	25	2.	2.017	2.038	-0.021	23	2.279	2.202	0.077	0.7	7
9	8	10.	1.568	1.621	-0.053	25	2.591	2.517	0.074	0.7	8
9	15	1.	1.924	2.021	-0.097	31	2.415	2.341	0.074	0.7	9
9	22	1.	2.121	2.021	0.100	26	2.661	2.595	0.066	0.6	10
9	29	2.	1.851	2.006	-0.155	15	2.149	2.109	0.040	0.4	11
10	6	2.	2.090	1.967	0.123	4	1.544	1.514	0.030	0.3	12
10	13	3.	1.826	1.873	-0.047	29	2.274	2.246	0.028	0.3	13
10	20	2.	2.149	2.109	0.040	2	2.079	2.057	0.022	0.2	14
11	3	6.	1.885	1.804	0.082	24	2.431	2.410	0.021	0.2	15
11	10	2.	2.097	2.130	-0.034	3	2.467	2.450	0.017	0.1	16
12	1	32.	1.716	2.042	-0.326	37	2.204	2.203	0.001	0.0	17
12	8	24.	1.398	1.801	-0.403	30	2.491	2.500	-0.009	-0.1	18
12	15	1.	2.009	2.126	-0.117	7	2.017	2.038	-0.021	-0.2	19
12	22	5.	2.314	2.337	-0.023	22	2.314	2.337	-0.023	-0.2	20
12	29	2.	2.279	2.202	0.077	17	2.097	2.130	-0.034	-0.3	21
1	5	2.	2.431	2.410	0.021	32	2.279	2.324	-0.045	-0.4	22
1	12	4.	2.591	2.517	0.074	14	1.826	1.873	-0.047	-0.5	23
1	19	6.	2.661	2.595	0.066	9	1.568	1.621	-0.053	-0.5	24
1	26	5.	2.111	2.210	-0.100	34	2.140	2.216	-0.076	-0.7	25
2	2	2.	2.428	2.527	-0.099	36	2.009	2.090	-0.082	-0.8	26
2	9	2.	2.274	2.246	0.028	10	1.924	2.021	-0.097	-1.0	27
2	16	2.	2.491	2.500	-0.009	28	2.428	2.527	-0.099	-0.9	28
2	23	3.	2.415	2.341	0.074	27	2.111	2.210	-0.100	-1.0	29
3	1	1.	2.279	2.324	-0.045	6	2.025	2.126	-0.101	-1.0	30
3	15	0.	2.140	2.216	-0.076	21	2.009	2.126	-0.117	-1.2	31
3	22	2.	2.305	2.213	0.093	12	1.851	2.006	-0.155	-1.5	32
4	12	4.	2.009	2.090	-0.082	1	1.820	2.124	-0.305	-3.1	33
4	21	2.	2.204	2.203	0.001	19	1.716	2.042	-0.326	-2.5	34
4	28	2.	2.004	1.898	0.106	20	1.398	1.801	-0.403	-3.3	35

Figure 7A.39

189

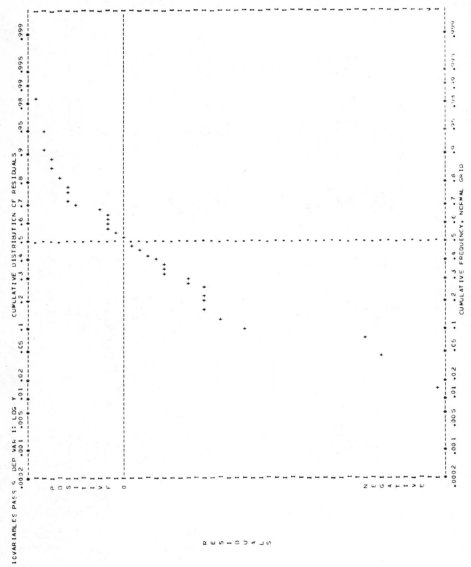

Figure 7A.40

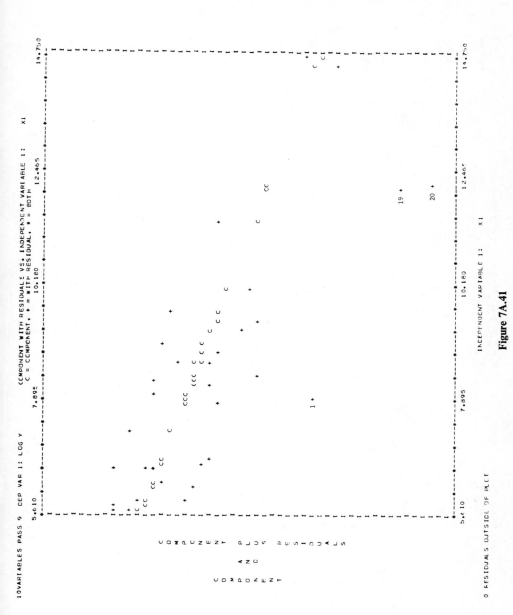

Figure 7A.41

191

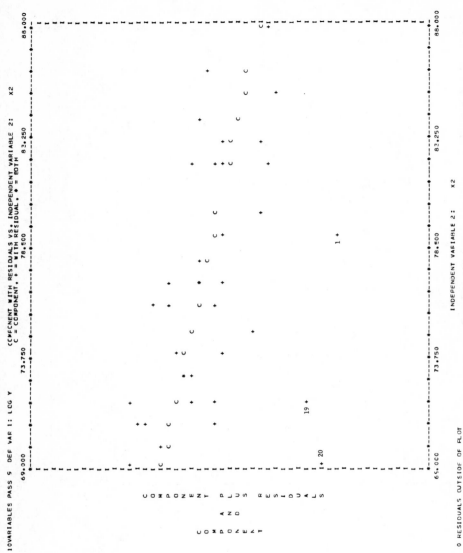

Figure 7A.42

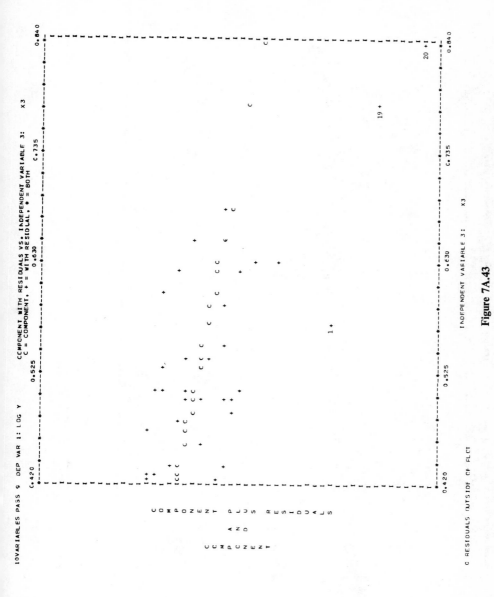

Figure 7A.43

193

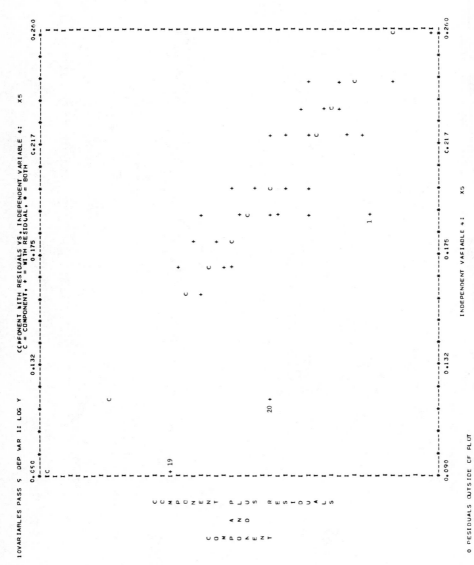

Figure 7A.44

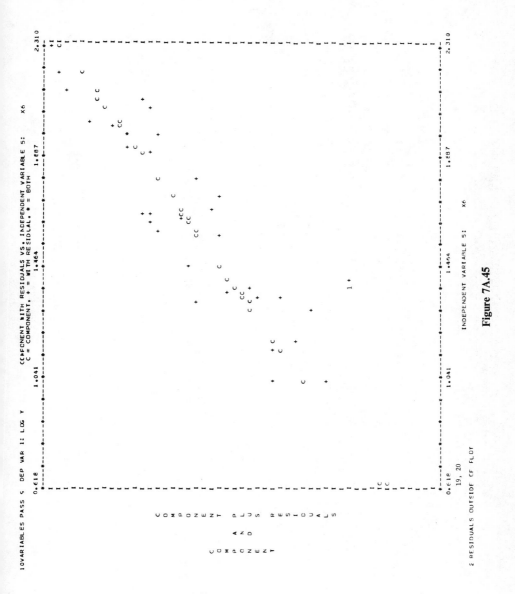

Figure 7A.45

195

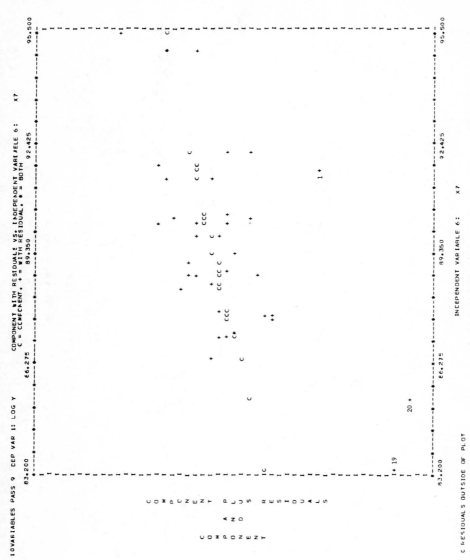

Figure 7A.46

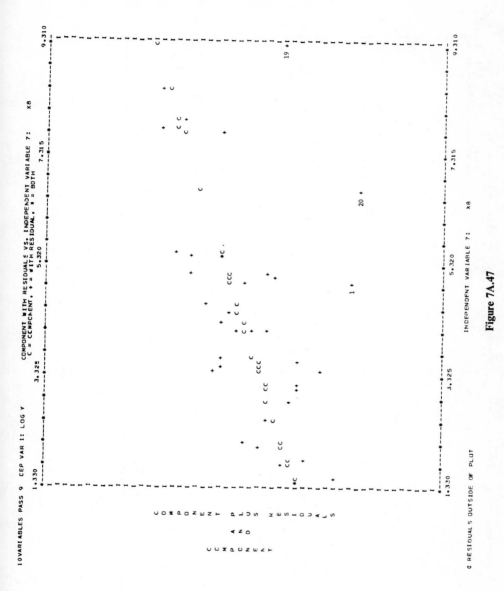

Figure 7A.47

197

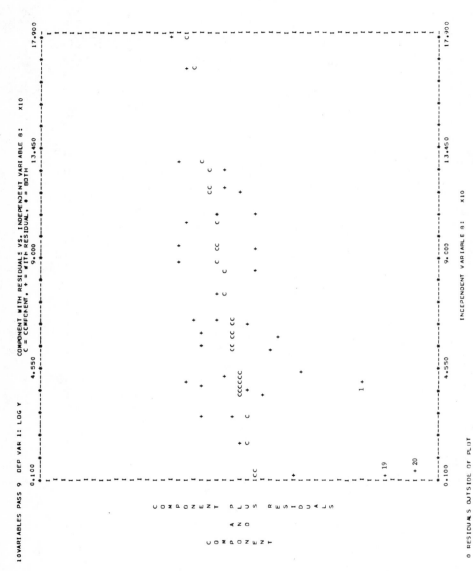

Figure 7A.48

LINEAR LEAST-SQUARES CURVE FITTING PROGRAM

6 VARIABLES PASS 2   DEP VAR 1: LOG Y   RESIDUAL ROOT MEAN SQUARE OF FITTED EQUATION: 0.11

STANDARD DEVIATION ESTIMATED FROM RESIDUALS OF NEIGHBORING OBSERVATIONS (OBSERVATIONS 1 TO 4 APART IN FITTED Y ORDER).

NC.	CUMULATIVE STD DEV	WSSD	OBSV.	OBSV.	DEL RESIDUALS	WSSD	DEL RESIDUALS	FITTED Y	OBSV.	SEG.
			ORDERED BY WSSD					ORDERED BY FITTED Y		
1	0.19	0.08	7	4	0.20	2.48	0.27	-0.41	26	1
2	0.13	0.21	30	17	0.06	1.07	0.08	-0.40	28	2
3	0.13	0.21	22	18	0.12	6.48	0.29	-0.37	27	3
4	0.12	0.29	10	13	0.08	5.01	0.03	-0.32	25	4
5	0.10	0.38	17	22	0.03	3.66	0.13	-0.26	29	5
6	0.12	0.45	27	29	0.25	7.96	0.15	-0.11	30	6
7	0.13	0.48	26	25	0.05	7.53	0.12	-0.07	24	7
8	0.13	0.53	22	19	0.22	0.38	0.03	-0.05	17	8
9	0.12	0.60	8	3	0.03	0.53	0.22	0.02	22	9
10	0.13	0.63	1	4	0.24	6.53	0.14	0.03	19	10
11	0.12	0.67	18	20	0.05	7.36	0.01	0.06	21	11
12	0.12	0.70	7	1	0.04	5.06	0.04	0.08	23	12
13	0.11	0.78	14	3	0.06	4.04	0.07	0.10	18	13
14	0.12	0.80	7	16	0.20	6.14	0.12	0.20	31	14
15	0.11	0.86	16	4	0.00	34.51	0.17	0.22	20	15
16	0.11	0.87	12	7	0.12	1.13	0.26	0.74	11	16
17	0.10	0.88	14	6	0.03	14.05	0.09	0.75	12	17
18	0.10	0.88	30	22	0.10	5.51	0.02	0.89	15	18
19	0.11	0.93	12	1	0.16	0.70	0.04	0.89	7	19
20	0.11	0.97	19	16	0.05	2.89	0.24	0.90	1	20
21	0.11	0.98	1	14	0.11	0.86	0.00	0.91	16	21
22	0.12	1.02	28	29	0.33	3.98	0.19	0.93	4	22
23	0.11	1.07	28	27	0.06	3.82	0.06	0.93	5	23
24	0.12	1.13	11	12	0.28	6.31	0.05	1.02	14	24
25	0.12	1.17	3	9	0.08	3.17	0.03	1.04	6	25
26	0.12	1.35	16	6	0.18	0.60	0.03	1.09	8	26
27	0.12	1.61	17	15	0.19	1.17	0.08	1.18	3	27
28	0.12	1.83	8	5	0.12	3.88	0.03	1.20	9	28
29	0.12	1.84	5	8	0.04	2.85	0.12	1.25	2	29
30	0.12	2.01	4	8	0.15	0.29	0.08	1.33	10	30
31	0.12	2.09	30	19	0.12	6.50	0.19	1.35	13	31

Figure 7B.1

199

LINEAR LEAST-SQUARES CURVE FITTING PROGRAM

STACK LOSS PASS 13   DEP VAR 1: S.LOSS

COMPONENT EFFECT OF EACH VARIABLE ON EACH OBSERVATION (IN UNITS OF Y)
(VARIABLES ORDERED BY THEIR RELATIVE INFLUENCE --- OBSERVATIONS ORDERED BY INFLUENCE OF MOST INFLUENTIAL VARIABLE)

| | | VARIABLES | | |
SEQ.	OBSV.	1 A.FLOW	2 W.TEMP	3 FLOWSQ
1	2	15.96	3.48	3.04
2	5	3.04	0.84	-0.22
3	6	3.04	1.37	-0.22
4	7	3.04	1.89	-0.22
5	8	3.04	1.89	-0.22
6	9	0.17	1.37	-0.35
7	10	0.17	-1.27	-0.35
8	11	0.17	-1.27	-0.35
9	12	0.17	-1.80	-0.35
10	13	0.17	-1.27	-0.35
11	14	0.17	-0.75	-0.35
12	20	-1.27	-0.22	-0.33
13	15	-5.57	-1.27	0.05
14	16	-5.57	-1.27	0.05
15	17	-5.57	-0.75	0.05
16	18	-5.57	-0.75	0.05
17	19	-5.57	-0.22	0.05

Figure 7C.1

200

LINEAR LEAST-SQUARES CURVE FITTING PROGRAM

STACK LOSS PASS 13   DEP VAR 1: S.LOSS   RESIDUAL ROOT MEAN SQUARE OF FITTED EQUATION: 1.12

STANDARD DEVIATION ESTIMATED FROM RESIDUALS OF NEIGHBORING OBSERVATIONS (OBSERVATIONS 1 TO 4 APART IN FITTED Y ORDER).

NO.	CUMULATIVE STD DEV	ORDERED BY WSSD				ORDERED BY WSSD		ORDERED BY FITTED Y		
		WSSD	OBSV.	OBSV.	DEL RESIDUALS	WSSD	DEL RESIDUALS	FITTED Y	OBSV.	SEQ.
1	3.04	0.0	13	10	3.00	0.0	1.00	7.68	16	1
2	2.03	0.0	8	7	1.00	0.22	0.53	7.68	15	2
3	1.35	0.0	11	10	0.0	0.0	0.0	8.21	18	3
4	1.77	0.0	13	11	3.00	0.22	0.47	8.21	17	4
5	1.42	0.0	18	17	0.0	28.18	0.25	8.74	19	5
6	1.35	0.0	16	15	1.00	3.61	1.83	12.49	12	6
7	1.24	0.22	5	6	0.53	2.51	4.36	12.66	20	7
8	1.14	0.22	13	14	0.47	0.0	3.00	13.02	13	8
9	1.07	0.22	16	17	0.47	0.0	0.0	13.02	11	9
10	1.22	0.22	11	14	2.53	0.22	2.53	13.02	10	10
11	1.15	0.22	15	17	0.53	3.52	0.89	13.55	14	11
12	1.10	0.22	16	18	0.47	6.75	0.53	15.66	9	12
13	1.21	0.22	10	14	2.53	0.22	0.53	18.13	5	13
14	1.16	0.22	15	18	0.53	0.22	1.47	18.65	6	14
15	1.12	0.22	12	10	0.47	0.0	1.00	19.18	8	15
16	1.08	0.22	12	11	0.47	142.38	0.23	19.18	7	16
17	1.04	0.22	6	7	0.47	0.22	0.47	36.95	2	17

Figure 7C.2

LINEAR LEAST-SQUARES CURVE FITTING PROGRAM

STACK LOSS PASS 28

EQUATION 1 OF A MULTI-EQUATION PROBLEM

DATA READ WITH SPECIAL FORMAT

FORMAT CARD 1 (A6, 2I1, 4X, F3.0, 3X, F6.2, F4.0, F6.2)

B ZERO = CALCULATED VALUE

DATA INPUT   3 INDEPENDENT VARIABLES   1 DEPENDENT VARIABLE(S)

OBSV.	SEQ.	1-11-21	2-12-22	3-13-23	4-14-24	5-15-25	6-16-26	7-17-27	8-18-28	9-19-29	10-20-30
1	0	80.000	27.000	0.0	27.000						
2	0	62.000	23.250	0.0	18.750						
3	0	58.000	23.000	0.0	15.000						
4	0	58.000	18.000	0.0	12.800						
5	0	50.000	18.800	0.0	8.000						
6	0	56.000	20.000	0.0	15.000						

DATA TRANSFORMATIONS

POSITION	CODE	OPERATION	CONSTANT -E0.	LOCATION	OMIT	VARIABLE	NAME
1	8	ADD CONSTANT		1	0	1	A.FLOW
2		NONE		2	0	2	W.TEMP
3	9	MULTIPLY POSITIONS, 1X 1		3	0	3	FLOWSQ
4		NONE		4	0	4	S.LOSS

DATA AFTER TRANSFORMATIONS   THE FITTED EQUATION HAS 3 INDEPENDENT VARIABLES, 1 DEPENDENT VARIABLE(S)

OBSV.	A.FLOW 1-11-21	W.TEMP 2-12-22	FLOWSQ 3-13-23	S.LOSS 4-14-24
1	2.00000D+01	2.70000D+01	4.00000D+02	3.70000D+01
2	2.00000D+00	2.32500D+01	4.00000D+00	1.87500D+01
3	-2.00000D+00	2.30000D+01	4.00000D+00	1.50000D+01
4	-2.00000D+00	1.80000D+01	4.00000D+00	1.28000D+01
5	-1.00000D+01	1.88000D+01	1.00000D+02	8.00000D+00
6	-4.00000D+00	2.00000D+01	1.60000D+01	1.50000D+01

SUMS OF VARIABLES
4.00000D+00 1.30050D+02 5.28000D+02 1.06550D+02

MEANS OF VARIABLES
6.66667D-01 2.16750D+01 8.80000D+01 1.77562D+01

ROOT MEAN SQUARES OF VARIABLES
1.02502D+01 3.38138D+00 1.57379D+02 1.00623D+01

SIMPLE CORRELATION COEFFICIENTS, RII.I FPIME)

1	1.000			
2	0.864	1.000		
3	0.815	0.681	1.000	
4	0.991	0.889	0.640	1.000

Figure 7C.3

LINEAR LEAST-SQUARES CURVE FITTING PROGRAM

STACK LOSS PASS 28   DEP VAR 1: S.LOSS

MIN Y = 8.000D+00   MAX Y = 3.700D+01   RANGE Y = 2.900D+01

Y = B(0) + B(1)X1 + B(2)X2 + B(3)X1*1SGRD
Y = STACK LOSS
X1 = AIR FLOW - MEAN VALUE OF AIR FLOW(60.)
X2 = COOLING WATER TEMPERATURE
1ST DAY OBSERVATIONS WITH FLOW GREATER THAN 60 OMITTED(1,3,4,21)
CONTIGUOUS LINED OUT OBSERVATIONS AVERAGED(5-8, 10-14 AND 15-19).

IND.VAR(I)	NAME	COEF.B(I)	S.E. COEF.	T-VALUE	R(I)SQRD	MIN X(I)	MAX X(I)	RANGE X(I)	REL.INF.X(I)
0		7.62999D+00							
1	A.FLOW	7.651C4D-01	1.81D-01	4.3	0.8429	-1.0C0D+01	2.000D+C1	3.000D+01	0.80
2	W.TEMP	4.1C159D-01	4.24D-01	1.0	0.7488	1.800D+01	2.700D+01	9.000D+00	0.13
3	FICWSQ	6.76552C-03	8.09D-03	0.8	0.6669	4.CC0D+CC	4.000D+C2	3.960D+02	0.09

NO. OF OBSERVATIONS   6
NC. OF IND. VARIABLES   3
RESIDUAL DEGREES OF FREEDCM   2
F-VALUE   61.8
RESIDUAL RCCT MEAN SQUARE   1.64351794
RESIDUAL MEAN SQUARE   2.70115122
RESIDUAL SUM OF SQUARES   5.40230243
TOTAL SUM CF SQUARES   506.25208333
MULT. CORREL. COEF. SQUARED   .5893

-------------ORDERED BY COMPUTER INPUT-------------

IDENT.	OBSV.	WSS DISTANCE	OBS. Y	FITTED Y	RESIDUAL
2	1	66.	37.000	36.555	0.045
5- 8	2	1.	13.757	18.871	-C.121
9	3	2.	15.030	15.690	-0.690
10-14	4	3.	12.803	13.610	-0.810
15-19	5	25.	8.030	8.439	-0.429
20	6	5.	15.000	12.985	2.015

-------ORDERED BY RESIDUALS-------

OBSV.	OES. Y	FITTED Y	CRDERED RESID.	STUD.RESID.	SEQ
6	15.000	12.985	2.015	1.4	1
1	37.000	36.955	0.045	0.5	2
2	18.750	18.871	-0.121	-0.1	3
5	8.000	8.439	-0.439	-0.9	4
3	15.000	15.690	-0.690	-0.6	5
4	12.800	13.610	-0.810	-1.3	6

Figure 7C.4

STACK LOSS PASS 28   DEP VAR 1: S.LOSS   SEARCH NUMBER 1

SELECTION OF VARIABLES FOR BASIC EQUATION

CP = 1.5, VARIABLE ADDED 1

CP = 2.7, VARIABLE ADDED 2

NUMBER OF OBSERVATIONS   6
NUMBER OF VARIABLES IN FULL EQUATION   3
NUMBER OF VARIABLES IN BASIC EQUATION   0
REMAINDER OF VARIABLES TO BE CONSIDERED   3

EQUATION	P	CP	VARIABLES IN EQUATION		
	4	4.0	FULL EQUATION		
			NO BASIC SET OF VARIABLES		
1	2	1.5	BASIC SET PLUS	A.FLOW 1	
2	3	2.7	BASIC SET PLUS	A.FLOW 1	W.TEMP 2
3	4	4.0	BASIC SET PLUS	A.FLOW 1	W.TEMP 2 FLOWSQ 3
4	3	2.9	BASIC SET PLUS	A.FLOW 1	FLOWSQ 3

SWEEP NUMBER 8. DELTA Z(K,K) = 0.

SEQ	EQUATION	P	ORDERED CP
1	1	2	1.5
2	2	3	2.7
3	4	3	2.9
4	3	4	4.0

Figure 7C.5

204

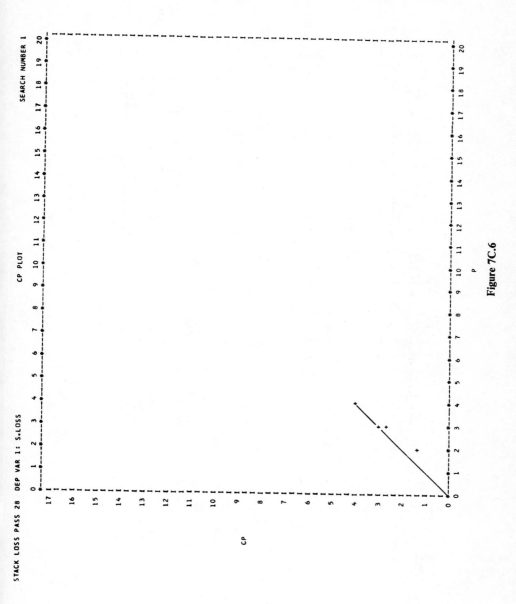

Figure 7C.6

205

LINEAR LEAST-SQUARES CURVE FITTING PROGRAM

STACK LOSS PASS 30   DEP VAR 1: S.LOSS   MIN Y = 8.000D+00   MAX Y = 3.700D+01   RANGE Y = 2.900D+01

$$Y = B(C) + B(1)X1$$
$$Y = STACK\ LOSS$$
$$X1 = AIR\ FLOW$$

1ST DAY OBSERVATIONS WITH FLOW GREATER THAN 60 OMITTED(1,3,4,21)
CONTIGUOUS LINED OUT OBSERVATIONS AVERAGED(5-8, 10-14 AND 15-19).

IND.VAR(I)	NAME	COEF.B(I)	S.E. COEF.	T-VALUE	R(I)SQRD	MAX X(I)	MIN X(I)	RANGE X(I)	REL.INF.X(I)
C		-4.12377D+01							
1	A.FLOW	5.72420D-01	6.71D-02	14.5	0.0	8.000D+01	5.000D+01	3.000D+01	1.01

NO. OF OBSERVATIONS	6
NO. OF IND. VARIABLES	1
RESIDUAL DEGREES OF FREEDOM	4
F-VALUE	210.2
RESIDUAL ROOT MEAN SQUARE	1.53734343
RESIDUAL MEAN SQUARE	2.36342461
RESIDUAL SUM OF SQUARES	9.45369924
TOTAL SUM OF SQUARES	506.2220E333
MULT. CORREL. COEF. SQUARED	.9813

--ORDERED BY COMPUTER INPUT--						--ORDERED BY RESIDUALS--					
IDENT.	OBSV.	WSS DISTANCE	OBS. Y	FITTED Y	RESIDUAL	OBSV.	OBS. Y	FITTED Y	ORDERED RESID.	STUD.RESID.	SEQ
2	1	150.	37.000	36.559	0.441	6	15.000	13.220	1.780	1.3	1
5- 8	2	1.	13.750	19.055	-0.305	5	8.000	7.385	0.615	0.5	2
9	3	3.	15.000	15.165	-0.165	1	37.000	36.559	0.441	0.8	3
10-14	4	3.	12.800	15.165	-2.365	3	15.000	15.165	-0.165	-0.1	4
15-19	5	46.	8.000	7.385	0.615	2	18.750	19.055	-0.305	-0.2	5
20	6	9.	13.000	13.220	1.780	4	12.800	15.165	-2.365	-1.7	6

Figure 7C.7

LINEAR LEAST-SQUARES CURVE FITTING PROGRAM

MIN Y = 8.966D+01   MAX Y = 9.761D+01   RANGE Y = 7.9500+00

PROCESS VARIABLE STUDY

$Y = B(0) + B(1)X1 + B(2)X2 + E(3)X3 + B(4)X4$

X1 IS THE OCTANE NUMBER OF THE GASOLINE PRODUCT
X1, X2 AND X3 ARE FEED COMPOSITION VARIABLES
X4 IS THE LOG OF AN OPERATING VARIABLE

IND.VAR(I)	NAME	CCEF.B(I)	S.E. COEF.	T-VALUE	R(I)SQRD	MIN X(I)	MAX X(I)	RANGE X(I)	REL.INF.X(I)
0		9.5E570D+01							
1	X1	-9.33672C-02	5.16D-03	18.1	0.4325	4.23CD+CC	7.554D+01	7.131D+01	0.84
2	X2	-1.25685D-01	3.17D-02	4.1	0.3567	0.0	1.076D+01	1.076D+01	0.18
3	X3	-2.51338D-02	1.28D-02	1.8	0.5734	4.000D+01	6.400D+01	2.400D+01	0.08
4	X4	1.95869D+00	3.20D-01	6.1	0.5265	1.200D+00	2.319D+00	1.119D+00	0.28

NO. OF OBSERVATIONS   82
NC. OF IND. VARIABLES   4
RESIDUAL DEGREES OF FREEDOM   77
F-VALUE   190.8
RESIDUAL ROOT MEAN SQUARE   0.43490772
RESIDUAL MEAN SQUARE   0.18514473
RESIDUAL SUM OF SQUARES   14.56414390
TOTAL SUM OF SQUARES   158.94124512
MULT. COFFEL. CDEF. SQUARED   .9CE4

IDENT.	OBSV.	WSS DISTANCE	OBS. Y	FITTED Y	RESIDUAL	OBSV.	OBS. Y	FITTED Y	ORDERED RESID.	STUD.RESID.	SEQ
9 7	1	1.	92.190	92.406	-0.216	21	92.480	91.508	0.972	2.3	1
9 8	2	0.	92.740	91.991	0.749	52	92.550	91.655	0.951	2.1	2
9 9	3	0.	91.880	92.140	-0.260	82	92.160	91.299	0.861	2.1	3
9 10	4	1.	92.800	92.271	0.529	2	92.740	91.991	0.749	1.7	4
9 11	5	1.	92.560	92.305	0.255	30	92.050	91.450	0.730	1.7	5
9 12	6	1.	92.610	92.262	0.348	78	91.860	91.191	0.669	1.6	6
9 13	7	1.	92.330	92.366	-0.036	39	92.050	91.519	0.571	1.3	7
9 14	8	3.	92.220	92.348	-0.128	67	91.110	90.54t	0.564	1.4	8
9 15	9	2.	91.960	92.302	-0.342	31	91.730	91.186	0.544	1.3	9
9 16	10	2.	92.170	92.388	-0.218	4	92.900	92.371	0.529	1.2	10
9 17	11	4.	92.750	92.602	-0.052	16	92.430	91.931	0.499	1.2	11
9 18	12	2.	92.890	92.776	0.114	17	92.770	92.282	0.488	1.1	12
9 19	13	2.	92.790	92.333	0.457	44	91.010	90.524	0.486	1.3	13
9 20	14	1.	92.550	92.305	0.245	13	92.790	91.833	0.467	1.1	14
9 21	15	2.	92.420	92.338	0.082	80	92.170	91.722	0.448	1.0	15
9 22	16	2.	92.430	91.531	0.499	81	91.560	91.117	0.443	1.1	16
9 23	17	2.	92.770	92.282	0.488	74	94.650	94.215	0.435	1.1	17
9 24	18	4.	92.600	92.714	-0.114	6	92.610	92.262	0.348	1.0	18
9 25	19	2.	92.300	92.714	0.106	45	91.900	91.557	0.343	0.8	19
9 29	20	0.	92.300	92.480	0.467	65	91.060	90.731	0.329	0.8	20
9 30	21	0.	92.480	91.508	0.972	75	97.610	97.328	0.282	0.8	21
10 1	22	0.	91.610	91.728	-0.118	5	92.560	92.305	0.255	0.6	22
10 22	23	1.	91.300	91.179	0.121	14	92.550	92.305	0.245	0.6	23
10 4	24	1.	91.330	91.813	-0.443	66	90.650	90.468	0.222	0.6	24
10 6	25	1.	91.250	91.713	-0.463	51	92.880	92.662	0.218	0.5	25
10 26	26	3.	90.760	91.038	-0.278	49	92.160	91.975	0.185	0.5	26
10 10	27	4.	90.900	90.823	0.077	37	91.900	91.727	0.173	0.4	27
10 11	28	5.	90.430	91.097	-0.667	53	91.350	91.194	0.156	0.4	28
10 17	29	2.	50.830	91.123	-0.293	29	91.900	91.300	0.121	0.3	30
10 19	30	3.	92.180	91.450	0.730	23	91.300	91.179			

Figure 7D.1

207

10 19	21	1.	91.730	91.186	0.544	12	92.890	92.776	0.114	0.3	31
10 22	32	1.	91.100	91.334	-0.224	46	91.920	91.809	0.111	0.3	32
10 23	33	0.	91.740	91.738	-0.058	19	92.300	92.194	0.106	0.2	33
10 24	34	1.	91.460	91.609	-0.149	79	91.610	91.520	0.090	0.2	34
10 25	35	2.	91.440	91.706	-0.266	15	92.420	92.338	0.082	0.2	35
10 26	36	2.	91.560	91.727	-0.120	27	90.900	90.823	0.077	0.2	36
10 27	37	2.	91.900	91.747	0.173	41	91.090	91.099	-0.009	-0.0	37
10 28	38	2.	91.610	91.519	0.137	7	92.330	92.366	-0.036	-0.0	38
10 30	39	3.	92.090	91.053	0.571	42	90.510	90.551	-0.041	-0.1	39
11 9	40	9.	90.640	91.099	-0.413	47	92.160	92.202	-0.042	-0.1	40
11 15	41	6.	91.090	90.551	-0.009	11	92.750	92.802	-0.052	-0.1	41
11 16	42	6.	90.510	90.510	-0.041	33	92.430	52.802	-0.058	-0.1	42
11 17	43	6.	90.240	90.524	-0.270	59	91.740	91.798	-0.074	-0.2	43
11 19	44	2.	91.900	91.809	0.486	18	89.980	90.054	-0.114	-0.3	44
11 20	45	1.	91.923	92.202	0.343	22	92.600	92.714	-0.118	-0.3	45
11 21	46	2.	92.160	91.655	0.111	36	91.610	91.728	-0.120	-0.3	46
11 23	47	2.	91.363	91.975	-0.042	50	91.560	91.680	-0.121	-0.3	47
11 24	48	3.	92.160	92.601	-0.305	8	92.680	92.801	-0.128	-0.3	48
11 25	49	4.	92.680	92.662	0.185	38	92.220	92.348	-0.137	-0.3	49
11 27	50	4.	92.883	91.699	0.121	64	91.610	91.747	-0.149	-0.3	50
11 28	51	3.	92.590	91.194	0.218	34	90.590	90.727	-0.165	-0.4	51
11 30	52	9.	91.350	91.115	0.891	76	97.080	97.245	-0.167	-0.4	52
12 1	53	2.	90.290	91.115	0.156	70	90.570	90.737	-0.216	-0.5	53
12 2	54	2.	90.710	90.895	-0.825	1	92.150	92.406	-0.218	-0.5	54
12 3	55	2.	90.430	90.836	-0.405	10	92.170	92.388	-0.234	-0.6	55
12 5	56	6.	89.870	90.295	-0.406	32	91.100	91.334	-0.260	-0.6	56
12 6	57	4.	89.980	90.645	-0.425	3	91.380	92.140	-0.266	-0.6	57
12 7	58	4.	60.000	90.667	-0.074	35	91.440	91.706	-0.270	-0.7	58
12 9	59	7.	89.660	90.740	-0.645	43	90.240	90.510	-0.278	-0.7	59
12 10	60	7.	90.080	90.955	-1.007	26	90.760	91.038	-0.285	-0.7	60
12 11	61	4.	90.670	90.727	-0.660	63	90.670	90.955	-0.293	-0.7	61
12 14	62	6.	90.590	90.731	-0.285	29	90.830	91.123	-0.305	-0.8	62
12 16	63	8.	91.060	90.468	-0.137	48	91.360	91.665	-0.331	-0.8	63
12 17	64	7.	90.690	90.546	0.329	68	90.320	90.651	-0.342	-0.8	64
12 18	65	6.	91.110	90.708	0.222	9	91.960	92.302	-0.348	-1.0	65
12 19	66	6.	90.320	90.737	0.564	69	92.360	90.708	-0.387	-1.0	66
12 20	67	11.	90.360	90.557	-0.331	71	94.170	94.557	-0.405	-1.0	67
12 21	68	13.	90.570	94.557	-0.348	55	90.710	91.115	-0.406	-1.0	68
12 22	69	16.	94.390	94.222	-0.167	57	90.430	90.836	-0.409	-1.0	69
12 23	70	18.	93.420	94.215	-0.367	72	94.390	94.799	-0.413	-1.0	70
12 24	71	16.	94.650	97.328	-0.409	40	90.640	91.053	-0.443	-1.1	71
12 25	72	156.	57.610	97.245	-0.802	58	85.870	90.955	-0.463	-1.3	72
12 26	73	145.	57.080	95.024	0.435	24	91.370	91.813	-0.485	-1.5	73
12 29	74	87.	95.120	911.191	0.282	25	91.250	91.713	-0.504	-1.6	74
1 4	75	3.	91.860	91.520	-0.165	56	90.410	90.895	-0.645	-2.0	75
1 7	76	0.	91.610	91.722	-0.504	77	95.120	95.624	-0.660	-1.9	76
1 10	77	1.	92.170	91.117	0.669	60	90.300	90.740	-0.667	-2.4	77
1 11	78	1.	92.160	91.299	0.090	62	90.080	91.097	-0.802		78
	79				0.448	28	90.430	94.222	-0.825		79
	80				0.443	73	93.420	91.115	-1.007		80
	81				0.861	54	90.290	91.115			81
	82					61	85.660	90.667			82

Figure 7D.1 (continued)

208

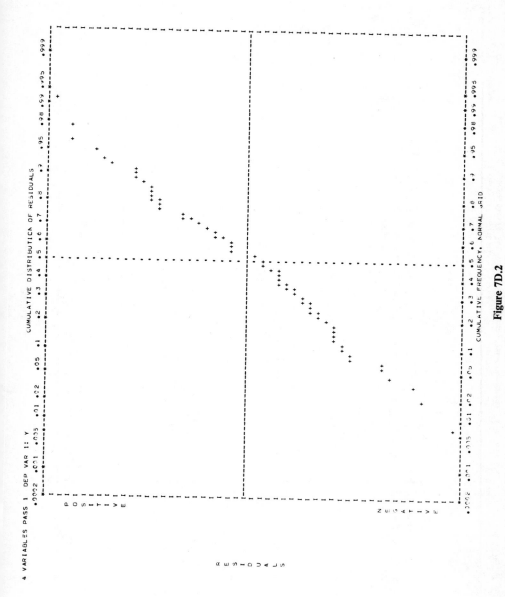

Figure 7D.2

209

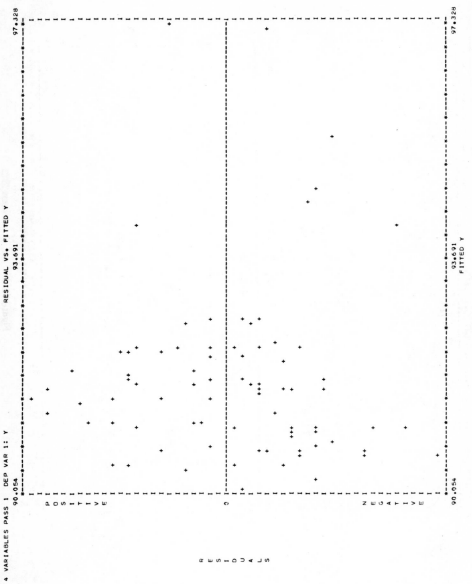

Figure 7D.3

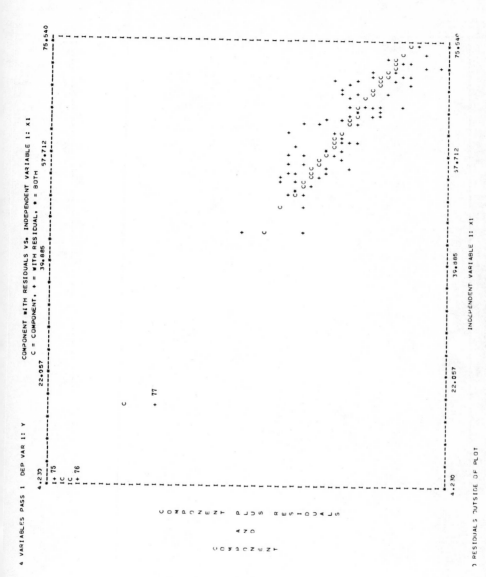

Figure 7D.4

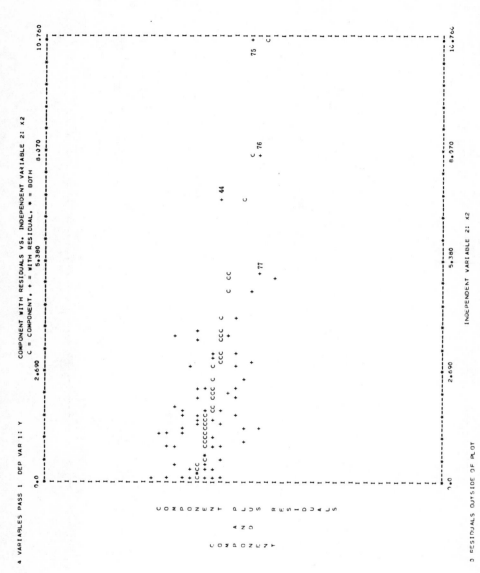

Figure 7D.5

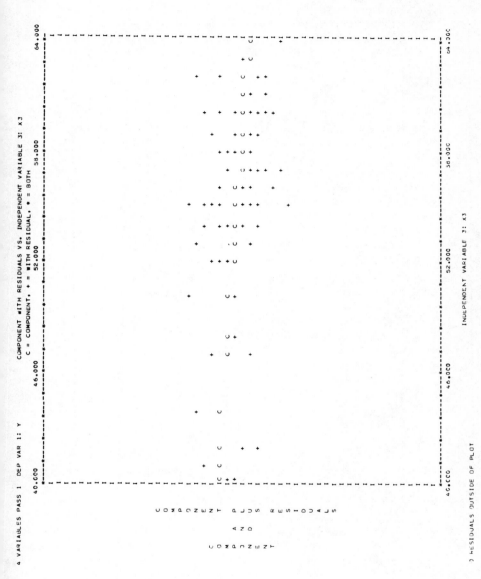

Figure 7D.6

213

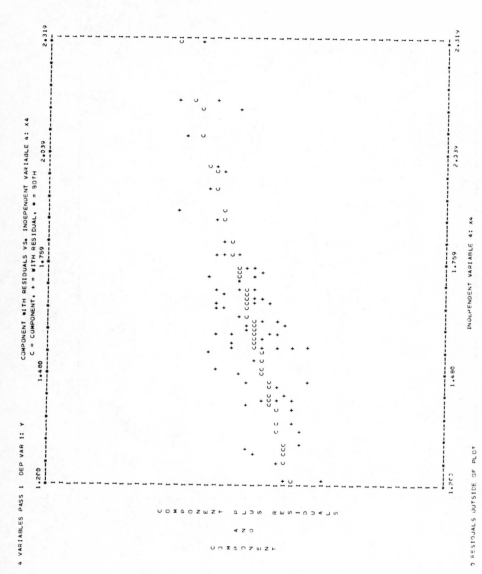

Figure 7D.7

LINEAR LEAST-SQUARES CURVE FITTING PROGRAM

4 VARIABLES PASS 2   DEP VAR 1: Y   MIN Y = 8.966D+01   MAX Y = 9.761D+01   RANGE Y = 7.950D+00

PROCESS VARIABLE STUDY
$Y = B(0) + B(1)X1 + B(2)X2 + B(3)X3 + B(4)X4$
$\quad\quad + B(5)X5 + B(6)X1X5 + B(7)X2X5$

Y IS THE OCTANE NUMBER OF THE GASOLINE PRODUCT
X1, X2 AND X3 ARE FEED COMPOSITION VARIABLES
X4 IS THE LOG OF AN OPERATING VARIABLE
X5 = 1 FOR OBSERVATIONS 75, 76, AND 77; 0 OTHERWISE.

IND.VAR(I)	NAME	COEF.B(I)	S.E. COEF.	T-VALUE	R(I)SQRD	MIN X(I)	MAX X(I)	RANGE X(I)	REL.INF.X(I)
0		9.66436D+01							
1	X1	-9.74058D-02	6.870D-03	14.2	0.6776	4.230D+00	7.554D+01	7.131D+01	0.87
2	X2	-1.31333D-01	3.560D-02	3.7	0.4872	0.0	1.076D+01	1.076D+01	0.18
3	X3	-3.00375D-02	1.44D-02	2.1	0.6048	4.000D+01	6.400D+01	2.400D+01	0.09
4	X4	1.82606D+00	3.32D-01	5.5	0.5583	1.200D+00	2.319D+00	1.119D+00	0.26
5	X5	-1.95752D+00	3.10D+00	0.6	0.9931	0.0	1.000D+00	1.000D+00	0.25
6	X1X5	1.85483D-02	1.08D-01	0.2	0.9511	0.0	1.711D+01	1.711D+01	0.04
7	X2X5	1.89139D-01	2.72D-01	0.7	0.9870	0.0	1.075D+01	1.076D+01	0.24

NO. OF OBSERVATIONS   82
NO. OF IND. VARIABLES   7
RESIDUAL DEGREES OF FREEDOM   74
F-VALUE   108.5
RESIDUAL ROOT MEAN SQUARE   0.4366802
RESIDUAL MEAN SQUARE   0.19069643
RESIDUAL SUM OF SQUARES   14.1115358C
TOTAL SUM OF SQUARES   158.94124512
MULT. CORREL. COEF. SQUARED   .9112

Figure 7D.8

215

CP VALUES FOR THE SELECTION OF VARIABLES
4 VARIABLE EXAMPLE PASS 2

NUMBER OF OBSERVATIONS   82
NUMBER OF VARIABLES IN FULL EQUATION   7
NUMBER OF VARIABLES IN BASIC EQUATION   4
REMAINDER OF VARIABLES TO BE CONSIDERED   3

P	CP		X1	X2	X3	X4
5	4.4	BASIC SET OF VARIABLES =	X1	X2	X3	X4
6	4.4	BASIC SET PLUS	-	X1X5		
6	5.4	BASIC SET PLUS	X5	-		
7	6.0	BASIC SET PLUS	X5	-	X2X5	
6	6.3	BASIC SET PLUS	-	-	X2X5	
7	6.4	BASIC SET PLUS	-	X1X5	X2X5	
7	6.4	BASIC SET PLUS	X5	X1X5	-	
8	8.0	BASIC SET PLUS	X5	X1X5	X2X5	

Figure 7D.9

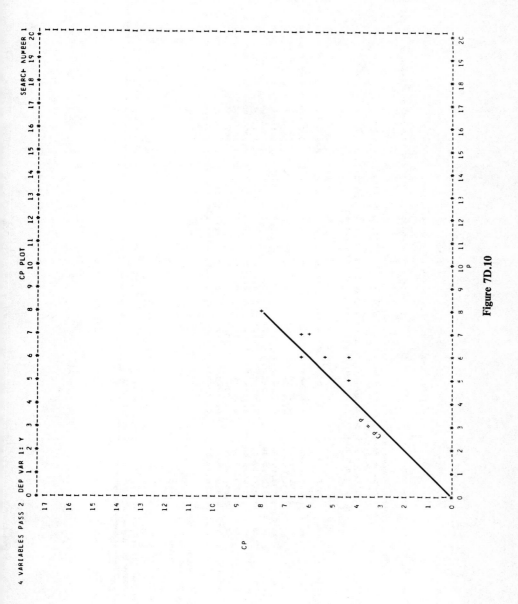

Figure 7D.10

LINEAR LEAST-SQUARES CURVE FITTING PROGRAM

11 VARIABLE PASS 1   DEP VAR 1: Y   MIN Y = 2.655D+00   MAX Y = 2.929D+00   RANGE Y = 2.738D-01

PRICE VOLUME STUDY.
Y = B(0) + B(1)X1 + --- + B(5)X5 + B(6)X6 + --- + B(9)X9 + B(10)X10
    + B(11)X11

WHERE Y = LOG VOLUME. X1 TO X5 = MONTHS 2 TO 6, X6 TO X9 = DAYS 2
TO 5, X10 = PRICE, X11 = PRICE DIFFERENTIAL TO COMPETITION.

IND.VAR(I)	NAME	COEF.B(I)	S.E. COEF.	T-VALUE	R(I)SQRD	MIN X(I)	MAX X(I)	RANGE X(I)	REL.INF.X(I)
0		2.95421D+00							
1	X1	1.40053D-02	7.34D-03	1.9	0.4971	0.0	1.000D+00	1.000D+00	0.05
2	X2	3.89073D-02	6.30D-03	6.2	0.3415	0.0	1.000D+00	1.000D+00	0.14
3	X3	2.92968D-02	6.56D-03	4.5	0.3445	0.0	1.000D+00	1.000D+00	0.11
4	X4	5.03914D-02	6.53D-03	7.7	0.3652	0.0	1.000D+00	1.000D+00	0.18
5	X5	1.81225D-02	8.11D-03	2.2	0.3366	0.0	1.000D+00	1.000D+00	0.07
6	X6	-7.07196D-02	6.24D-03	11.3	0.3915	0.0	1.000D+00	1.000D+00	0.26
7	X7	2.13640D-03	6.16D-03	0.3	0.3889	0.0	1.000D+00	1.000D+00	0.01
8	X8	6.69655D-02	6.22D-03	10.8	0.3842	0.0	1.000D+00	1.000D+00	0.24
9	X9	2.98530D-02	6.30D-03	3.7	0.5327	0.0	1.000D+00	1.000D+00	0.08
10	X10	-4.81546D-03	8.48D-04	5.7		2.731D+01	3.778D+01	1.047D+01	0.18
11	X11	-1.13827D-02	1.96D-03	5.8	0.2011	-1.470D+00	8.500D+00	9.970D+00	0.41

NO. OF OBSERVATIONS   124
NO. OF IND. VARIABLES   11
RESIDUAL DEGREES OF FREEDOM   112
F-VALUE   66.6
RESIDUAL ROOT MEAN SQUARE   0.02173644
RESIDUAL MEAN SQUARE   0.00047247
RESIDUAL SUM OF SQUARES   0.05291696
TOTAL SUM OF SQUARES   0.39883716
MULT. CORREL. COEF. SQUARED   .8673

Figure 7E.1

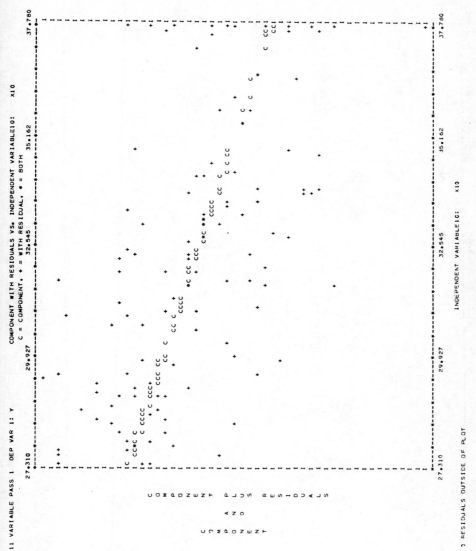

Figure 7E.2

219

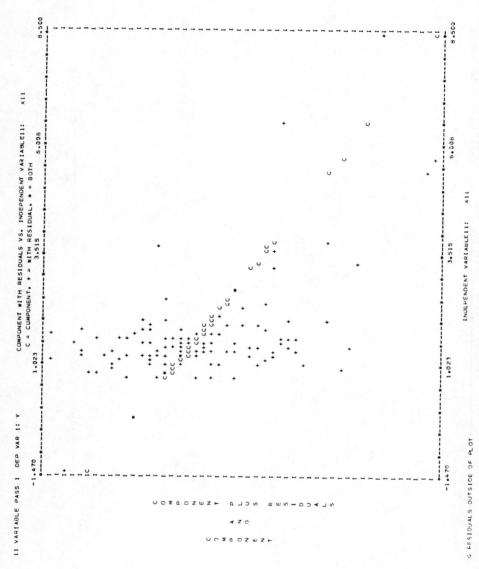

Figure 7E.3

220

LINEAR LEAST-SQUARES CURVE FITTING PROGRAM

11 VARIABLE PASS 2   DEP VAR 1: Y   MIN Y = 2.655D+00   MAX Y = 2.929D+00   RANGE Y = 2.738D-01

PRICE VOLUME STUDY.
$Y = B(0) + B(1)X1 + --- + B(5)X5 + B(6)X6 + --- + B(9)X9 + B(10)X10$
$+ B(11)X11 + B(12)X10*X12 + B(13)X11*X12 + B(14)X11*X13$

WHERE Y = LOG VOLUME, X1 TO X5 = MONTHS 2 TO 6, X6 TO X9 = DAYS 2
TO 5, X10 = PRICE, X11 = PRICE DIFFERENTIAL TO COMPETITION, X12 AND X13
ARE INDICATOR VARIABLES (X12 = 1 FOR SPECIAL SALES CONDITIONS, X13 = 1
FOR OUTER OBSERVATIONS OF X11, OTHERWISE X12 AND X13 = 0).
ARE ALL VARIABLES INFLUENTIAL? SEE CP SEARCH.

IND.VAR(I)	NAME	COEF.B(I)	S.E. COEF.	T-VALUE	K(I)SQRD	MIN X(I)	MAX X(I)	RANGE X(I)	REL.INF.X(I)
0		3.00676D+00							
1	X1	8.85186D-03	7.73D-03	1.1	0.5667	0.0	1.000D+00	1.000D+00	0.03
2	X2	3.76399D-02	6.24D-03	6.0	0.3593	0.0	1.000D+00	1.000D+00	0.14
3	X3	2.23533D-02	7.33D-03	3.1	0.4981	0.0	1.000D+00	1.000D+00	0.08
4	X4	4.97240D-02	6.49D-03	7.7	0.3857	0.0	1.000D+00	1.000D+00	0.18
5	X5	1.21861D-02	8.70D-03	1.4	0.4494	0.0	1.000D+00	1.000D+00	0.04
6	X6	-7.25085D-02	6.14D-03	11.8	0.3991	0.0	1.000D+00	1.000D+00	0.26
7	X7	-3.20170D-04	6.17D-03	0.1	0.4221	0.0	1.000D+00	1.000D+00	0.00
8	X8	6.59421D-02	6.11D-03	10.8	0.3942	0.0	1.000D+00	1.000D+00	0.24
9	X9	2.17640D-02	6.18D-03	3.5	0.3882	0.0	1.000D+00	1.000D+00	0.08
10	X10	-6.26533D-03	1.29D-03	4.9	0.8072	2.731D+01	3.773D+01	1.047D+01	0.24
11	X11	-1.43105D-02	4.24D-03	3.4	0.8363	-1.470D+00	8.500D+00	9.970D+00	0.52
12	X10X12	1.64685D-04	3.37D-04	0.5	0.8553	0.0	3.773D+01	3.778D+01	0.02
13	X11X12	6.25129D-03	3.75D-03	1.7	0.7725	0.0	8.500D+00	8.500D+00	0.19
14	X11X13	1.03211D-04	3.39D-03	0.0	0.7728	-1.470D+00	8.500D+00	9.970D+00	0.00

NO. OF OBSERVATIONS   124
NO. OF IND. VARIABLES   14
RESIDUAL DEGREES OF FREEDOM   109
F-VALUE   55.3
RESIDUAL ROOT MEAN SQUARE   0.02125809
RESIDUAL MEAN SQUARE   0.00045191
RESIDUAL SUM OF SQUARES   0.04925778
TOTAL SUM OF SQUARES   0.39883716
MULT. CORREL. COEF. SQUARED   .8765

Figure 7E.4

```
CP VALUES FOR THE SELECTION OF VARIABLES
       11 VARIABLE EXAMPLE PASS 2

NUMBER OF OBSERVATIONS                    125
NUMBER OF VARIABLES IN FULL EQUATION       14
NUMBER OF VARIABLES IN BASIC EQUATION       9
REMAINDER OF VARIABLES TO BE CONSIDERED     5


P    CP
10   9.2   BASIC SET OF VARIABLES = X2   X3   X4   X6
                                    X8   X9   X10  X11
                                    X11X12


11   8.8   BASIC SET PLUS  -   X5   -
12   9.2   BASIC SET PLUS X1   X5   -
11   10.1  BASIC SET PLUS X1   -    -
12   10.8  BASIC SET PLUS  -   X5   X7
11   11.2  BASIC SET PLUS  -   -    X7
13   11.2  BASIC SET PLUS X1   X5   X7
15   15.0  BASIC SET PLUS X1   X5   X7   X10X12  X11X13
```

Figure 7E.5

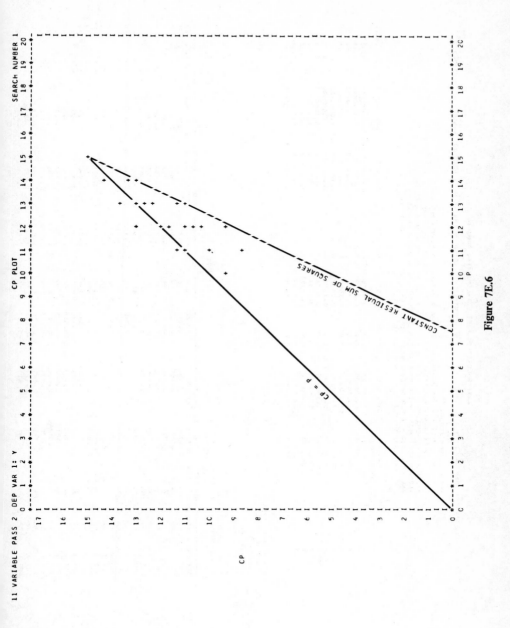

Figure 7E.6

223

LINEAR LEAST-SQUARES CURVE FITTING PROGRAM

11 VARIABLE PASS 3   DEP VAR 1: Y   MIN Y = 2.655D+00   MAX Y = 2.925D+00   RANGE Y = 2.738D-01

PRICE VOLUME STUDY.
Y = B(0) + B(1)X1 + --- + B(5)X5 + E(6)X6 + B(7)X8 + B(8)X9 + B(5)X10
    + B(10)X11 + B(11)X11*X12
WHERE Y = LOG VOLUME, X1 TO X5 = MONTHS 2 TO 6, X6 TO X9 = DAYS 2 TO 5,
X10 = PRICE, X11 = PRICE DIFFERENTIAL TO COMPETITION, X12 = INDICATOR
VARIABLE (X12 = 1 FOR SPECIAL SALES CONDITIONS, X12 = 0 OTHERWISE).
RESPONSE OF DAY 3, X7, SIMILAR TO DAY 1.

IND.VAR(I)	NAME	COEF.B(I)	S.E. COEF.	T-VALUE	R(I)SQRD	MIN X(I)	MAX X(I)	RANGE X(I)	REL.INF.X(I)
0		2.99275D+00							
1	X1	5.27264D-03	7.28D-03	1.3	0.5235	0.0	1.000D+00	1.000D+00	0.03
2	X2	3.77353D-02	6.10D-03	6.2	0.3445	0.0	1.000D+00	1.000D+00	0.14
3	X3	2.38057D-02	6.62D-03	3.6	0.3996	0.0	1.000D+00	1.000D+00	0.09
4	X4	4.94681D-02	6.32D-03	7.8	0.3668	0.0	1.000D+00	1.000D+00	0.18
5	X5	1.37219D-02	7.58D-03	1.7	0.3612	0.0	1.000D+00	1.000D+00	0.05
6	X6	-7.21530D-02	5.17D-03	14.0	0.1742	0.0	1.000D+00	1.000D+00	0.26
7	X8	6.63940D-02	5.16D-03	12.9	0.1697	0.0	1.000D+00	1.000D+00	0.24
8	X9	2.19949D-02	5.24D-03	4.2	0.1704	0.0	1.000D+00	1.000D+00	0.08
9	X10	-5.80895D-03	8.89D-04	6.5	0.6039	2.731D+01	3.778D+01	1.047D+01	0.22
10	X11	-1.47852D-02	2.24D-03	6.6	0.4269	-1.470D+00	8.500D+00	9.970D+00	0.54
11	X11X12	7.49159D-03	2.62D-03	2.9	0.5449	0.0	8.500D+00	8.500D+00	0.23

NO. OF OBSERVATIONS   124
NO. OF IND. VARIABLES   11
RESIDUAL DEGREES OF FREEDOM   112
F-VALUE   72.1
RESIDUAL ROOT MEAN SQUARE   0.2099518
RESIDUAL MEAN SQUARE   0.0C440C80
RESIDUAL SUM OF SQUARES   0.04926932
TOTAL SUM OF SQUARES   0.35283716
MULT. CORREL. COEF. SQUARD   .8762

		---ORDERED BY COMPUTER INPUT---				---ORDERED BY RESIDUALS---				
IDENT.	OBSV.	WSS DISTANCE	OBS. Y	FITTED Y	RESIDUAL	OBSV.	OBS. Y	FITTED Y	ORDERED RESID.	STUD.RESID.	SEQ
SET 1	48	2.	2.741	2.752	-0.011	76	2.747	2.703	0.045	2.6	1
SET 1	50	2.	2.803	2.766	0.037	183	2.855	2.811	0.044	2.2	2
SET 1	52	2.	2.807	2.798	0.009	59	2.855	2.813	0.042	2.1	3
SET 1	53	8.	2.691	2.706	-0.015	137	2.754	2.756	0.038	1.9	4
SET 1	55	1.	2.759	2.786	-0.027	50	2.803	2.766	0.037	1.8	5
SET 1	57	1.	2.810	2.791	0.019	69	2.841	2.805	0.035	1.7	6
SET 1	58	7.	2.864	2.857	0.007	184	2.910	2.877	0.034	1.7	7
SET 1	59	2.	2.855	2.813	0.042	148	2.872	2.839	0.034	1.7	8
SET 1	60	8.	2.690	2.720	-0.029	181	2.822	2.789	0.033	1.7	9
SET 1	62	2.	2.812	2.796	0.016	130	2.776	2.749	0.027	1.3	10
SET 1	64	2.	2.804	2.796	0.008	172	2.790	2.763	0.027	1.3	11
SET 1	65	8.	2.841	2.865	-0.025	170	2.929	2.903	0.026	1.3	12
SET 1	66	2.	2.829	2.826	0.003	108	2.803	2.779	0.025	1.3	13
SET 1	67	9.	2.688	2.732	-0.045	144	2.774	2.750	0.024	1.2	14
SET 1	69	2.	2.841	2.805	0.035	213	2.856	2.832	0.023	1.2	15
SET 1	71	2.	2.787	2.803	-0.016	87	2.838	2.815	0.023	1.2	16
SET 1	72	8.	2.981	2.873	0.008	85	2.815	2.792	0.023	1.1	17
SET 1	73	3.	2.847	2.828	0.019	228	2.778	2.756	0.022	1.1	18
SET 1	74	9.	2.725	2.734	-0.009	198	2.920	2.899	0.021	1.0	19
SET 1	76	8.	2.747	2.703	0.045	155	2.650	2.629	0.021	1.0	20
SETII	78	8.	2.701	2.717	-0.016	73	2.847	2.828	0.019	1.0	21
SET 1	79	9.	2.787	2.813	-0.026	57	2.810	2.791	0.019	0.9	22

SET 1	80	2.	2.772	2.783	219	-0.010	2.899	2.881	0.018	0.9	23
SET 1	81	9.	2.661	2.694	136	-0.023	2.866	2.850	0.016	0.8	24
SET 1	83	1.	2.747	2.778	62	-0.030	2.812	2.796	0.016	0.8	25
SET 1	85	2.	2.815	2.792	116	-0.023	2.723	2.707	0.016	0.8	26
SET 1	86	8.	2.838	2.815	100	-0.201	2.861	2.835	0.015	0.8	27
SET 1	87	2.	2.853	2.815	188	-0.020	2.829	2.814	0.016	0.8	28
SET 1	92	4.	2.757	2.777	186	-0.020	2.753	2.738	0.015	0.7	29
SET 1	53	10.	2.822	2.843	95	-0.021	2.719	2.705	0.015	0.7	30
SET 1	94	2.	2.765	2.799	193	-0.015	2.750	2.735	0.014	0.7	31
SET 1	95	11.	2.713	2.705	104	0.011	2.779	2.764	0.014	0.7	32
SET 1	97	4.	2.782	2.772	113	0.011	2.792	2.779	0.013	0.6	33
SET 1	99	10.	2.763	2.835	115	-0.007	2.814	2.802	0.012	0.6	34
SET 1	100	11.	2.851	2.790	97	0.016	2.782	2.772	0.011	0.6	35
SET 1	101	5.	2.801	2.697	101	0.011	2.757	2.746	0.011	0.5	36
SET 1	102	4.	2.688	2.764	221	-0.005	2.829	2.819	0.011	0.5	37
SET 1	104	12.	2.779	2.763	216	0.014	2.807	2.798	0.010	0.5	38
SET 1	106	7.	2.758	2.779	52	-0.015	2.881	2.873	0.009	0.5	39
SET 1	107	14.	2.811	2.826	191	-0.015	2.807	2.796	0.008	0.4	40
SET 1	108	2.	2.803	2.684	72	0.025	2.881	2.873	0.008	0.4	41
SET 1	109	8.	2.055	2.755	64	-0.026	2.804	2.796	0.008	0.4	42
SET 1	111	2.	2.773	2.779	111	0.008	2.773	2.765	0.008	0.4	43
SET 1	113	2.	2.792	2.779	214	0.013	2.747	2.740	0.007	0.3	44
SET 1	114	9.	2.848	2.846	58	0.002	2.864	2.857	0.007	0.3	45
SET 1	116	10.	2.814	2.802	149	0.012	2.912	2.905	0.006	0.3	46
SET 1	118	3.	2.723	2.707	156	0.016	2.901	2.896	0.006	0.3	47
SET 1	120	6.	2.759	2.783	139	-0.024	2.752	2.787	0.006	0.2	48
SET 1	121	4.	2.785	2.783	135	0.002	2.898	2.894	0.004	0.2	49
SET 1	122	11.	2.870	2.881	190	-0.011	2.811	2.807	0.004	0.2	50
SET 1	123	3.	2.839	2.835	134	0.004	2.859	2.825	0.004	0.2	51
SET 1	125	9.	2.746	2.742	177	-0.004	2.862	2.858	0.004	0.2	52
SET 1	127	3.	2.787	2.816	202	-0.028	2.834	2.831	0.004	0.2	53
SET 1	128	9.	2.803	2.821	123	-0.018	2.746	2.742	0.004	0.2	54
SET 1	129	10.	2.687	2.685	122	-0.003	2.839	2.835	0.004	0.2	55
SET 1	130	3.	2.776	2.749	163	0.027	2.908	2.905	0.003	0.1	56
SET 1	132	3.	2.806	2.818	66	-0.012	2.829	2.826	0.003	0.1	57
SET 1	134	6.	2.829	2.825	128	0.004	2.829	2.885	0.003	0.1	58
SET 1	135	4.	2.858	2.854	114	0.004	2.867	2.844	0.003	0.1	59
SET 1	136	11.	2.450	2.894	120	0.016	2.848	2.841	0.002	0.1	60
SET 1	137	3.	2.756	2.850	192	0.038	2.785	2.783	0.002	0.1	61
SET 1	139	9.	2.794	2.787	200	0.000	2.833	2.831	0.002	0.0	62
SET 1	141	11.	2.736	2.750	207	-0.014	2.761	2.760	0.001	0.0	63
SETI 1	142	37.	2.767	2.811	171	-0.014	2.766	2.766	0.000	0.0	64
SET 1	143	18.	2.774	2.838	169	-0.005	2.858	2.858	0.000	0.0	65
SET 1	144	4.	2.833	2.750	165	0.024	2.766	2.768	-0.000	-0.0	66
SET 1	148	11.	2.831	2.837	158	0.006	2.755	2.757	-0.001	-0.1	67
SET 1	149	4.	2.872	2.839	86	0.034	2.858	2.859	-0.001	-0.1	68
SET 1	151	10.	2.912	2.505	160	-0.006	2.831	2.833	-0.002	-0.1	69
SET 1	153	11.	2.731	2.767	204	-0.036	2.827	2.838	-0.003	-0.1	70
SET 1	155	3.	2.829	2.826	197	0.002	2.850	2.830	-0.003	-0.1	71
SET 1	156	9.	2.850	2.829	199	0.021	2.847	2.854	-0.003	-0.2	72
SET 1	157	4.	2.901	2.896	157	0.006	2.851	2.851	-0.004	-0.2	73
SET 1	158	4.	2.647	2.851	167	-0.004	2.852	2.857	-0.004	-0.2	74
SET 1	160	10.	2.735	2.757	143	-0.301	2.823	2.838	-0.004	-0.2	75
SET 1	162	3.	2.831	2.833	106	-0.302	2.253	2.763	-0.005	-0.2	76
SET 1	163	10.	2.827	2.837	226	-0.010	2.758	2.892	-0.005	-0.3	77
SET 1	164	5.	2.905	2.905	223	0.003	2.816	2.822	-0.005	-0.3	78
SET 1	165	10.	2.646	2.768	146	-0.015	2.831	2.837	-0.006	-0.3	79
SETI 1	167	5.	2.852	2.857	99	-0.004	2.763	2.770	-0.007	-0.4	81
					211		2.827	2.835	-0.008	-0.4	82

Figure 7E.7

SET11	169	8.	2.874	2.675	-0.000	102	2.688	2.697	-0.009	-0.5	83
SET 1	170	10.	2.929	2.903	0.026	74	2.725	2.734	-0.009	-0.5	84
SET 1	171	5.	2.858	2.858	-0.000	162	2.827	2.837	-0.010	-0.5	85
SET 1	172	11.	2.790	2.763	0.027	218	2.820	2.830	-0.010	-0.5	86
SET 1	176	4.	2.737	2.790	-0.053	227	2.840	2.850	-0.010	-0.5	87
SET 1	177	5.	2.862	2.858	0.004	80	2.772	2.783	-0.010	-0.5	88
SET 1	178	12.	2.794	2.814	-0.020	121	2.870	2.881	-0.011	-0.5	89
SET 1	179	4.	2.708	2.720	-0.012	48	2.741	2.752	-0.011	-0.5	90
SET 1	181	7.	2.822	2.789	-0.033	132	2.806	2.818	-0.012	-0.6	91
SET 1	183	13.	2.855	2.811	0.044	179	2.708	2.720	-0.012	-0.6	92
SET 1	184	14.	2.910	2.877	0.034	142	2.797	2.811	-0.014	-0.7	93
SET 1	186	7.	2.753	2.738	0.015	94	2.785	2.799	-0.014	-0.7	94
SET 1	188	6.	2.829	2.814	0.015	141	2.736	2.750	-0.014	-1.2	95
SET 1	190	12.	2.811	2.807	0.004	53	2.691	2.706	-0.015	-0.7	96
SET 1	191	6.	2.887	2.878	-0.009	164	2.811	2.861	-0.015	-0.8	97
SET 1	192	13.	2.833	2.831	0.002	107	2.846	2.826	-0.015	-0.8	98
SET 1	193	5.	2.750	2.735	-0.014	71	2.767	2.803	-0.016	-0.8	99
SET 1	195	11.	2.827	2.813	-0.018	78	2.701	2.717	-0.016	-0.9	100
SET 1	197	12.	2.923	2.830	-0.003	195	2.795	2.813	-0.018	-0.9	101
SET 1	198	5.	2.650	2.699	0.021	127	2.803	2.821	-0.018	-0.9	102
SET 1	199	6.	2.761	2.654	-0.003	178	2.794	2.814	-0.020	-1.0	103
SET 1	200	12.	2.834	2.760	0.001	92	2.757	2.777	-0.020	-1.0	104
SET 1	202	5.	2.835	2.831	0.004	93	2.822	2.843	-0.021	-1.1	105
SET 1	204	11.	2.883	2.838	-0.003	205	2.880	2.902	-0.022	-1.1	106
SET 1	205	6.	2.808	2.902	-0.022	118	2.759	2.783	-0.024	-1.1	107
SET 1	206	12.	2.766	2.860	0.000	65	2.817	2.855	-0.025	-1.2	108
SET 1	207	5.	2.812	2.766	-0.052	129	2.841	2.842	-0.025	-1.2	109
SET 1	209	11.	2.827	2.841	-0.029	212	2.879	2.904	-0.025	-1.3	110
SET 1	211	5.	2.879	2.835	-0.025	79	2.787	2.813	-0.026	-1.3	111
SET 1	212	3.	2.356	2.924	0.023	55	2.759	2.786	-0.027	-1.3	112
SET 1	213	9.	2.379	2.832	0.040	125	2.787	2.816	-0.028	-1.4	113
SET 1	214	3.	2.356	2.740	-0.010	109	2.655	2.684	-0.029	-1.4	114
SET 1	216	9.	2.747	2.819	-0.010	209	2.812	2.841	-0.029	-1.4	115
SET 1	218	4.	2.829	2.630	0.018	60	2.690	2.720	-0.030	-1.5	116
SET 1	219	3.	2.820	2.891	-0.031	225	2.754	2.824	-0.030	-1.5	117
SET 1	220	9.	2.393	2.839	0.011	83	2.747	2.778	-0.031	-1.5	118
SET 1	221	4.	2.808	2.746	-0.005	220	2.808	2.839	-0.033	-1.6	119
SET 1	223	10.	2.757	2.822	-0.030	81	2.661	2.594	-0.036	-1.6	120
SET 1	225	3.	2.815	2.824	-0.006	151	2.731	2.767	-0.045	-1.8	121
SET 1	226	10.	2.794	2.592	-0.010	67	2.688	2.732	-0.052	-2.2	122
SET 1	227	4.	2.840	2.850	-0.010	206	2.808	2.860	-0.052	-2.6	123
SET 1	228	11.	2.778	2.756	0.022	176	2.737	2.790	-0.053	-2.7	124

Figure 7E.7 (continued)

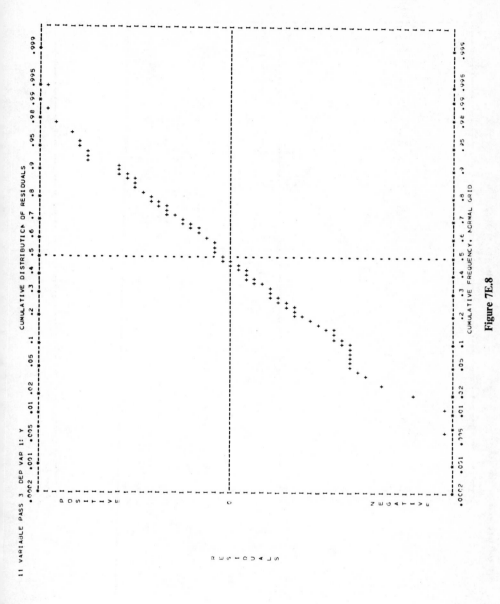

Figure 7E.8

227

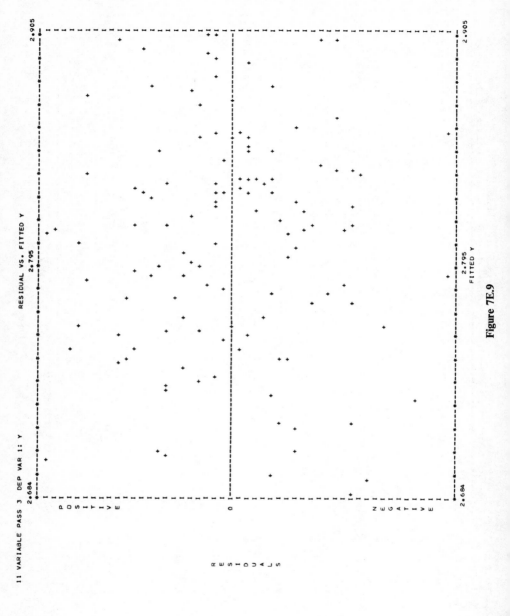

Figure 7E.9

11 VARIABLE PASS 3   DEP VAR 1: Y   COMPONENT WITH RESIDUALS VS. INDEPENDENT VARIABLE 9:   X10
                                    C = COMPONENT, + = WITH RESIDUAL, * = BOTH

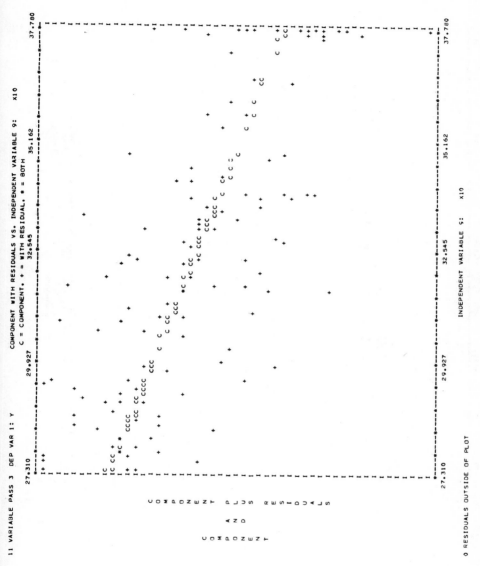

Figure 7E.10

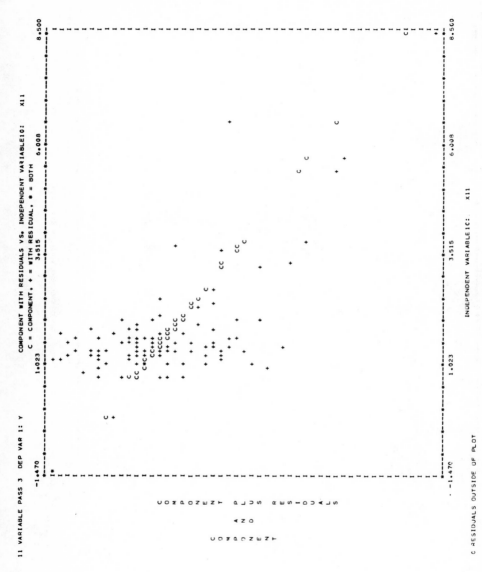

Figure 7E.11

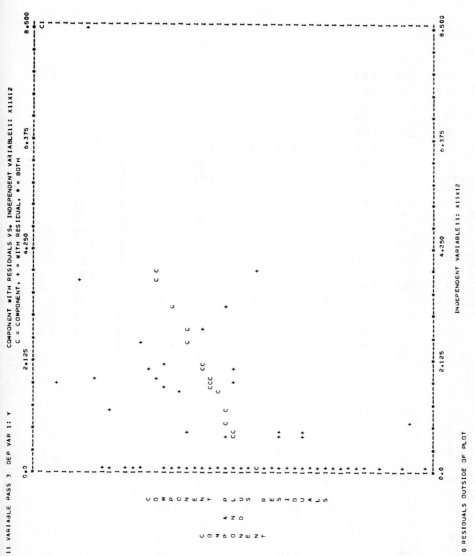

Figure 7E.12

231

CRITICAL VALUES FOR STUDENTIZED RESIDUAL TO TEST FOR SINGLE OUTLIER

N	1	2	3	4	5	6	8	10	15	25
					$\alpha = 0.10$					
5	1.87									
6	2.00	1.89								
7	2.10	2.02	1.90							
8	2.18	2.12	2.03	1.91						
9	2.24	2.20	2.13	2.05	1.92					
10	2.30	2.26	2.21	2.15	2.06	1.92				
12	2.39	2.37	2.33	2.29	2.24	2.17	1.93			
14	2.47	2.45	2.42	2.39	2.36	2.32	2.19	1.94		
16	2.53	2.51	2.50	2.47	2.45	2.42	2.34	2.20		
18	2.58	2.57	2.56	2.54	2.52	2.50	2.44	2.35		
20	2.63	2.62	2.61	2.59	2.58	2.56	2.52	2.46	2.11	
25	2.72	2.72	2.71	2.70	2.69	2.68	2.66	2.63	2.50	
30	2.80	2.79	2.79	2.78	2.77	2.77	2.75	2.73	2.66	2.13
35	2.86	2.85	2.85	2.85	2.84	2.84	2.82	2.81	2.77	2.55
40	2.91	2.91	2.90	2.90	2.90	2.89	2.88	2.87	2.84	2.72
45	2.95	2.95	2.95	2.95	2.94	2.94	2.93	2.93	2.90	2.82
50	2.99	2.99	2.99	2.99	2.98	2.98	2.98	2.97	2.95	2.89
60	3.06	3.06	3.05	3.05	3.05	3.05	3.05	3.04	3.03	3.00
70	3.11	3.11	3.11	3.11	3.11	3.11	3.10	3.10	3.09	3.07
80	3.16	3.16	3.16	3.15	3.15	3.15	3.15	3.15	3.14	3.12
90	3.20	3.20	3.19	3.19	3.19	3.19	3.19	3.19	3.18	3.17
100	3.23	3.23	3.23	3.23	3.23	3.23	3.23	3.22	3.22	3.21
					$\alpha = 0.05$					
5	1.92									
6	2.07	1.93								
7	2.19	2.08	1.94							
8	2.28	2.20	2.10	1.94						
9	2.35	2.29	2.21	2.10	1.95					
10	2.42	2.37	2.31	2.22	2.11	1.95				
12	2.52	2.49	2.45	2.39	2.33	2.24	1.96			
14	2.61	2.58	2.55	2.51	2.47	2.41	2.25	1.96		
16	2.68	2.66	2.63	2.60	2.57	2.53	2.43	2.26		
18	2.73	2.72	2.70	2.68	2.65	2.62	2.55	2.44		
20	2.78	2.77	2.76	2.74	2.72	2.70	2.64	2.57	2.15	
25	2.89	2.88	2.87	2.86	2.84	2.83	2.80	2.76	2.60	

N = number of observations.
p = number of independent variables (including count for intercept if fitted).

N	\multicolumn{10}{c}{p}
	1	2	3	4	5	6	8	10	15	25
30	2.96	2.96	2.95	2.94	2.93	2.93	2.90	2.88	2.79	2.17
35	3.03	3.02	3.02	3.01	3.00	3.00	2.98	2.97	2.91	2.64
40	3.08	3.08	3.07	3.07	3.06	3.06	3.05	3.03	3.00	2.84
45	3.13	3.12	3.12	3.12	3.11	3.11	3.10	3.09	3.06	2.96
50	3.17	3.16	3.16	3.16	3.15	3.15	3.14	3.14	3.11	3.04
60	3.23	3.23	3.23	3.23	3.22	3.22	3.22	3.21	3.20	3.15
70	3.29	3.29	3.28	3.28	3.28	3.28	3.27	3.27	3.26	3.23
80	3.33	3.33	3.33	3.33	3.33	3.33	3.32	3.32	3.31	3.29
90	3.37	3.37	3.37	3.37	3.37	3.37	3.36	3.36	3.36	3.34
100	3.41	3.41	3.40	3.40	3.40	3.40	3.40	3.40	3.39	3.38

$\alpha = 0.01$

N	1	2	3	4	5	6	8	10	15	25
5	1.98									
6	2.17	1.98								
7	2.32	2.17	1.98							
8	2.44	2.32	2.18	1.98						
9	2.54	2.44	2.33	2.18	1.99					
10	2.62	2.55	2.45	2.33	2.18	1.99				
12	2.76	2.70	2.64	2.56	2.46	2.34	1.99			
14	2.86	2.82	2.78	2.72	2.65	2.57	2.35	1.99		
16	2.95	2.92	2.88	2.84	2.79	2.73	2.58	2.35		
18	3.02	3.00	2.97	2.94	2.90	2.85	2.75	2.59		
20	3.08	3.06	3.04	3.01	2.98	2.95	2.87	2.76	2.20	
25	3.21	3.19	3.18	3.16	3.14	3.12	3.07	3.01	2.78	
30	3.30	3.29	3.28	3.26	3.25	3.24	3.21	3.17	3.04	2.21
35	3.37	3.36	3.35	3.34	3.34	3.33	3.30	3.28	3.19	2.81
40	3.43	3.42	3.42	3.41	3.40	3.40	3.38	3.36	3.30	3.08
45	3.48	3.47	3.47	3.46	3.46	3.45	3.44	3.43	3.38	3.23
50	3.52	3.52	3.51	3.51	3.51	3.50	3.49	3.48	3.45	3.34
60	3.60	3.59	3.59	3.59	3.58	3.58	3.57	3.56	3.54	3.48
70	3.65	3.65	3.65	3.65	3.64	3.64	3.64	3.63	3.61	3.57
80	3.70	3.70	3.70	3.70	3.69	3.69	3.69	3.68	3.67	3.64
90	3.74	3.74	3.74	3.74	3.74	3.74	3.73	3.73	3.72	3.70
100	3.78	3.78	3.78	3.77	3.77	3.77	3.77	3.77	3.76	3.74

Source: R. E. Lund, "Tables for an Approximate Test for Outliers in Linear Models," *Technometrics*, **17**, No. 4 (1975), pp. 473–476. Reprinted with permission of the author and of the journal.

N = number of observations.
p = number of independent variables (including count for intercept if fitted).

CHAPTER 8

Selection of Variables in Nested Data

8.1 Background of Example

As an example of cautious least-square fitting and selection of influential variables, we take the problem of estimating gasoline yields from various characteristics of the crude and a characteristic of the gasoline produced. The data in Table 8.1 appeared first in an article by N. H. Prater in the *Petroleum Refiner*, and again in a chapter on "Simple and Multiple Regression Analyses" by Hader and Grandage.

The data were obtained in a laboratory study of the distillation properties of various crude oils with respect to their yield of gasoline. The four independent variables measured were:

x_1, crude oil gravity, °API x_3, crude oil ASTM 10% point, °F
x_2, crude oil vapor pressure, psi x_4, gasoline ASTM end point, °F

Thus x_1, x_2, and x_3 are properties of the crude, and x_4 is a more or less independent property of the gasoline produced. The response, y, is gasoline yield, as percentage of crude. The crude oil ASTM 10% point and the gasoline ASTM end point are measurements of crude oil and gasoline volatility. Both measure the temperature at which a given amount of liquid has been vaporized. The end point is the temperature at which *all* of the liquid has been vaporized. As gasoline is distilled from a given crude, the volume of gasoline increases, as does its end point.

The purpose of this work was to obtain an equation for estimating gasoline yields, given the end point of the gasoline desired and the properties of available crude.

TABLE 8.1
PRATER'S DATA ON CRUDE OIL PROPERTIES AND GASOLINE YIELDS

Crude Oil			Gasoline	
Gravity °API x_1	Vapor Pressure, psi x_2	ASTM 10% Point, °F x_3	ASTM End Point, °F x_4	Yield, % y
38.4	6.1	220	235	6.9
40.3	4.8	231	307	14.4
40.0	6.1	217	212	7.4
31.8	0.2	316	365	8.5
40.8	3.5	210	218	8.0
41.3	1.8	267	235	2.8
38.1	1.2	274	285	5.0
50.8	8.6	190	205	12.2
32.2	5.2	236	267	10.0
38.4	6.1	220	300	15.2
40.3	4.8	231	367	26.8
32.2	2.4	284	351	14.0
31.8	0.2	316	379	14.7
41.3	1.8	267	275	6.4
38.1	1.2	274	365	17.6
50.8	8.6	190	275	22.3
32.2	5.2	236	360	24.8
38.4	6.1	220	365	26.0
40.3	4.8	23!	395	34.9
40.0	6.1	217	272	18.2
32.2	2.4	284	424	23.2
31.8	0.2	316	428	18.0
40.8	3.5	210	273	13.1
41.3	1.8	267	358	16.1
38.1	1.2	274	444	32.1
50.8	8.6	190	345	34.7
32.2	5.2	236	402	31.7
38.4	6.1	220	410	33.6
40.0	6.1	217	340	30.4
40.8	3.5	210	347	26.6
41.3	1.8	267	416	27.8
50.8	8.6	190	407	45.7

235

8.2   Fitting Equation

Pass 1 is the least-squares fit of the gasoline yields as a linear function of the four independent variables:

$$Y = b_0 + b_1 x_1 + b_2 x_2 + b_3 x_3 + b_4 x_4.$$

This corresponds to the treatment of the data by Prater* and by Hader and Grandage. Computer printouts of the results are shown in Figures 8A.3–8A.10.

8.3   Recognition of Nested Data

In the component and component-plus-residual plots of x_1, crude oil gravity, x_2, crude oil vapor pressure, and x_3, crude oil ASTM 10% point, Figures 8A.7–8A.9, we see evidence of nesting. There are a number of observations at each combination of values of x_1, x_2, and x_3. The plot of x_4, gasoline ASTM end point, appears to be relatively evenly distributed. Rearranging the data in order of decreasing x_1, as in Table 8.2, reveals that we are dealing with *ten* crudes rather than thirty-two, and that gasoline ASTM end point, x_4, is indeed "nested" within crudes. An alternative way of recognizing nesting, and one we used before the advent of the component and component-plus-residual plots, is to sort the data on each variable on a computer or a card sorter and list the observations after each sort to look for matching x-sets. Since the previous authors analyzed the data (as we have in Pass 1) as though there were thirty-two independent observations with constant variance, we will have an opportunity to see how their conclusions differ from ours.

8.4   Data Identification and Entry

The data were identified for convenience by crude number and by observation number within crudes, and then entered on special forms for linear least-squares fitting by the computer. Indicator variables were included in order to separate and quantify the differing responses of the various crudes. Details concerning data entry and the various computer passes discussed in this chapter are given in Appendix 8A.

8.5   Fitting Equation to Nested Data

A number of computer passes were made to determine whether some other equations would fit the nested data better. The effect of gasoline end point on gasoline yield can be seen in Figure 8.1. A least-squares line has been drawn for each crude.

* There are a number of numerical errors in the Prater paper.

TABLE 8.2
PRATER'S DATA ARRANGED IN ORDER OF DECREASING X_1

Identi-fication Number	Crude Oil			Gasoline	
	Gravity, °API	Vapor Pressure, psi	ASTM 10% Point, °F	ASTM End Point, °F	Yield, %
	x_1	x_2	x_3	x_4	y
1-1	50.8	8.6	190	205	12.2
2				275	22.3
3				345	34.7
4				407	45.7
2-1	41.3	1.8	267	235	2.8
2				275	6.4
3				358	16.1
4				416	27.8
3-1	40.8	3.5	210	218	8.0
2				273	13.1
3				347	26.6
4-1	40.3	4.8	231	307	14.4
2				367	26.8
3				395	34.9
5-1	40.0	6.1	217	212	7.4
2				272	18.2
3				340	30.4
6-1	38.4	6.1	220	235	6.9
2				300	15.2
3				365	26.0
4				410	33.6
7-1	38.1	1.2	274	285	5.0
2				365	17.6
3				444	32.1
8-1	32.2	5.2	236	267	10.0
2				360	24.8
3				402	31.7
9-1	32.2	2.4	284	351	14.0
2				424	23.2
10-1	31.8	0.2	316	365	8.5
2				379	14.7
3				428	18.0

237

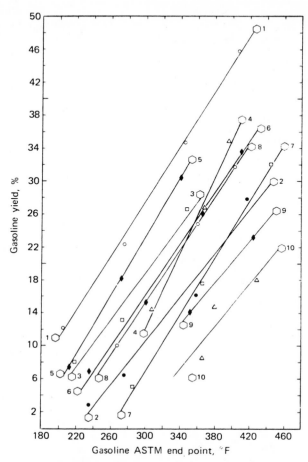

Figure 8.1   Gasoline yields from ten crudes.

Use of indicator (dummy) variables

Pass 2 was made under the assumption that a correlation between gasoline end point and yield could be represented by a set of ten straight lines with a common slope (each line representing a different crude). The equation used was:

$$Y = b_0 + b_1 x_1 + b_2 x_2 + \cdots + b_{10} x_{10},$$

where y = gasoline yield,

x_1 = gasoline end point—all crudes have common b_1 slope,

$x_2, \ldots, x_{10}$ = indicator or "dummy" variables for crudes 2–10, to separate and offset by crude type (see Chapter 4, page 56).

The constant term, b_0, is an estimate of y for Crude 1 at zero end point, and the coefficient of each of the other crude variables indicates the magnitude by which that crude systematically differs from Crude 1. Thus, with $i = 2$–10, when a given x_i (indicator variable) is 1, the value of the associated b_i becomes the offset of the ith crude from Crude 1.

Crude	i, Where $x_i = 1$, All Other $x_i = 0$	Slope Intercept
1		$Y_1 = b_1 x_1 + b_0$
2	2	$Y_2 = b_1 x_1 + b_0 + b_2 x_2$
3	3	$Y_3 = b_1 x_1 + b_0 + b_3 x_3$
4	4	$Y_4 = b_1 x_1 + b_0 + b_4 x_4$
.	.	.
.	.	
.	.	
10	10	$Y_{10} = b_1 x_1 + b_0 + b_{10} x_{10}$

The control and transformation card entry form for Pass 2 is shown in Figure 8A.11. The computer printouts of Pass 2 are shown in Figures 8A.12–8A.14. As can be seen in Figure 8A.12, the fitted values for points 1 and 3 of ⌐rude 4—observations *41* and *43*—have the largest residuals, suggesting that ᴉe slope for Crude 4 may differ from the slopes for other crudes.

Test for significance of added variables, "individual" versus ʿcommon" slopes

Pass 3 was run to determine whether the data could be represented significantly better by straight lines with individual slopes. The equation used was:

$$Y = b_0 + b_1 x_1 + b_2 x_2 + \cdots + b_{10} x_{10} + b_{11} x_{11} + \cdots + b_{19} x_{19}.$$

Variables $x_2, \ldots, x_{10}$ are indicator variables as before. Variables $x_{11}, \ldots, x_{19}$ are cross-product terms to allow each crude to have its own slope, for example, $x_{11} = x_1 x_2$, $x_{12} = x_1 x_3$. Transformation card entries were made as shown in Figures 8A.1 and 8A.11 to include variables representing the cross products. The mean end point of all the data ($332°F$, obtained from Pass 1) was subtracted from each observation in order to obtain more meaningful t-values by reducing the correlations between independent variables.

The constant term, b_0, is then the value of Y for Crude 1 at $332°F$, and Y is calculated for the various crudes as shown in the table. The results of the fit of the full equation are shown in Figure 8A.15.

Crude	i, Where $x_i = 1$, All Other $x_i = 0$	Slope	Intercept
1		$Y_1 = b_1 x_1$ $\qquad + b_0$	
2	2	$Y_2 = b_1 x_1 + b_{11} x_1 x_2 + b_0 + b_2 x_2$	
3	3	$Y_3 = b_1 x_1 + b_{12} x_1 x_3 + b_0 + b_3 x_3$	
4	4	$Y_4 = b_1 x_1 + b_{13} x_1 x_4 + b_0 + b_4 x_4$	
.	.	.	
.	.	.	
.	.	.	
10	10	$Y_{10} = b_1 x_1 + b_{19} x_1 x_{10} + b_0 + b_{10} x_{10}$	

The hypothesis that the additional individual slope coefficients do not improve the fit was tested by the F ratio of the mean square due to the extra variables divided by the residual mean square (Brownlee, Section 13.8, page 443, and Davies, 1960, Section 11B.32).

Source of Variation	Number of Coefficients	Degrees of Freedom	Residual Sum of Squares	Residual Mean Square
Common slope	$p \qquad = 10$	21	74	
Individual slope	$p + q = 19$	12	30	2.5
Extra variables	$q = 9$	9	44	4.9

$$F(9, 12) = \frac{\text{RMS}_q}{\text{RMS}_{p+q}} = \frac{4.9}{2.5} = 1.9.$$

From F-tables, $F(0.95, 9, 12)$ is 2.8. Thus, if we use the 95% level as our criterion of significance, the calculated F-ratio value is nonsignificant. The data are therefore considered to be compatible with the hypothesis that the additional coefficients do not improve the fit.

8.6   Fitting Equation Among Sets of Nested Data

From the computer printout of Pass 2, Figure 8A.12, we see that b_1 (the rate of change of gasoline yield per degree change in gasoline end point) was 0.1587. The standard error of this coefficient is 0.0057. Hence the slope might well be reported as 0.159 ± 0.006.

This average within-crude slope was used to adjust the ten crudes to the same end-point value, 332°F. This is done graphically in Figure 8.2, and the resulting gasoline yields are given in Figure 8A.16 as variable 4. This figure presents the data used to correlate the yield of gasoline at this constant end point with gravity, vapor pressure, and 10% point of the crudes, that is, with x_1, x_2, and x_3.

With three independent variables there are 2^3 equations to be considered. All seven combinations of variables were used, as shown in the table on page 243.

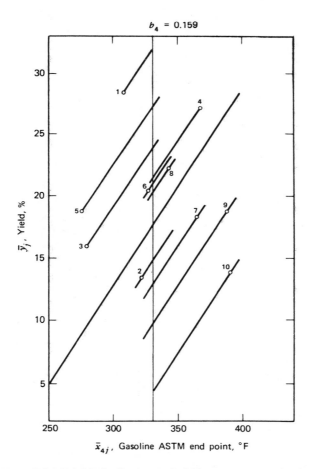

Figure 8.2 Graphical adjustment of yields to constant end point.

TABLE 8.3

SUMMARY OF PASSES 4–10: YIELD OF 332°F END-POINT GASOLINE VERSUS GRAVITY, VAPOR PRESSURE, AND 10% POINT OF CRUDE

	Pass 4 (x_1, x_2, x_3)				Pass 5 (x_1, x_2)				Pass 6 (x_1, x_3)				Pass 7 (x_2, x_3)			
	Value	b_i	$s(b_i)$	t_i	Value	b_i	$s(b_i)$	t_i	Value	b_i	$s(b_i)$	t_i	Value	b_i	$s(b_i)$	t_i
Degrees of freedom	6				7				7				7			
F-value	70				30				100				85			
R^2	0.971				0.89				0.966				0.961			
RMS	3.10				10.10				3.15				3.63			
RSS	18.6				70.7				22.1				25.4			
SSFE	630.6				578.5				627.1				623.8			
C_p	4.0				19.0				3.1				4.2			
Variable																
x_1, gravity		0.22	0.16	1.5		0.50	0.23	2.1		0.20	0.15	1.4				
x_2, vapor pressure		0.54	0.51	1.0		2.27	0.51	4.4*						0.47	0.55	0.9
x_3, 10% point		−0.16	0.04	4.1*						−0.19	0.02	8.9*		−0.18	0.04	5.0*

	Pass 8 (x_1)				Pass 9 (x_2)				Pass 10 (x_3)			
	Value	b_i	$s(b_i)$	t_i	Value	b_i	$s(b_i)$	t_i	Value	b_i	$s(b_i)$	t_i
Degrees of freedom	8				8				8			
F-value	10				35				175			
R^2	0.58				0.82				0.957			
RMS	33.76				14.5				3.52			
RSS	270.1				116.0				28.1			
SSFE	379.1				533.2				621.1			
C_p	81.0				31.0				3.0			
Variable												
x_1, gravity		1.13	0.34	3.3*								
x_2, vapor pressure						2.93	0.48	6.1*				
x_3, 10% point										−0.21	0.02	13.3*

* Significant at the 99% level.

	Crude Properties		
Pass	Gravity	Vapor Pressure	10% Point
4	x_1	x_2	x_3
5	x_1	x_2	—
6	x_1	—	x_3
7	—	x_2	x_3
8	x_1	—	—
9	—	x_2	—
10	—	—	x_3

Results are summarized in Table 8.3.

8.7 Selecting Variables Based on Total Error, Recognizing Bias and Random Error

Since there are a number of combinations of variables from which to choose, Mallows' graphical method of plotting p versus C_p is used to review all the alternatives. The statistic C_p was defined in Chapter 6 as follows:

$$C_p = \frac{\text{RSS}}{s^2} - (N - 2p),$$

where RSS = residual sum of squares,
s^2 = best estimate of variance (normally the residual mean square of the equation with all variables present),
N = number of observations,
p = number of constants to be estimated; $p_{\max} = K + 1$ with b_0 present, $p_{\max} = K$ with b_0 absent.

For illustration, values for p and C_p have been calculated from computer Passes 4–10 and are given in Table 8.4. In practice, only Pass 4, the full equation, need be run in the curve fitting program, with the option set to search for the candidate equations.

The C_p versus p plot of all of the equations is shown in Figure 8.3.

As all equations close to the zero-bias line, $C_p = p$, contain x_3, this is clearly the most important variable. The equation with x_1 and x_3 is on the line and so is judged to have no bias, or at least no more than the full equation in x_1, x_2, and x_3. The equation with x_3 alone has more bias (A to B) and less random error (B to C) than the equation with x_1 and x_3. Both have sensibly the same total squared error (bias plus random).

TABLE 8.4

C_p OF EQUATIONS USED IN PASSES 4–10

$N = 10, s^2 = 3.10*$

Pass	Variables	p	RSS	$\dfrac{\text{RSS}}{s^2}$	$N - 2p$	C_p
4	x_1, x_2, x_3	4	18.6	6.0	2	4.0
5	x_1, x_2	3	70.7	23.0	4	19.0
6	x_1, x_3	3	22.1	7.1	4	3.1
7	x_2, x_3	3	25.4	8.2	4	4.2
8	x_1	2	270.1	87.0	6	81.0
9	x_2	2	116.0	37.0	6	31.0
10	x_3	2	28.1	9.0	6	3.0
	None	1	649.2	209.0	8	201.0

* s^2 from Pass 4.

The value of each coefficient, as affected by the inclusion of other variables, is as follows:

Pass	b_1	b_2	b_3
4	0.22	0.54	−0.16
5	0.50	2.27	—
6	0.20	—	−0.19
7	—	0.47	−0.18
8	1.13	—	—
9	—	2.93	—
10	—	—	−0.21

Coefficient b_3 is relatively stable, regardless of the presence of other variables. With b_3 present, b_1 is relatively stable, b_2 more erratic. We do not have a good estimate of the influence of x_2, but it appears that we have determined clearly the influences of x_1 and x_3. Combining this information with that from the C_p plots, we conclude that the x_3 equation is the most economical one for minimizing total error (bias plus random), and the x_1, x_3 equation the most economical for minimizing bias. We have chosen the x_3 equation from Pass 10 for our prediction equation.

8.8   Properties of Final Equation

From Pass 10, b_3, the rate of change of gasoline yield per degree of the 10% point of the crude, was found to be -0.212 ± 0.016. It is rather im-

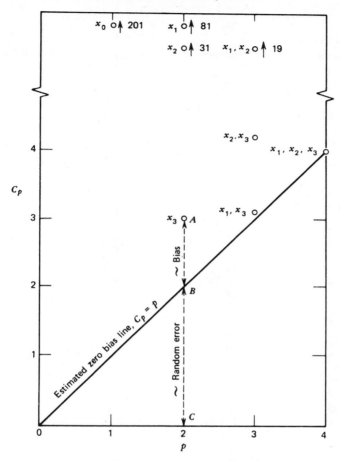

Figure 8.3 C_p versus p plot.

portant to remember that the standard error of the coefficient is based on only 8 residual degrees of freedom and that b_3 is based on only ten crudes. The equation can be expected to hold only for crudes like these, and *not* for dissimilar ones.

The residual mean square of Pass 10 (3.52) is really composed of two parts. One is due to the fact that each adjusted y is the average of several within-crude values and hence includes some within-crude variation, and the other is due to whatever random factors vary from crude to crude. We have an estimate of the "within-crude" component from Pass 2; it is 3.53 with 21

TABLE 8.5
Comparison of Observed and Predicted Yields
(Prater Data)

Identification Number	Gasoline Yields, % on Crude		
	Observed	d_w (Pass 2)	d_a (Pass 10)
1-1	12.2	−0.2	2.0
2	22.3	−1.2	
3	34.7	0.1	
4	45.7	1.3	
2-1	2.8	3.2	0.8
2	6.4	0.4	
3	16.1	−3.0	
4	27.8	−0.6	
3-1	8.0	1.8	−2.0
2	13.1	−1.8	
3	26.6	0.0	
4-1	14.4	−3.1	−0.3
2	26.8	−0.3	
3	34.9	3.4	
5-1	7.4	−1.3	2.9
2	18.2	0.0	
3	30.4	1.3	
6-1	6.9	1.2	−3.0
2	15.2	−0.9	
3	26.0	−0.4	
4	33.6	0.1	
7-1	5.0	−0.6	0.3
2	17.6	−0.7	
3	32.1	1.3	
8-1	10.0	−0.1	−0.4
2	24.8	−0.1	
3	31.7	0.2	
9-1	14.0	1.2	−0.8
2	23.2	−1.2	
10-1	8.5	−1.2	0.5
2	14.7	2.8	
3	18.0	−1.6	

degrees of freedom. We can get a rough estimate of the "among-crude" component as follows:

$$\text{RMS (Pass 10)} = s^2_{\text{among}} + \frac{s^2_{\text{within}}}{n},$$

where n is the average number of runs per crude, or $32/10$ or 3.2. Thus:

$$3.52 = s^2_{\text{among}} + \frac{3.53}{3.2},$$

whence
$$s^2_{\text{among}} = 3.52 - 1.10 = 2.42.$$

All this work can be summarized in one equation *plus* the two separate variance-component estimates just given. The "nested" equation is

$$Y = 70.84 - 0.212x_3 + 0.159(x_4 - 332).$$

A detailed view of how this equation represents the thirty-two data points from which it was derived can be observed in the last two columns of Table 8.5, headed d_w and d_a. The d_w-values represent *within*-crude variations, and the d_a *among*-crude variations. The total deviation between an observation and its value predicted by the equation is the sum of a d_w and a d_a.

The estimated variance of Y for the equation given above, at $x_4 = 332°F$ end point, is:

$$\text{Est. Var. } (Y) = s^2 \left[\frac{1}{N} + \frac{(x_3 - \bar{x}_3)^2}{\Sigma(x_{3i} - \bar{x}_3)^2} \right]$$

(see Chapter 2, page 12). For $x_3 = \bar{x}_3$, Est. Var. $(\bar{Y})$ is s^2/N.

For this final equation, the effective variance is 3.52. Since there are ten crudes,

$$\text{Est. Var. } (\bar{Y}) = \frac{3.52}{10} = 0.352.$$

The resultant 95% confidence bounds at the center of the data are then

$$\bar{Y} \pm \{[2F(0.95, 2, N - 2)][\text{Est. Var. } (\bar{Y})]\}^{\frac{1}{2}} = \bar{y} \pm [2(4.46)(0.35)]^{\frac{1}{2}}$$
$$= \bar{y} \pm 1.77.$$

8.9 Comparison of Equations

It is interesting to compare both the fit and the usefulness of the equation just given with those of the former equation, which ignored nesting. As seen in Table 8.6, the new equation is simpler (only half the number of variables are required) and the fit is better (the residual mean square is one-third smaller). In addition, we now have estimates of two components of variance (among and within crudes). The earlier equation, ignoring these components, produced a confidence interval for Y that was 40% too small.

TABLE 8.6
COMPARISON OF EQUATIONS, FORMER AND PRESENT
(Prater Data)

Criterion	Equation	
	Former	Present
Number of independent variables	4	2
Overall fit		
C_p	4.0	3.0
Residual mean square	5.0	3.5
Components of variance		
Among crudes (9 df)	—	2.42
Within crudes (19 df)	—	3.53
95% confidence interval for Y at		
center of data	1.02	1.77

APPENDIX 8A

DATA PREPARATION AND COMPUTER PRINTOUTS OF EXAMPLE

Figure 8A.1 shows (1) the data entry form used for card keypunching; (2) the information given on the format card to tell the computer the locations and dimensions of the fields used for identification, observation numbers, sequence numbers, and variables on the data cards; and (3) a summary of the transformation card entries for the first six computer passes. Details of the control cards and transformation entry forms used for the first and second passes are shown in Figures 8A.2 and 8A.11.

The order of the cards for the first three passes run in sequence was as follows:

Pass 1   Control card (1 in column 48)
          Format card
          Transformation cards (1)
          Information cards (4)
          Variable name cards (1)
          Data cards
          END card (the word "END" in columns 1–3)

Pass 2   Control card (1 in columns 42 and 48)
             Format card
             Transformation cards (5)
             Information cards (6)
             Variable name cards (5)
             Data cards
             END card (the word "END" in columns 1–3)

Pass 3   Control card (2 in columns 42 and 48)
             Transformation card (5)
             Information cards (3)

Passes 2 and 3 used the same format; otherwise, new format, data, and END cards would have been required for each pass.

Printouts of the various computer passes are given in the following figures:

Pass	Figures
1	8A.3 –8A.10
2	8A.11–8A.14
3	8A.15
4	8A.16

ENTER WEIGHTING FACTOR, IF ANY, AS LAST ENTRY.
"END" CARD MUST BE LAST CARD.

IDENT.	OBSV. NO.	SEQ. NO.	$X_{1-11-21}$	$X_{2-12-22}$	$X_{3-13-23}$	$X_{4-14-24}$	$X_{5-15-25}$	$X_{6-16-26}$	$X_{7-17-27}$	$X_{8-18-28}$	$X_{9-19-29}$	$X_{10-20-30}$	USER'S OPTION
CRUDE			GRAVITY	VAPOR PRS	10% PT	END PT	2,3,4,5,6 ← CRUDE → 7,8,9,10					YIELD	
1	1,1		5,0,8	8.6	1,9,0	2,0,5						1,2,2	
	1,2					2,7,5						1,2,23	
	1,3					3,4,5						3,4,7	
	1,4					4,0,7						4,5,7	
2	2,1		4,1,3	1,.8	2,6,7	2,3,5	1					2,8	
	2,2					2,7,5						6,4	
	2,3					3,5,8						1,6,1	
	2,4					4,1,6						2,7,8	
3	3,1		4,0,8	3,.5	2,1,0	2,1,8	1					8,0	
	3,2					2,7,3						1,3,1	
	3,3					3,4,7						2,6,6	
4	4,1		4,0,3	4,.8	23,1	3,07	1					1,4,4	
	4,2					3,6,7						2,6,8	
	4,3					3,9,5						3,4,9	
5	5,1		4,0,0	6,.1	2,1,7	2,1,2	1					7,4	
	5,2					2,7,2						1,8,2	
	5,3					3,4,0						3,0,4	
6	6,1		3,8,4		2,2,0	2,3,5		1				6,9	
6	6,2					3,0,0						1,5,2	

FORMAT CARD INFORMATION FOR PASSES 1 TO 6

```
              *a  *b      *c          *d  *e
Card:        (A6, I4, I2, 4F6.2, 26F1.0, 4X, F6.1)
Columns:     12345678901234567890123456789012345567890---
```

* Explanation of Symbols Used

	Number with this type of Spacing	Type of Field	Number of Columns Each	Number of Columns to Right of Decimal Point
a	1	A = letter or number	6	0
b	1	I = integer	4	0
c	4	F = field	6	2
d	4	X = exclude	1	0
e	Statement must start and end with parenthesis.			

Figure 8A.1

Passes 1, 2 and 3

					Variable		
					Pass	Pass	Pass
Position	Name	C	ARG	LOC	1	2	3
1	Gravity	00		1	1		
2	Vapor Pressure	00		2	2		
3	10% Point	00		3	3		
4	End Point	00		4	4	1	
4	End Point-Mean E.P.	08*	-332.	4			1
5	(End Point-Mean E.P.)^2	09**	0404	5			
6							
7							
8							
9							
10							
11							
	Crude						
12	2	00		12		2	2
13	3	00		13		3	3
14	4	00		14		4	4
15	5	00		15		5	5
16	6	00		16		6	6
17	7	00		17		7	7
18	8	00		18		8	8
19	9	00		19		9	9
20	10	00		20		10	10
21							
	End Point x Crude						
22	End Point x 2	09**	0412	22			11
23	End Point x 3	09	0413	23			12
24	End Point x 4	09	0414	24			13
25	End Point x 5	09	0415	25			14
26	End Point x 6	09	0416	26			15
27	End Point x 7	09	0417	27			16
28	End Point x 8	09	0418	28			17
29	End Point x 9	09	0419	29			18
30	End Point x 10	09	0420	30			19
31	Gasoline Yield, %	00		40	5	11	20

   * 08 denotes operation: X ± K
  ** 09 denotes operation: X(N)·X(M)

Figure 8A.1 (*continued*).

251

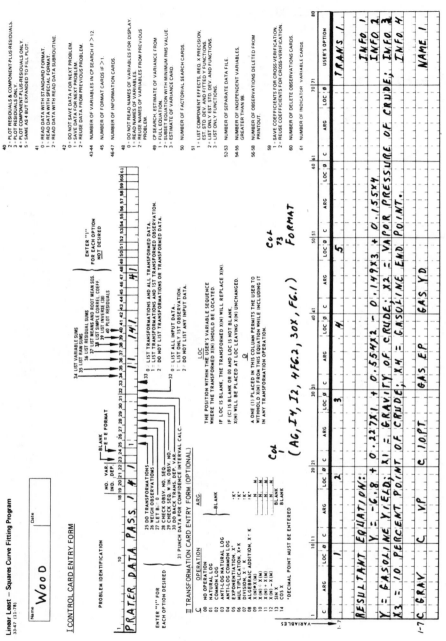

Figure 8A.2

PRATER DATA PASS 1

DATA INPUT   4 INDEPENDENT VARIABLES   1 DEPENDENT VARIABLE(S)

OBSV.	SEQ.	1-11-21	2-12-22	3-13-23	4-14-24	5-15-25	6-16-26	7-17-27	8-18-28	9-19-29	10-20-30
11	0	50.800	8.600	190.000	205.000	12.200					
12	0	50.800	8.600	190.000	275.000	22.300					
13	0	50.800	8.600	190.000	345.000	34.700					
14	0	50.800	8.600	190.000	407.000	45.700					
21	0	41.300	1.800	267.000	235.000	2.800					
22	0	41.300	1.800	267.000	275.000	6.400					
23	0	41.300	1.800	267.000	358.000	16.100					
24	0	41.300	1.800	267.000	416.000	27.800					
31	0	40.800	3.500	210.000	218.000	8.000					
32	0	40.800	3.500	210.000	273.000	13.100					
33	0	40.800	3.500	210.000	347.000	26.600					
41	0	40.300	4.800	231.000	307.000	14.400					
42	0	40.300	4.800	231.000	367.000	26.800					
43	0	40.300	4.800	231.000	395.000	34.900					
51	0	40.000	6.100	217.000	212.000	7.400					
52	0	40.000	6.100	217.000	272.000	18.200					
53	0	40.000	6.100	217.000	340.000	30.400					
61	0	38.400	6.100	220.000	235.000	6.900					
62	0	38.400	6.100	220.000	300.000	15.200					
63	0	38.400	6.100	220.000	365.000	26.000					
64	0	38.400	6.100	220.000	410.000	33.600					
71	0	38.100	1.200	274.000	285.000	5.000					
72	0	38.100	1.200	274.000	365.000	17.600					
73	0	38.100	1.200	274.000	444.000	32.100					
82	0	32.200	5.200	236.000	360.000	24.800					
81	0	32.200	5.200	236.000	267.000	10.000					
83	0	32.200	5.200	236.000	402.000	31.700					
91	0	32.200	2.400	284.000	351.000	14.000					
92	0	32.200	2.400	284.000	424.000	23.200					
101	0	31.800	0.200	316.000	365.000	8.500					
102	0	31.800	0.200	316.000	379.000	14.700					
103	0	31.800	0.200	316.000	428.000	18.000					

SUMS OF VARIABLES
1.25600D+03 1.33800D+02 7.72800D+03 1.06270D+04 6.29100D+02

MEANS OF VARIABLES
3.92500D+01 4.18125D+00 2.41500D+02 3.32094D+02 1.96594D+01

ROOT MEAN SQUARES OF VARIABLES
5.63543D+00 2.61983D+00 3.75140D+01 6.97560D+01 1.07224D+01

SIMPLE CORRELATION COEFFICIENTS, R(I,I PRIME)

1	1.000				
2	0.621	1.000			
3	-0.700	-0.906	1.000		
4	-0.322	-0.298	0.412	1.000	
5	0.246	0.384	-0.315	0.712	1.000

Figure 8A.3

LINEAR LEAST-SQUARES CURVE FITTING PROGRAM

CRATER DATA PASS 1   DEP VAR 1: GAS YD   MIN Y = 2.800D+00   MAX Y = 4.570D+01   RANGE Y = 4.290D+01

RESULTANT EQUATION:

Y = -6.8 + 0.227X1 + 0.554X2 - 0.149X3 + 0.155X4

Y = GASOLINE YIELD; X1 = GRAVITY OF CRUDE; X2 = VAPOR PRESSURE OF CRUDE;
X3 = 10 PERCENT POINT OF CRUDE; X4 = GASOLINE END POINT.

IND.VAR(I)	NAME	COEF.B(I)	S.E. COEF.	T-VALUE	R(I)SQRD	MIN X(I)	MAX X(I)	RANGE X(I)	REL.INF.x X(I)
0		-6.6201D+00							
1	C GRAV	2.2724C-01	9.09D-02	2.3	0.4922	3.1800+C1	5.080D+01	1.9900+01	0.10
2	C VP	5.5372C-01	3.70D-01	1.5	0.8284	2.0000-01	8.600+00	8.400D+00	0.11
3	C 10PT	-1.4562C0-01	2.52D-01	5.1	0.8662	1.900D+02	3.160D+02	1.260D+02	C.44
4	GAS EP	1.5465CD-01	6.45D-03	24.0	0.2034	2.050D+02	4.440D+02	2.390D+02	0.86

NO. OF OBSERVATIONS   32
NO. OF IND. VARIABLES   4
RESIDUAL DEGREES OF FREEDOM   27
F-VALUE   171.7
RESIDUAL ROOT MEAN SQUARE   2.2344438E
RESIDUAL MEAN SQUARE   4.99273937
RESIDUAL SUM OF SQUARES   134.6039€29C
TOTAL SUM OF SQUARES   3564.07718750
MULT. CORREL. COEF. SQUARED   .9622

IDENT.	OBSV.	WSS DISTANCE	OBS. Y	FITTED Y	RESIDUAL		OBSV.	OBS. Y	MIN X(I)	FITTED Y	RESIDUAL	ORDERED RESID.	STUD.RESID.	SEQ
1	11	92.	12.200	12.777	-0.577		53	30.400	30.400	25.779	4.621	4.621	2.1	1
1	12	30.	22.300	23.602	-1.302		43	34.9CC	34.9CC	31.539	3.361	3.361	1.6	2
1	13	15.	34.700	34.428	0.272		52	18.2CC	18.2CC	15.262	2.938	2.938	1.4	3
1	14	41.	45.700	44.016	1.684		102	14.700	14.700	11.876	2.824	2.824	1.4	4
1	21	46.	2.800	-0.022	2.822		21	2.800	2.800	-0.022	2.822	2.822	1.4	5
2	22	19.	6.430	6.164	0.236		73	3€.100	3€.100	30.194	1.906	1.906	0.9	6
2	23	7.	16.100	19.000	-2.900		14	45.700	45.700	44.016	1.654	1.654	0.9	7
2	24	37.	27.800	27.570	-0.170		83	31.7CC	31.7CC	30.255	1.445	1.445	0.7	8
3	31	67.	8.000	6.700	1.300		51	7.400	7.400	5.983	1.417	1.417	0.7	9
3	32	21.	13.100	15.206	-2.1C6		31	8.000	8.000	6.700	1.300	1.300	0.7	10
3	33	6.	26.600	26.650	-C.050		82	24.8CC	24.8CC	23.760	1.040	1.040	0.5	11
4	41	4.	14.400	17.930	-3.530		81	10.300	10.300	9.377	0.623	0.623	C.3	12
4	42	6.	26.800	27.209	-0.405		91	14.300	14.300	13.640	0.360	0.360	0.2	13
4	43	19.	34.937	31.539	3.361		13	34.700	34.700	34.428	0.272	0.272	0.1	14
5	51	72.	7.400	5.983	1.417		22	6.400	6.400	5.164	0.236	0.236	0.1	15
5	52	20.	18.233	15.262	2.938		33	26.6CC	26.6CC	26.650	-0.050	-0.050	-0.0	16
5	53	3.	30.400	25.779	4.621		24	27.800	27.800	27.970	-0.170	-C.170	-0.1	17
6	61	47.	6.900	6.728	0.172		72	17.6CO	17.6CO	17.976	-0.376	-0.376	-0.2	18
6	62	7.	15.230	18.780	-3.580		42	26.800	26.800	27.209	-0.4C9	-C.4C9	-0.2	19
6	63	31.	26.000	28.633	-2.833		11	12.200	12.200	12.777	-0.577	-0.577	-0.2	20
6	64	7.	33.600	35.792	-2.192		71	5.000	5.000	5.604	-0.604	-0.604	-C.3	21
7	71	16.	5.000	5.604	-0.604		101	8.5CC	8.5CC	9.710	-1.210	-1.210	-0.6	22
7	72	10.	17.600	17.976	-0.376		12	22.300	22.300	23.602	-1.302	-1.302	-0.7	23
7	73	65.	32.100	30.194	1.9C6		103	18.000	18.000	19.453	-1.453	-1.453	-0.7	24
8	81	4.	24.803	23.760	1.040		92	23.200	23.200	24.929	-1.729	-1.729	-0.8	25
8	82	21.	10.030	9.377	0.623		61	6.900	6.900	8.728	-1.828	-1.828	-0.9	26
8	83	24.	31.700	30.255	1.445		32	13.1CC	13.1CC	15.206	-2.106	-2.106	-1.1	27
9	91	11.	14.030	13.640	0.360		64	33.600	33.600	35.792	-2.192	-2.192	-1.1	28
9	92	45.	23.203	24.525	-1.729		63	26.000	26.000	28.833	-2.833	-2.633	-1.3	29
10	101	32.	8.500	9.710	-1.210		23	16.100	16.100	19.000	-2.900	-2.900	-1.4	30
10	102	37.	14.700	11.876	2.824		41	14.400	14.400	17.930	-3.530	-3.530	-1.6	31
10	103	70.	18.000	19.453	-1.453		62	15.200	15.200	18.780	-3.580	-3.580	-1.7	32

Figure 8A.4

254

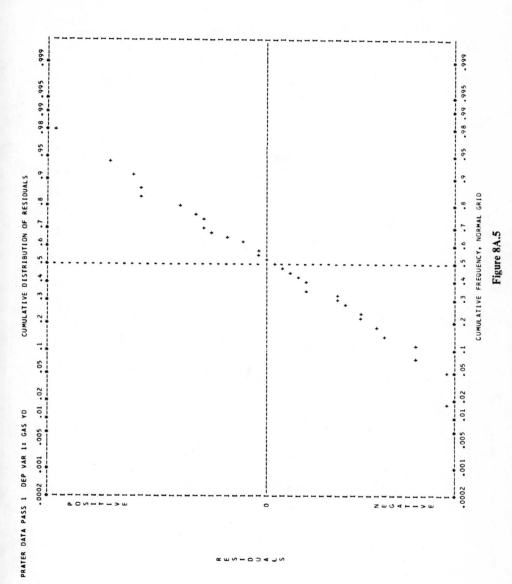

PRATER DATA PASS 1 DEP VAR 1: GAS YD   CUMULATIVE DISTRIBUTION OF RESIDUALS

CUMULATIVE FREQUENCY, NORMAL GRID

Figure 8A.5

255

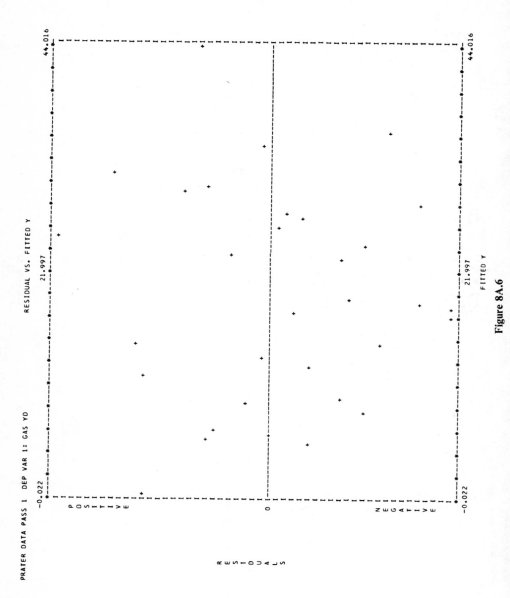

Figure 8A.6

256

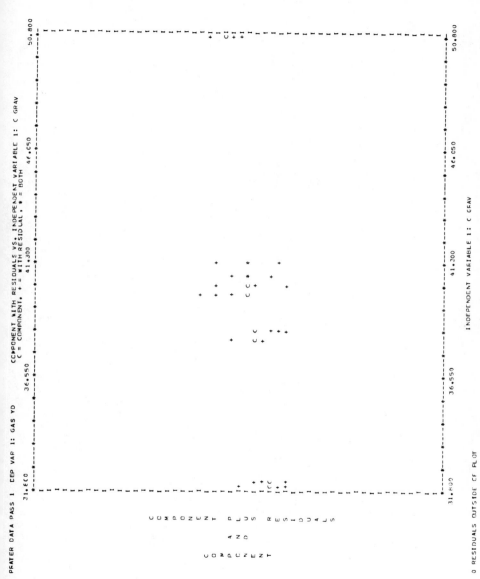

PRATER DATA PASS 1 DEP VAR 1: GAS YD

COMPONENT WITH RESIDUALS VS. INDEPENDENT VARIABLE 1: C GRAV
C = COMPONENT, + = WITH RESIDUAL, * = BOTH

INDEPENDENT VARIABLE 1: C GRAV

O RESIDUALS OUTSIDE OF PLOT

Figure 8A.7

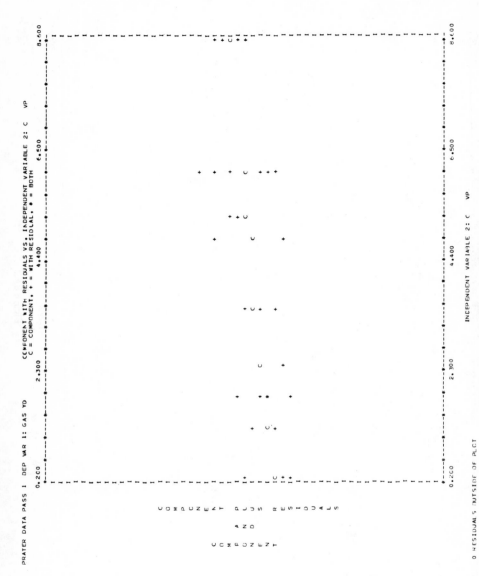

Figure 8A.8

PRATER DATA PASS 1   DEF VAR 1: GAS YD   COMPONENT WITH RESIDUALS VS. INDEPENDENT VARIABLE 3: C 10PT
                                             C = COMPONENT, + = WITH RESIDUAL, * = BOTH

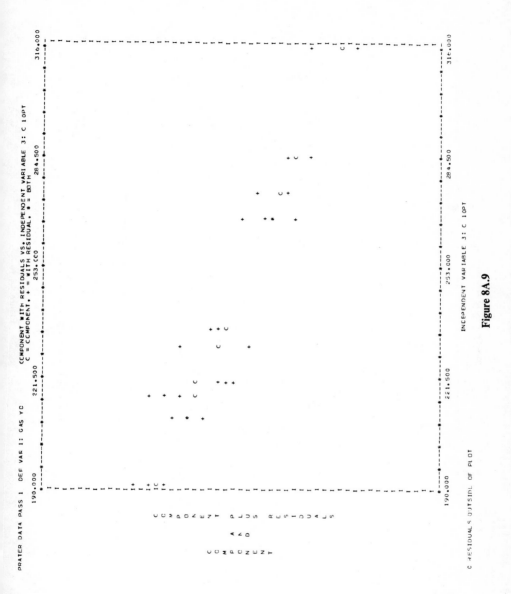

INDEPENDENT VARIABLE 3: C 10PT

Figure 8A.9

259

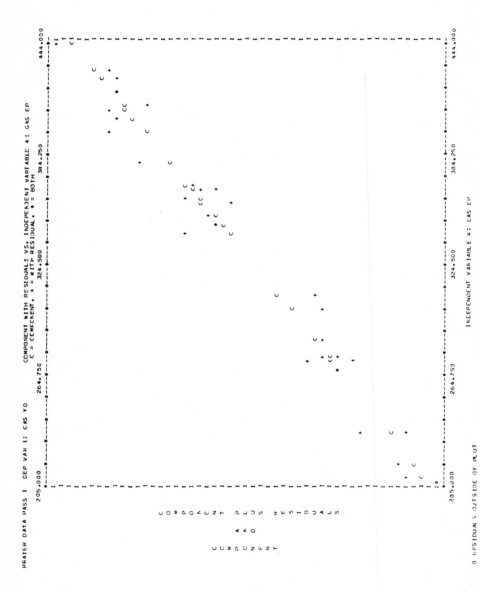

Figure 8A.10

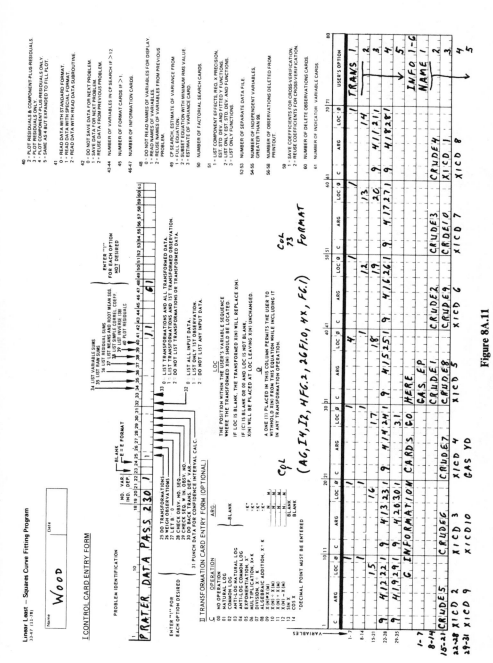

Figure 8A.11

261

PRATER DATA PASS 2   DEP VAR 1: GAS YD   LINEAR LEAST-SQUARES CURVE FITTING PROGRAM

MIN Y = 2.600D+00   MAX Y = 4.570D+01   RANGE Y = 4.290D+01

RESULTANT EQUATION:
Y = -20.2 + 0.159X1 - 17.5X2 - 8.3X3 - 11.0X4 - 4.8X5 - 11.4X6

Y = GASOLINE YIELD; X1 = GASOLINE END POINT; X2 --- X10 = CRUDES 2 TO 10
Y FOR CRUDE 1 = B(0)+ B(1)X1; Y FOR CRUDE 2 = (B(0)+ B(2)X2) + B(1)X1;
Y FOR CRUDE 3 = (B(0)+ B(3)X3) + B(1)X1; ETC.

IND.VAR(I)	NAME	COEF.B(I)	S.E. COEF.	T-VALUE	R(I)SQRD	MIN X(I)	MAX X(I)	RANGE X(I)	REL.INF.X(I)
0	GAS EP	-2.017630D+01	5.720D-03	27.8	0.2843	2.050D+02	4.440D+02	2.390D+02	0.88
1	CRUDE2	-1.951350D+01	1.33D+00	13.2	0.4304	0.0	1.0000+00	1.0000+00	0.88
2	CRUDE3	-8.274150D+00	1.44D+00	5.7	0.3776	0.0	1.0000+00	1.0000+00	0.19
3	CRUDE4	-1.103030D+01	1.46D+00	7.5	0.3920	0.0	1.0000+00	1.0000+00	0.26
4	CRUDE5	-4.767350D+00	1.45D+00	3.3	0.3804	0.0	1.0000+00	1.0000+00	0.11
5	CRUDE6	-1.139520D+01	1.33D+00	8.5	0.4326	0.0	1.0000+00	1.0000+00	0.27
6	CRUDE7	-1.948630D+01	1.47D+00	13.2	0.4000	0.0	1.0000+00	1.0000+00	0.45
7	CRUDE8	-1.211390D+01	1.45D+00	8.4	0.3815	0.0	1.0000+00	1.0000+00	0.28
8	CRUDE9	-2.274400D+01	1.69D+00	13.5	0.3404	0.0	1.0000+00	1.0000+00	0.53
10	CRDE10	-2.811330D+01	1.51D+00	18.6	0.4312	0.0	1.0000+00	1.0000+00	0.66

NO. OF OBSERVATIONS   32
NO. OF IND. VARIABLES   10
RESIDUAL DEGREES OF FREEDOM   21
F-VALUE   98.9
RESIDUAL ROOT MEAN SQUARE   1.87885466
RESIDUAL MEAN SQUARE   3.53009485
RESIDUAL SUM OF SQUARES   74.13109186
TOTAL SUM OF SQUARES   3564.07718750
MULT. CORREL. COEF. SQUARED   .9752

---------ORDERED BY COMPUTER INPUT---------

IDENT.	OBSV. WS DISTANCE	OBSV. Y	FITTED Y	RESIDUAL
1	12	12.300	12.476	-0.176
1	13	22.300	23.487	-1.187
1	14	22.400	34.598	-0.102
1	46	45.700	44.439	1.261
2	21	2.800	-0.376	3.176
2	95	6.400	5.973	0.427
2	135	16.100	19.148	-3.048
2	76	27.200	28.354	-0.554
2	122	8.000	6.165	1.835
3	31	13.100	14.895	-1.795
3	47	26.600	26.641	-0.041
4	24	14.400	17.536	-3.136
4	39	26.800	27.060	-0.260
4	42	34.900	31.504	3.396
4	63	7.400	8.720	-1.320
5	114	18.200	18.243	-0.043
5	37	30.400	29.037	1.363
5	12	6.900	6.743	0.157
6	101	15.200	16.060	-0.860
6	41	26.000	26.377	-0.377
6	42	33.680	28.354	5.288
7	77	17.600	18.286	-0.686
7	101	20.200	20.659	-0.659
7	183	32.300	23.687	-1.187
8	46	10.300	10.103	-0.065
8	7C	31.700	21.532	-0.168
8	75	14.000	12.806	1.194
9	137	23.200	24.394	-1.194
9	195	8.510	9.659	-1.659
10	101	14.700	11.881	2.819
10	204	18.000	19.655	-1.655

---------ORDERED BY RESIDUALS---------

OBS. Y	FITTED Y	ORDERED RESID.	STUD.RESID.	SEQ
34.900	31.504	3.356	2.0	1
2.800	-0.376	3.176	2.0	2
14.700	11.881	2.819	1.8	3
2.000	6.165	1.835	1.2	4
30.400	29.037	1.363	0.9	5
45.700	44.439	1.261	0.8	6
32.800	30.826	1.274	0.8	7
44.700	44.439	1.261	0.8	8
14.000	12.806	1.194	0.8	9
14.000	12.806	1.157	0.8	10
6.900	6.400	0.427	0.3	11
31.700	31.532	0.168	0.1	12
34.700	34.598	0.102	0.1	13
33.600	33.520	0.080	0.1	14
26.600	26.641	-0.041	-0.0	15
18.200	18.243	-0.043	-0.0	16
24.800	24.865	-0.065	-0.0	17
10.000	10.103	-0.103	-0.0	18
12.200	12.376	-0.176	-0.1	19
26.800	27.060	-0.260	-0.2	20
26.000	26.377	-0.377	-0.2	21
27.800	28.354	-0.554	-0.4	22
15.600	16.260	-0.686	-0.5	23
9.000	9.659	-0.659	-0.5	24
23.400	23.687	-0.187	-0.8	25
22.300	23.394	-1.194	-0.9	26
7.400	8.720	-1.320	-0.9	27
18.000	19.659	-1.659	-1.1	28
13.100	14.895	-1.795	-1.2	29
16.400	19.148	-3.043	-1.9	30
14.400	17.536	-3.136	-2.1	31

Figure 8A.12

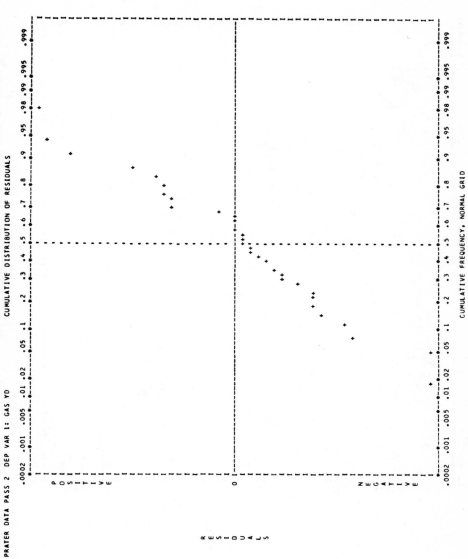

PRATER DATA PASS 2 DEP VAR 1: GAS YD   CUMULATIVE DISTRIBUTION OF RESIDUALS

Figure 8A.13

263

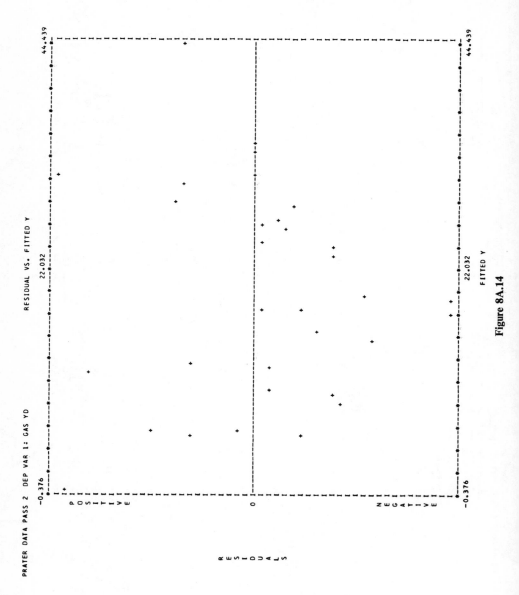

PRATER DATA PASS 2   DEP VAR 1: GAS YD   RESIDUAL VS. FITTED Y

Figure 8A.14

264

LINEAR LEAST-SQUARES CURVE FITTING PROGRAM

PRATER DATA PASS 3   DEP VAR 1: GAS YD

MIN Y = 2.800D+00   MAX Y = 4.570D+01   RANGE Y = 4.290D+01

Y = GASOLINE YIELD;
X1 = GASOLINE END PCINT - 332 ; X2 --- X10 = CRUDE 2 THRCUGH 10;
X11 = (END POINT - 332)(CRUDE 21); X19 = (END POINT - 332)(CRUDE 10)

IND.VAR(I)	NAME	COEF.R(I)	S.E. COEF.	T-VALUE	R(I)SQRD	MIN X(I)	MAX X(I)	RANGE X(I)	REL.INF.X(I)
0	GAS EP	3.272960+01	1.05D-12	15.9	0.8484	-1.27CD+02	1.120D+02	2.390D+02	0.93
1	CRUDE2	-1.66858D-01	1.160D+00	15.5	0.4622	0.0	1.0000+00	1.0000+00	0.42
2	CRUDE3	-1.7561D-01	1.540+00	5.5	0.4085	0.0	1.0000+00	1.0000+00	0.21
3	CRUDE4	-9.12714D+00	1.380+00	9.4	0.5128	0.0	1.0000+00	1.0000+00	0.30
4	CRUDE5	-1.29302D+01	1.600+00	2.4	0.6356	0.0	1.0000+00	1.0000+00	0.09
5	CRUDE6	-3.7611BD+00	1.150D+00	10.1	0.6352	0.0	1.0000+00	1.0000+00	0.30
6	CRUDE6	-1.16137D+01	1.320D+00	15.2	0.4572	0.0	1.0000+00	1.0000+00	0.27
7	CRUDE7	-2.00631D+01	1.250D+00	9.8	0.4692	0.0	1.0000+00	1.0000+00	0.47
8	CRUDE8	-1.22281D+01	2.21D+00	9.6	0.4079	0.0	1.0000+00	1.0000+00	0.29
9	CRD09	-2.11241D+01	2.350+00	11.3	0.7238	0.0	1.0000+00	1.0000+00	0.49
10	CRD20	-3.65641D+01	2.540+00	12.0	0.8314	-9.700D+01	8.400D+01	1.810D+02	0.62
11	X1CD2	-3.01449D+00	2.05D+00	2.3	0.4765	-1.140D+02	1.500D+01	1.290D+02	0.13
12	X1CD3	-2.193D+00	2.71D-02	2.3	0.6148	-2.500D+01	1.500D+01	8.400D+02	0.06
13	X1CD4	-1.28233D-02	2.05D-02	0.6	0.3922	-1.2CD+02	6.300D+01	8.280D+01	0.13
14	X1CD5	-1.33198D-02	1.600-02	0.8	0.6478	0.0	8.000D+01	1.750D+02	0.04
15	X1CD6	-3.55536D-03	1.940-02	0.3	0.4361	-9.700D+01	1.780D+01	1.590D+02	0.05
16	X1CD7	-6.38203D-03	3.250-02	0.1	0.4762	-9.700D+01	1.120D+02	1.350D+02	0.02
17	X1CD8	-4.C9302D-02	3.5ED-02	1.1	0.3180	-6.500D+01	7.800D+01	9.200D+01	0.09
18	X1CD9	-3.7B557D-02			0.7172	0.0	9.200D+01	9.600D+01	0.08
19	X1CD10				0.8270	0.0	9.600D+01		

NO. OF OBSERVATIONS   32
NC. OF IND. VARIABLES   19
RESIDUAL DEGREES CF FREEDOM   12
F-VALUE   73.6
RESIDUAL ROOT MEAN SQUARE   1.5857891
RESIDUAL MEAN SQUARE   2.52242910
RESIDUAL SUM CF SQUARES   30.32291923
TOTAL SUM CF SQUARES   3564.07718750
MULT. CORREL. COEF. SCLARED   .9915

| | | ORDERED BY INPUT | | | | ORDERED BY OBSV.Y | | ORDERED BY FITTED.Y | | RESIDUALS ORDERED BY | | |
IDENT.	OBSV.	WSS.DISTANCE	OBSV.Y	FITTED.Y	RESIDUAL	OBSV.Y	FITTED.Y	OBS.Y	FITTED.Y	ORDERED RESID.	STUD.RESID.	SEQ
1-1	102	107	12.200	11.539	0.661	14.700	102	14.700	12.228	2.472	2.9	1
1-2	13	45.	22.300	23.219	-0.919	27.800	24	27.800	26.168	1.632	1.9	2
1-3	14	71.	34.700	34.899	-0.199	2.800	21	2.800	1.604	1.196	1.4	3
1-4	21	212	45.300	45.244	0.456	8.000	31	5.925	0.799	1.175	1.4	4
2-1	22	142	6.400	1.604	1.032	34.600	33	25.801	0.687	0.799	1.0	5
2-2	112	112	16.100	18.296	-0.632	34.900	41	34.213	0.677	0.661	0.8	6
2-3	112	185	27.800	26.168	-2.196	6.900	61	6.223	0.661	0.508	0.6	7
3-1	31	75.	8.000	8.000	1.075	12.200	64	33.092	0.508	0.456	0.5	8
3-2	33	38.	13.100	14.925	-1.873	33.600	73	45.244	0.456	0.347	0.5	9
3-3	41	71.	26.800	25.801	-0.799	45.300	41	31.753	0.343	0.320	0.5	10
4-1	42	78.	14.400	14.090	0.320	32.100	81	14.657	0.320	0.065	0.1	11
4-2	43	112	26.800	27.807	-0.687	5.000	52	14.080	0.065	-0.029	0.0	12
5-1	51	173.	7.400	7.407	-0.007	14.400	91	31.635	-0.029	0.012	0.0	13
5-2	53	53.	30.400	34.213	0.012	31.700	92	9.971	-0.006	-0.006	0.0	14
5-3	14.	153.	6.900	18.138	-1.677	5.000	53	18.188	-0.095	-0.307	0.0	15
6-1	61	60.	15.200	6.923	0.183	18.200	51	14.000	-0.183	-0.095	-0.1	16
6-2	62	116.	26.000	16.223	0.343	23.200	82	23.200	-0.199	-0.183	-0.2	17
6-3	63	163.	24.657	26.183	-0.347	30.400	63	7.407	-0.307	-0.199	-0.2	18
7-1	72	250.	17.600	19.290	0.347	24.600	103	26.183	-0.549	-0.307	-0.4	19
7-2	73	76.	24.130	24.695	0.065	26.800	13	36.899	-0.632	-0.343	-0.4	20
8-1	82	104.	10.002	9.971	0.029	16.400	22	18.549	-0.690	-0.690	-0.8	21
8-2	81	112.	14.700	14.000	0.0	7.032	103	7.032	-0.632	-0.919	-1.1	22
9-1	91	167.	23.500	23.219	0.0	17.600	72	18.290	-0.919	-0.919	-1.1	23
9-2	92	262.	8.500	14.000	-1.922	23.219	12	23.219	-1.003	-1.003	-1.5	24
10-1	101	248.	13.100	12.422	-1.922	15.200	62	16.203	-0.690	-1.003	-0.8	25
10-2	101	261.	14.700	12.428	2.472	26.800	42	23.219	-0.919	-1.873	-1.4	26
10-3	103	342.	19.000	18.549	-0.549	13.100	32	27.807	-1.003	-2.196	-2.0	27
						8.500	101	14.973	-1.873		-1.7	28
						16.100	23	10.422	-1.922			29
								18.296	-2.196			30
												31
												32

Figure 8A.15

265

LINEAR LEAST-SQUARES CURVE-FITTING PROGRAM

PRATER DATA PASS 4

DATA INPUT   3 INDEPENDENT VARIABLES   1 DEPENDENT VARIABLE(S)

OBSV.	SEQ.	1-11-21	2-12-22	3-13-23	4-14-24	5-15-25	6-16-26	7-17-27	8-18-28	9-19-29	10-20-30
100	0	50.800	8.600	190.000	32.550						
200	0	41.300	1.800	267.000	15.036						
300	0	40.800	3.500	210.000	24.275						
400	0	40.300	4.800	231.000	21.515						
500	0	40.000	6.100	217.000	27.782						
600	0	38.400	6.100	220.000	21.154						
700	0	38.100	1.200	274.000	13.063						
800	0	32.200	5.200	236.000	20.436						
900	0	32.200	2.400	284.000	9.806						
1000	0	31.800	0.200	316.000	4.436						

DATA TRANSFORMATIONS

POSITION	CODE	OPERATION
1		NONE
2		NONE
3		NONE
4		NONE

CONSTANT	LOCATION	OMIT	VARIABLE	NAME
	1	0	1	GRAV.
	2	0	2	VAP PR
	3	0	3	10 PCT
	4	0	4	YIELD

THE FITTED EQUATION HAS   3 INDEPENDENT VARIABLES,   1 DEPENDENT VARIABLE(S)

DATA AFTER TRANSFORMATIONS

OBSV.	GRAV. 1-11-21	VAP PR 2-12-22	10 PCT 3-13-23	YIELD 4-14-24	5-15-25	6-16-26	7-17-27	8-18-28	9-19-29	10-20-30
100	5.08000D+01	8.60000D+00	1.90000D+02	3.25500D+01						
200	4.13000D+01	1.80000D+00	2.67000D+02	1.50360D+01						
300	4.08000D+01	3.50000D+00	2.10000D+02	2.42750D+01						
400	4.03000D+01	4.80000D+00	2.31000D+02	2.15190D+01						
500	4.00000D+01	6.10000D+00	2.17000D+02	2.77820D+01						
600	3.84000D+01	6.10000D+00	2.20000D+02	2.11540D+01						
700	3.81000D+01	1.20000D+00	2.74000D+02	1.30630D+01						
800	3.22000D+01	5.20000D+00	2.36000D+02	2.04360D+01						
900	3.22000D+01	2.40000D+00	2.84000D+02	9.80600D+00						
1000	3.18000D+01	2.00000D-01	3.16000D+02	4.43600D+00						

SUMS OF VARIABLES
3.85900D+02 3.99000D+01 2.44500D+03 1.90057D+02

MEANS OF VARIABLES
3.85900D+01 3.99000D+00 2.44500D+02 1.90057D+01

ROOT MEAN SQUARES OF VARIABLES
5.71809D+00 2.62444D+00 3.91869D+01 8.49313D+00

SIMPLE CORRELATION COEFFICIENTS, R(I,I PRIME)

1	1.000			
2	0.613	1.000		
3	-0.712	-0.897	1.000	
4	0.764	0.906	-0.978	1.000

Figure 8A.16

266

CHAPTER 9

Nonlinear Least Squares, a Complex Example

9.1 Introduction

In this chapter we use both linear and nonlinear least-squares programs to arrive at our final equation. The data give the amounts of heat released over six different time periods by fourteen samples of cement of differing chemical compositions. The equation is *nonlinear* in time, while two of its coefficients are *linear* and *quadratic* functions of composition.

Indicator variables are used first in a nonlinear equation to allow for systematic differences between individual cements while ascertaining the relationship between observations of those cements over a period of time. The dependence of the coefficients of the indicator variables on composition is then investigated by the C_p-search technique in the linear least-squares program. The results are returned to the nonlinear iterative program to fit the final equation involving both time and composition.

In the end, statistics of the fit and interpretation of the results are compared with those obtained by others.

Statistical problems discussed include:
1. Using an equation which is nonlinear in time.
2. Handling biased data caused by using differences instead of raw observations.
3. Developing potential equation forms.
4. Finding outliers.
5. Using indicator variables to separate the effects of different samples.
6. Choosing between common and individual rate equations.
7. Finding high correlation between a variable that can be measured immediately and one that requires a long time before it can be measured.

267

8. Handling composition variables when the components must add to unity.

9. Selecting influential variables.

10. Studying the component effect of each sample on each variable.

11. Determining whether "far out" observations control the form of the equation or only extend the range of applicability.

12. Checking on the possibility of nested data.

13. Specifying components of error to measure lack-of-fit.

14. Judging the reasonableness of the final equation.

9.2 Background of Example

In the construction of massive structures such as dams, the amount and the rate of heat evolved during the hardening of cement become extremely important. A large rise in temperature during hardening may result in excessive initial expansion and later, when the cement is eventually cooled to the surrounding temperatures, in contraction and cracking. Because of structure thickness, it is often necessary to build cooling ducts or refrigeration coils within the structure to carry away excess heat. However, an inordinate number of cooling ducts weakens the structure, and refrigeration over a long period of time increases costs. Thus there is a considerable incentive to determine the effect of cement composition on the amount and the rate at which heat evolves during hardening.

In a study of this problem, Woods, Steinour, and Starke of the Riverside Cement Company made experimental cements covering a wide range of compositions. Table 9.1 shows both the analysis of the oxides and the calculated compounds of the kilned "clinkers" used to make each cement.

Samples prepared from each cement were aged for 3, 7, 28, 90, 180, and 365 days under uniform conditions. The cumulative heat given off by each sample was determined indirectly by measuring the amount of heat evolved when the samples were dissolved in acid. It was assumed that Hess's law prevailed, namely, that the change in the heat content of a system in passing from one state to another is independent of the path. Using this assumption, the experimenters determined (1) the acid heat of solution of each cement in the initial state (unreacted with water), (2) the acid heat of solution of each sample after reaction with water and aging for various lengths of time, and (3) the difference between the first and the second quantity to obtain the amount of heat evolved during hardening over each period of time. The calculated cumulative heats of hardening are given in Table 9.2.

The experimenters used *linear* least squares to fit a separate equation for each of the six time periods, with four composition proportions as independent variables. A single equation, including time as a variable, would have

TABLE 9.1
CLINKER ANALYSIS AND CALCULATED CLINKER COMPOUNDS

		Clinker Analysis, Weight Per Cent					
Analysis	Cement	Silica, SiO_2	Alumina, Al_2O_3	Ferric Oxide, Fe_2O_3	Lime, CaO	Magnesia, MgO	Total
1	22	27.68	3.76	1.98	64.97	2.48	100.87
2	23	25.96	3.48	5.06	63.15	2.32	99.97
3	92	21.86	5.75	2.77	65.02	5.04	100.44
4	88	24.60	5.85	2.80	64.18	2.40	99.83
5	96	25.04	3.86	2.11	66.57	2.36	99.94
6	85	22.32	6.17	2.85	66.47	2.43	100.24
7	94	20.93	4.64	5.74	66.26	2.08	99.65
8	24	23.54	4.83	7.21	62.03	2.24	99.85
9	89	21.96	4.65	6.06	64.07	2.32	99.06
10	90	21.44	8.81	1.19	66.64	2.48	100.64
11	25	22.48	5.00	7.46	62.72	2.24	99.90
12	95	21.34	6.07	2.93	67.03	2.56	99.93
13	91	21.94	5.57	2.68	67.71	2.44	100.34
14	70	25.72	4.12	6.06	61.05	2.08	99.03

		Published Calculated Clinker Compounds, Weight Per Cent					
Analysis	Cement	$4CaO\text{-}Al_2O_3\text{-}Fe_2O_3$	$3CaO\text{-}Al_2O_3$	$3CaO\text{-}SiO_2$	$2CaO\text{-}SiO_2$	MgO	Total
1	22	6	7	26	60	2.5	101.5
2	23	15	1	29	52	2.3	99.3
3	92	8	11	56	20	5.0	100.0
4	88	8	11	31	47	2.4	99.4
5	96	6	7	52	33	2.4	100.4
6	85	9	11	55	22	2.4	99.4
7	94	17	3	71	6	2.1	99.1
8	24	22	1	31	44	2.2	100.2
9	89	18	2	54	22	2.3	98.3
10	90	4	21	47	26	2.5	100.5
11	25	23	1	40	34	2.2	100.2
12	95	9	11	66	12	2.6	100.6
13	91	8	10	68	12	2.4	100.4
14	70	18	1	17	61	2.1	99.1

TABLE 9.2
CUMULATIVE HEAT FROM HARDENING CEMENT WITH 40% WATER,
CALORIES PER GRAM

	Period of Aging					
Cement	3 Days	7 Days	28 Days	90 Days	180 Days	365 Days
22	51.2	53.1	72.5	81.2	78.5	85.5
23	40.6	45.7	61.0	71.9	74.3	76.0
92	74.8	84.7	92.2	100.6	104.3	110.4
88	53.2	63.1	73.7	77.7	87.6	90.6
96	71.9	78.8	88.8	92.6	95.9	103.5
85	80.6	90.3	101.0	103.7	109.2	109.8
	74.0	88.6	101.7	104.5	107.0	110.3
94	77.3	87.5	92.4	100.5	102.7	108.0
24	42.1	46.8	61.5	70.0	72.5	71.6
89	64.7	77.9	85.8	90.1	93.1	97.0
90	89.5	96.6	108.8	114.1	115.9	122.7
25	58.8	62.8	74.2	78.1	83.8	83.1
95	84.8	94.5	103.7	107.5	113.3	115.4
91	84.0	91.3	99.9	106.2	109.4	116.3
70	28.3	36.3	49.8	—	—	62.6

been more compact, more precise, more comprehensive, and hence more informative.

Hald (Section 20.3) and Draper and Smith (Chapter 6 and Appendix B) have used observations of thirteen of the fourteen cements at 180 days to illustrate the standard arithmetic of "multiple regression." No contribution to understanding the interrelationships of composition, time, and heat of hardening was claimed or expected.

Let us review in detail how the data were taken. Although ultimately we may wish to have our equations in terms of the cumulative heats of hardening, *measurements* were made in terms of the acid heat of solution of each aged sample. These values in turn were subtracted from the *initial* acid heat of solution of the dry cement from which the samples were made. This gave an estimate of the amount of heat given off by the exothermic reactions of aging and hardening. However, as this subtraction was repeated at each of the six time periods, any random fluctuation in a cement's *initial* measurement affected all six differences equally. Thus, the six differences for each cement are not statistically independent, although the seven measurements are. Least-squares fitting is mathematically simpler and easier to interpret if we fit an equation directly to the statistically independent observed heats of acid solution. This equation, in turn, can be modified to calculate the desired cumulative heats of hardening. We will pursue this course of action.

TABLE 9.3
Observed Acid Heat of Solution, Calories per Gram

NO.	CEMENT	0 DAYS	3 DAYS	7 DAYS	28 DAYS	90 DAYS	180 DAYS	365 DAYS
1	22	588.7	537.5	535.6	516.2	507.5	510.2	503.2
2	23	575.7	535.1	530.0	514.7	503.8	501.4	499.7
3	92	623.4	548.8	538.9	531.4	523.0	519.3	513.2
3	92	624.1						
4	88	595.5	542.3	532.4	521.8	517.8	507.9	504.9
5	96	603.2						
5	96	604.5	532.6	525.7	515.7	511.9	508.6	501.0
6	85	618.8	538.2	528.5	517.8	515.1	509.6	509.0
6	85		544.8	530.2	517.1	514.3	511.8	508.5
7	94	610.0	532.7	522.5	517.6	509.5	507.3	502.0
7	94	611.7						
8	24	573.0	530.9	526.2	511.5	503.0	500.5	501.4
9	89	599.0	534.3	521.1	513.2	508.9	505.9	502.0
10	90	633.4	543.9	536.8	524.6	519.3	517.5	510.7
11	25	585.9	527.1	523.1	511.7	507.8	502.1	502.8
12	95	625.6	540.8	531.1	521.9	518.1	512.3	510.2
13	91	624.0	540.0	532.7	524.1	517.8	514.6	507.7
14	70	564.7	536.4	528.4	514.9			502.1

Before we try to fit an equation to the observed values, let us see how much can be learned about their measurement. We realize from the start that this may be only an academic exercise because of the small number of replicates. Nonetheless, such a search is often worthwhile. The observed values are given in Table 9.3.

9.3 Replicates

Clinker mixtures were prepared from a finely ground commercial kiln feed by adding technical oxides and carbonates to bring the oxides to the desired values. Each mixture was fired in a furnace, cooled, and pulverized. Before storage, a portion of each was withdrawn for chemical analysis. There were no replicate clinker mixtures or replicate chemical analyses.

The fourteen cements were prepared by grinding the same amount of gypsum (3.2%) with each clinker mixture. Gypsum, $CaSO_4 \cdot 2H_2O$, is used commercially to delay setting, increase strength, and improve dimensional stability.

Duplicate initial heat of solution measurements were made on three cements to estimate the precision of the calorimeter. The results are shown in the table.

DUPLICATE INITIAL HEAT OF SOLUTION MEASUREMENTS

Observation	Cements		
	92	94	96
1	623.4	610.0	603.2
2	624.1	611.7	604.5
Difference	0.7	1.7	1.3
d^2	0.49	2.89	1.69

$\Sigma\, d^2/2 = 5.07/2 = 2.535$
Variance $= 2.535/3 = 0.845$
Standard deviation $= 0.92$ with 3 df

A portion of each cement was then mixed with water, and the resulting cement paste put into ten glass vials. All ten vials were then placed in a constant-temperature cabinet until required for the heat of solution measurements after aging 3, 7, 28, 90, 180, and 365 days.

To test the cumulative precision of the preparation and curing of the cement pastes, their preparation for admission to the calorimeter, and the heat of solution measurements, a second set of cement paste samples was prepared at a different time from Cement 85. The heat of solution measurements of these duplicate samples are given in the table. The 3-day difference

HEAT OF SOLUTION MEASUREMENTS ON DUPLICATE AGED SAMPLES

Batch	Days					
	3	7	28	90	180	365
1	538.2	528.5	517.8	515.1	509.6	509.0
2	544.8	530.2	517.1	514.3	511.8	508.5
Difference	6.6	1.7	−0.7	−0.8	2.2	−0.5
d^2	43.56	2.89	0.49	0.64	4.84	0.25
$\Sigma\, d^2 =$	52.67					

is so much larger than the differences for the other five periods that we use Cochran's g-test to judge whether all six differences could come from the same random normal source:

$$g(1, 6) = \frac{d^2 \max}{\Sigma d^2} = \frac{43.56}{52.67} = 0.827.$$

From Dixon and Massey (Table A-17) we find the value

$$g(0.95, 1, 6) = 0.781.$$

Thus at the 95% confidence level, the difference in the pair of duplicate observations at 3 days is *not* compatible with the others.

From the five acceptable pairs, we compute

$$\Sigma d^2/2 = 9.11/2 = 4.55,$$

$$\text{Variance} = 4.55/5 = 0.910, \text{ and the}$$

$$\text{Standard deviation} = 0.95 \text{ with 5 df.}$$

The standard deviation of 0.95 for the duplicate aged sample determinations and the standard deviation of 0.92 for the duplicate initial heat of solution measurements are remarkably close.

Although one of the 3-day observations of Cement 85 is an outlier, we will not know which it is until we determine the relationship between the various heat of solution measurements and time.

9.4 Potential Equations

The usual trial plots of the heat of solution, z, versus time, t, were made: z vs. t, z vs. $\log t$, $\log z$ vs. t, and $\log z$ vs. $\log t$. Only the last gave tolerably straight lines, and even these showed upward curvature at small t and large z. When a constant, E, was subtracted from z, we were able to obtain straight-line plots on log-log paper [equivalent to $\log (z - E)$ vs. $\log (t)$]. Two such plots are shown in Figure 9.1. A different value of E was required for each cement. The constant E appears to represent the acid heat of solution after hardening for an infinite length of time, that is, the ultimate or equilibrium heat of solution. (Cement used in aqueducts by the ancient Romans still contains unreacted cells or granules; water has not broken through all of the crystalline barriers.)

The resulting equation, $\log (z_t - E) = A - B \log (t)$, is nonlinear in that the value of E is unknown and there is no transformation which will linearize

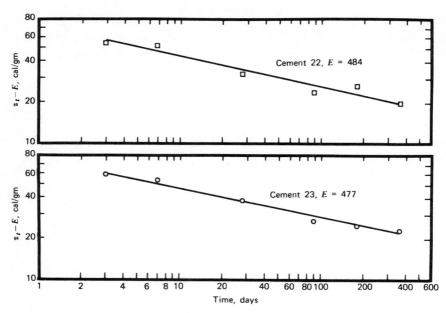

Figure 9.1 Acid heat of solution minus constant versus time.

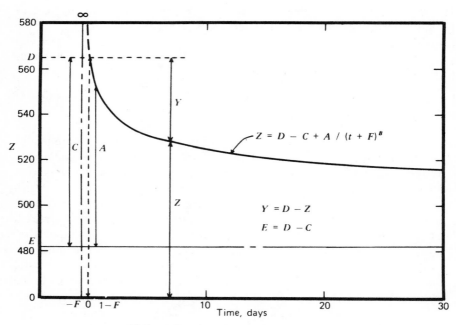

Figure 9.2 Heat of solution versus time.

274

it in all coefficients. Since we have observations at time zero, a time correction, F, may also be needed. Hence a convenient form of the equation is:

(9.1) $$Z - E = \frac{A}{(t + F)^B},$$

where t = time from moment of adding water, measured in days,

Z = estimated heat evolved when sample is dissolved in acid at time t (heat of acid solution measured in calories per gram), and z = observed value of same quantity,

E = heat of acid solution after infinite time, that is, the ultimate heat of solution,

A = a constant characteristic of the sample, the remaining heat of hardening at time $(1 - F)$,

B = an exponent, and

F = time correction.

Since

(9.2) $$E = D - C,$$

where D = initial heat of solution at time zero, and C = cumulative heat of hardening after infinite time, that is, the ultimate heat of hardening, then

(9.3) $$Z = D - C + \frac{A}{(t + F)^B}.$$

At $t = 0$ by definition, $Z = D$, the initial heat of solution, and

$$D = D - C + \frac{A}{(0 + F)^B},$$

so that

(9.4) $$F = \left(\frac{A}{C}\right)^{1/B}.$$

Substituting for F in 9.3 gives the equation for the heat of solution in terms of four constants: A, B, C, and D,

(9.5) $$Z = D - C + \frac{A}{[t + (A/C)^{1/B}]^B}.$$

The equation can also be written in terms of B, C, D, and F:

(9.6) $$Z = D - C\left[1 - \left(\frac{F}{t + F}\right)^B\right].$$

Both forms of the equation will be helpful in interpreting the final fit of the data. The basic equation, 9.3, is illustrated in Figure 9.2.

The cumulative heat of hardening, Y, has been defined as

$$Y = D - Z.$$

Substituting $D - Y$ for Z in equations 9.3, 9.5, and 9.6, we have:

(9.7)
$$Y = C - \frac{A}{(t + F)^B},$$

(9.8)
$$Y = C - \frac{A}{[t + (A/C)^{1/B}]^B},$$

and

(9.9)
$$Y = C\left[1 - \left(\frac{F}{t + F}\right)^B\right].$$

We will return to these equations later.

9.5 Nonlinear Fit—Observations of Individual Cements versus Time

The data from each of the fourteen cements were used separately to fit an equation of the form 9.5. The results of each fit are given in Table 9.4.

TABLE 9.4
ACID HEAT OF SOLUTION:
NONLINEAR ESTIMATION OF COEFFICIENTS—INDIVIDUAL EQUATIONS

Cement Number	Constant, A	Exponent, B	Ultimate Heat of Hardening, C	Initial Heat of Solution, D	Number of Observations	Degrees of Freedom	Residual Sum of Squares
22	72.41	0.221	104.74	588.67	7	3	56.35
23	79.13	0.209	100.44	575.66	7	3	17.13
92	95.90	0.103	161.34	623.74	8	4	8.24
88	104.63	0.102	147.46	595.50	7	3	19.66
96	86.45	0.098	149.23	603.84	8	4	13.99
85	64.23	0.506	112.51	618.81	13	9	35.70
94	75.21	0.116	145.62	610.85	8	4	14.33
24	63.69	0.309	83.91	572.96	7	3	20.82
89	56.62	0.431	99.80	599.01	7	3	11.98
90	69.46	0.166	147.33	633.40	7	3	8.83
25	54.31	0.192	101.80	585.90	7	3	10.84
95	56.12	0.245	128.18	625.60	7	3	6.83
91	90.19	0.097	165.21	623.98	7	3	6.71
70	69.89	0.315	73.58	564.68	5	1	0.98
Totals					105	49	232.39

The Nonlinear Least-Squares Curve-Fitting Program, NONLINWOOD, was used to estimate the coefficients of the equation. This program, as mentioned in Chapter 2, is a modification of the University of Wisconsin's GAUSHAUS program, which utilizes the Marquardt "maximum neighborhood" nonlinear estimation technique. The program is available from a number of computer libraries (see Chapter 1). Pass 1 and Pass 10 of the cement problem are included with the program as test problems. Details are given in the User's Manual.

9.6 Fit with Indicator Variables

In order to determine whether all of the cements can be represented by an equation with a single value of B, we write:

$$(9.10) \qquad Z = D' - C' + \frac{A'}{(t + F')^{B}},$$

where Z = estimated acid heat of solution at time t (in days),

$D' = D_1v_1 + D_2v_2 + \cdots + D_kv_k$ ($v_1, \ldots, v_k$ are indicator variables to separate the initial heats of solution of the k cements, $D_1, \ldots, D_k$),

$C' = C_1v_1 + C_2v_2 + \cdots + C_kv_k$,

$A' = A_1v_1 + A_2v_2 + \cdots + A_kv_k$, and

$F' = (A'/C')^{1/B}$.

Thus, there is a D, C, and A coefficient for each of the fourteen cements and one B coefficient—43 coefficients in all.

The starting values of the coefficients were guessed from plots such as Figure 9.1, but taken purposely wide of their marks to test the ability of the computer program to converge: 0.8 for B, 100 for each C, 80 for each A, and the average initial heat of solution for each cement's D coefficient. Computer printouts of the result of fitting this equation are shown in Figures 9A.1–9A.4.

Only nine iterations were required to obtain the final fit. The criterion used for ending the search for the "best" values of the coefficients was that the relative change in each coefficient from the previous iteration be less than 0.000001. The resulting change in the sum of squares of the 105 residuals from the previous iteration was less than 0.01.

The method appears to be rather robust in regard to the approximation of the starting values. The starting value of 0.8 for the exponent B shifted to 0.22; the starting value of 100 for C (the ultimate heat of hardening) shifted to a range of values from 83 to 137; and the starting value of 80 for A shifted to a range of 52–78. The values chosen for D (the initial acid heat o. solution) remained essentially unchanged.

The multiple correlation coefficient squared, a measure of how well the fitting equation accounts for the total variation of the observed values of the dependent variable, is 0.9976:

$$R_z^2 = 1 - \frac{\text{Residual sum of squares}}{\text{Total sum of squares}} = 1 - \frac{299}{122,985}$$

$$= 1 - 0.00243 = 0.9976.$$

The residual versus cumulative frequency plot indicates that the distribution of the residuals is roughly normal. (The center portion is close to a straight line.) Individual hand plots of the residuals of each cement versus time show that the fit is not biased for individual cements and that, at time periods above zero, the residuals are apparently randomly distributed.

The residual versus fitted Z plot looks peculiar at first sight, in that there is a larger spread of residuals at the lower fitted Z-values than at the higher values—the initial heats of solutions. In these equations the observations on aged samples have little influence on the fit of the observations at time zero. We rationalized this by supposing that the steepness of the curve was making back-extrapolation uninformative. But as we show later, when the data on composition are used, first weight percent and then mol percent, the residuals become more evenly dispersed over all values of fitted Y.

The statistics of the fit of equation 9.5 with an individual exponent for each cement, and the fit of equation 9.10 with a common exponent, are shown in the table.

EFFECT OF COMMON VERSUS INDIVIDUAL EXPONENTS

B	Number of Coefficients	Degrees of Freedom	Residual Sum of Squares	Residual Mean Square
Common	$p = 43$	62	299.3	4.8
Individual	$p + q = 56$	49	232.4	4.7
	$q = 13$	13	66.9	5.2

$$F(13, 49) = \frac{\text{RMS}_q}{\text{RMS}_{p+q}} = \frac{5.2}{4.7} = 1.1.$$

From F-tables, $F(0.95, 12, 40)$ is 2.0 and $F(0.95, 15, 60)$ is 1.8; so the individual exponents did not significantly improve the fit. Therefore, we conclude that all the cements can be represented by an equation with a single exponent.

Similar tests were made on coefficients D, C, and A. In each case, the use of individual coefficients for each cement significantly improved the fit over that

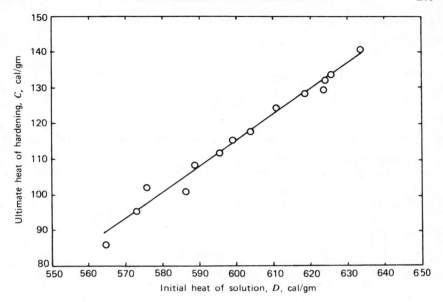

Figure 9.3 Ultimate heat of hardening versus initial heat of solution.

obtained with common coefficients. Hence, individual coefficients are needed for D, C, and A.

Earlier we had observed that one of the duplicate observations of sample 85 at 3 days was undoubtedly an outlier, but were unable to identify it. From the list of ordered residuals, Figure 9A.2, it is now apparent that the second observation is the culprit. It has the largest residual of all of the observations.

In Pass 2, Figures 9A.5–9A.7, the same equation is used as in Pass 1, but observation *85-3-2* is omitted. The residual sum of squares is reduced considerably, from 299.35 to 253.10, a decrease of 46.25 units. From the squared difference of the duplicate observations of *85-3*, we find that 43.56/2 or only 21.78 units can be directly attributed to removing the second observation. The remaining 24.47 units (46.25–21.78) is the indirect effect of the deletion. R_z^2 is increased from 0.9976 to 0.9979.

Observation *85-3-2* will be omitted in the remaining analysis.

A plot of coefficients C_i versus D_i ($i = 1, \ldots, 14$), Figure 9.3, reveals an unexpected *linear* relationship between the initial heat of solution of each cement and its ultimate heat of hardening. As seen in Figure 9A.8, the equation $C = GD - H$ represents the data with an R_c^2 of 0.98. Using

$$(9.11) \qquad C' = GD' - H,$$

we can rewrite the nonlinear equation as

(9.12)
$$Z = H + (1 - G)D' + \frac{A'}{(t + F')^B},$$

where now

$$F' = \left(\frac{A'}{GD' - H}\right)^{1/B}$$

and B, G, and H are *common* to all cements.

The nonlinear fit of equation 9.12 is shown in Figure 9A.9. The residual root mean square is 1.99 with 73 degrees of freedom, and R_z^2 is 0.9976. The residual versus cumulative frequency plot and the residual versus fitted Y plot (not shown) are similar to the plots of the nonlinear fit with individual C coefficients.

EFFECT OF COMMON G AND H VERSUS INDIVIDUAL C COEFFICIENTS

Equation with	Number of Coefficients	Degrees of Freedom	Residual Sum of Squares	Residual Mean Square
Common G and H	$p = 31$	73	290.7	4.0
Individual C	$p + q = 43$	61	253.1	4.1
	$q = 12$	12	37.6	3.1

$$F = \frac{\text{RMS}_q}{\text{RMS}_{p+q}} = \frac{3.1}{4.1} = 0.8; \quad F(0.95, 12, 60) = 1.9$$

The statistics of the fit of both equations are shown in the table. The new equation with *twelve* fewer coefficients fits as well as the old one. Thus, we conclude that equation 9.12 can be used to calculate the ultimate heat of hardening from the fitted value of the initial heat of solution.

9.7   Fit with Composition

Before composition variables can replace the indicator variables representing the different cements in equation 9.12, we will need to determine the *forms* of the equations and to estimate the starting values of their coefficients.

In Figure 9A.9 we saw how the indicator variables were used in the nonlinear equation to estimate coefficients D and A for each cement. We now try to find the dependence of these coefficients on the composition of the clinker from which each cement was made.

The proportions (or fractions) found by chemical analysis of the oxides of each clinker should total 100%. Since we do not know which of the oxides of each clinker is in error, we correct each proportionally by dividing by the clinker's total weight per cent and multiplying by 100.

We have chosen equations of the type first given by Scheffé (1958; also see Gorman and Hinman, 1962) to represent the dependence of the heat of solution on the five composition variables. These equations are symmetrical in all components. They do not arbitrarily omit one component and hence do not conceal real effects. The equations are polynomials of the second order whose terms have understandable meanings.

In order to reduce the covariances between the various coefficients, we subtract from each oxide the minimum value observed for it in all the samples. The resulting composition is reproportioned so that the sum of all five oxide proportions is again 100%. This is done by dividing each adjusted value by 100 minus the sum of the minimum observations. The minimum values and the resulting divisor are as follows:

Oxide	Minimum Concentration, %
Silica, SiO_2	21.00
Alumina, Al_2O_3	3.48
Ferric oxide, Fe_2O_3	1.18
Lime, CaO	61.60
Magnesia, MgO	2.08
Sum	$= \overline{89.34}$
Divisor $= (100 - $ Sum $) = 10.66$	

The transformed composition variables then are:

$$x_1 = (\% \text{ silica} \quad - 21.00)/10.66$$
$$x_2 = (\% \text{ alumina} \quad - 3.48)/10.66$$
$$x_3 = (\% \text{ ferric oxide} - 1.18)/10.66$$
$$x_4 = (\% \text{ lime} \quad - 61.60)/10.66$$
$$x_5 = (\% \text{ magnesia} \quad - 2.08)/10.66$$

Reproportioning in composition space is somewhat analogous to the usual technique of subtracting the mean from variables before taking their cross products and squared terms.

Because of the limited number of cements (fourteen), the number of cross products we can fit at one time is limited. Magnesia (x_5) theoretically remains unreacted, so we thought it safe to study later the potential influence of its cross products. The following set of equations can be fitted using *linear* least squares:

$$(9.13) \qquad D, A = b_1 x_1 + b_2 x_2 + b_3 x_3 + b_4 x_4 + b_5 x_5$$
$$+ b_{12} x_1 x_2 + b_{13} x_1 x_3 + b_{14} x_1 x_4$$
$$+ b_{23} x_2 x_3 + b_{24} x_2 x_4 + b_{34} x_3 x_4.$$

Computer printouts of the fit of these equations are given in Figures 9A.10–9A.16.

Since there is no b_0 term, the basic equations must contain the five corrected oxides. Thus, after fitting the above equations, C_p searches are made to determine which cross-product terms are influential. The following statistics are obtained:

Dependent Variable	Full Equations		Result of C_p Search		
	R_y^2	RRMS	Most Influential Cross Products	p	C_p
D	0.998	2.2	x_1x_3, x_2x_4	7	3.4
A	0.990	1.6	x_1x_2, x_1x_4	7	6.4

The fit of each equation is relatively good. With D, the search indicates that the value of C_p can be reduced from 11 with the full equation to 3.4 by using only the cross products x_1x_3 and x_2x_4. With A, C_p can be reduced from 11 to 6.4 by using cross products x_1x_2 and x_1x_4.

Plots of p (the number of coefficients in each equation) versus C_p are shown in Figures 9A.13 and 9A.16. Each plot represents the results of calculating the fit of 2^6 or 64 combinations of cross products added to the basic equation. Only equations with C_p's less than the full equation ($C_p = 11 = p$) are represented. Although we are not compelled to select the variable combinations with the lowest C_p-values, in this case (as seen in Figures 9A.17 and 9A.18) they appear reasonable. We have for the selected equations:

Dependent Variable	R_y^2	RRMS
D	0.997	1.55
A	0.978	1.57

The fit of the two dependent variables and the subsequent C_p searches took 6 seconds of computer time.

The cross products involving x_5—x_1x_5, x_2x_5, x_3x_5, x_4x_5—are now added to the reduced linear equations to determine whether any will provide a fit with a lower C_p-value. None does.

We are now ready to replace the indicator variable estimates of D' and A' in equation 9.12 with those of composition. The fourteen coefficients in the composition equations:

$$(9.14) \quad D = d_1x_1 + d_2x_2 + d_3x_3 + d_4x_4 + d_5x_5 + d_{13}x_1x_3 + d_{24}x_2x_4$$

and

$$(9.15) \quad A = a_1x_1 + a_2x_2 + a_3x_3 + a_4x_4 + a_5x_5 + a_{12}x_1x_2 + a_{14}x_1x_4$$

and the earlier estimates of B, G, and H are used to start the nonlinear iterations.

Results of this fit (Pass 9) are shown in Figure 9A.19. After eight iterations the relative change in each coefficient was less than 0.000001. The residual root mean square is 1.98, measured with 87 degrees of freedom. This matches closely the value of 1.99 found with the previous nonlinear fit using indicator variables. Replacing the *specific* indicator variables with the more *general* composition variables only reduced R_z^2 from 0.9976 to 0.9972. The cumulative distribution plot (Figure 9A.20) is similar to that obtained with the indicator variables (Figure 9A.6). The residual versus fitted Z plot (Figure 9A.21), however, shows a more nearly symmetrical dispersion of residuals at the larger Z-values (time = 0) than that obtained with the indicator variables (Figure 9A.7).

In order to determine with security just how much lack of fit there is in this equation, we would need some measure of the cumulative error caused by variations in kiln feed preparation, furnace firing, chemical analysis, gypsum addition, water addition, storage conditions, storage time, and calorimeter measurements. Unfortunately there were no replicate clinker samples. We have one replicate cement sample (85) from which to estimate the cumulative error of water addition, storage conditions, and storage time, and three replicate calorimeter measurements, but clearly these are not sufficient to provide the needed error estimates.

We have come to the end of the path on which we started. We do not think that further improvement is possible by empirical fitting. But now that a good fit has been obtained, we may hope for further theoretical developments by those able to make them.

In order to simplify the use of the equation given by Pass 9, Figure 9A.19, let us convert the coded composition to the composition actually observed by adding back the minimum observed value of each variable and multiplying by 10.66. The components of the nonlinear equation now contain only these variables:

$$t = \text{time measured in days,}$$
$$x_1 = \% \text{ silica in the clinker,}$$
$$x_2 = \% \text{ alumina,}$$
$$x_3 = \% \text{ ferric oxide,}$$
$$x_4 = \% \text{ lime, and}$$
$$x_5 = \% \text{ magnesia.}$$

In the *heat of solution* equation:

(9.16)
$$Z = H + (1 - G)D + \frac{A}{(t + F)^B},$$

and in the corresponding *cumulative heat of hardening* equation:

(9.17)
$$Y = -H + GD - \frac{A}{(t + F)^B},$$

where

$$F = \left(\frac{A}{GD - H}\right)^{1/B},$$

the coefficients B, G, and H, common to all cements, remain unchanged:

$$B = 0.20, \quad G = 0.726, \quad \text{and} \quad H = 320.$$

The decoded equations for D and A are:

(9.18)   Initial heat of solution, $D =$

$$\begin{aligned}
&4.2 \ \times \ (\% \text{ silica}) \\
&-56.4 \ \times \ (\% \text{ alumina}) \\
&+14.4 \ \times \ (\% \text{ ferric oxide}) \\
&+6.4 \ \times \ (\% \text{ lime}) \\
&+12.1 \ \times \ (\% \text{ magnesia}) \\
&-0.46 \times \ (\% \text{ silica} \times \% \text{ ferric oxide}) \\
&+1.00 \times \ (\% \text{ alumina} \times \% \text{ lime})
\end{aligned}$$

and

(9.19)   Sample constant, $A =$

$$\begin{aligned}
&50.0 \ \times \ (\% \text{ silica}) \\
&-28.7 \ \times \ (\% \text{ alumina}) \\
&-13.4 \ \times \ (\% \text{ ferric oxide}) \\
&+8.0 \ \times \ (\% \text{ lime}) \\
&-7.9 \ \times \ (\% \text{ magnesia}) \\
&+0.77 \times \ (\% \text{ silica} \times \% \text{ alumina}) \\
&-0.98 \times \ (\% \text{ silica} \times \% \text{ lime}).
\end{aligned}$$

A useful relationship is obtained from $C = GD - H$, that is, Ultimate heat of hardening $= 0.726(\text{Initial heat of solution}) - 320$, or simply

(9.20) $C = 0.726(D - 441).$

Using this relationship, we find

(9.21)   Ultimate heat of hardening, $C = -320$

$$\begin{aligned}
&+3.0 \ \times \ (\% \text{ silica}) \\
&-40.9 \ \times \ (\% \text{ alumina}) \\
&+10.4 \ \times \ (\% \text{ ferric oxide}) \\
&+4.6 \ \times \ (\% \text{ lime}) \\
&+8.8 \ \times \ (\% \text{ magnesia}) \\
&-0.33 \times \ (\% \text{ silica} \times \% \text{ ferric oxide}) \\
&+0.73 \times \ (\% \text{ alumina} \times \% \text{ lime}).
\end{aligned}$$

The component effects of each of the oxides on the initial heat of solution, D, and the sample constant, A, are given in Tables 9.5 and 9.6. In both tables we see that Cement 90 was highly influential in determining the effect of alumina, x_2, and alumina interactions, while Cement 92 was most

TABLE 9.5

COMPONENT EFFECT TABLE

INITIAL HEAT OF SOLUTION

IDENT.	SILICA PCT	B1X1	FERRIC OXIDE PCT	B3X3	CROSS PRODUCT X1X3	B6X1X3	U, TOTAL	ALUMINA PCT	B2X2	LIME PCT	B4X4	CROSS PRODUCT X2X4	B7X2X4	V, TOTAL	MAGNESIA PCT	B5X5	INITIAL HEAT OF SOLUTION
14 70	26.0	108.7	6.1	88.4	159.	-72.9	124.2	4.2	-234.7	61.6	392.8	256.	256.3	414.5	2.1	25.4	564.2
8 24	23.6	98.7	7.2	104.3	170.	-78.1	124.9	4.8	-272.8	62.1	395.9	301.	300.3	423.4	2.2	27.1	575.4
2 23	26.0	108.7	5.1	73.1	131.	-60.3	121.5	3.5	-196.3	63.2	402.5	220.	219.8	426.0	2.3	28.1	575.6
11 25	22.5	94.2	7.5	107.9	168.	-77.1	125.0	5.0	-282.3	62.8	400.1	314.	314.0	431.8	2.2	27.1	584.0
1 22	27.4	114.9	2.0	28.4	54.	-24.7	118.5	3.7	-210.2	64.4	410.4	240.	239.9	440.1	2.5	29.7	588.4
4 88	24.6	103.2	2.8	40.5	69.	-31.7	112.0	5.9	-330.5	64.3	409.7	377.	376.5	455.7	2.4	29.1	596.7
9 89	22.2	92.8	6.1	88.4	136.	-62.2	119.0	4.7	-264.8	64.7	412.2	304.	303.4	450.8	2.3	28.3	598.1
5 96	25.1	104.9	2.1	30.5	53.	-24.3	111.1	3.9	-217.8	66.6	424.5	257.	257.1	463.7	2.4	28.6	603.4
7 94	21.0	87.9	5.8	83.2	121.	-55.5	115.7	4.7	-262.6	66.5	423.7	310.	309.4	470.5	2.1	25.3	611.4
6 85	22.3	93.2	2.8	41.1	63.	-29.0	105.3	6.2	-347.2	66.3	422.6	408.	407.9	483.3	2.4	29.3	617.9
13 91	21.9	91.5	2.7	38.6	58.	-26.8	103.3	5.6	-313.1	67.5	430.0	375.	374.4	491.3	2.4	29.4	624.0
3 92	21.8	91.1	2.8	39.8	60.	-27.5	103.4	5.7	-322.9	64.7	412.5	371.	370.4	460.0	5.0	60.7	624.1
12 95	21.4	89.4	2.9	42.4	63.	-28.7	103.0	6.1	-342.6	67.1	427.4	407.	407.2	492.0	2.6	31.0	626.1
10 90	21.3	89.2	1.2	17.1	25.	-11.6	94.7	8.8	-493.7	66.2	422.0	580.	579.3	507.5	2.5	30.8	633.0

TABLE 9.6

COMPONENT EFFECT TABLE

SAMPLE-CONSTANT A

IDENT.	SILICA PCT	B1X1	ALUMINA PCT	B2X2	FERRIC OXIDE PCT	B3X3	LIME PCT	B4X4	MAGNESIA PCT	B5X5	X1X2	B6₁X1X2	X1X4	B7X1X4	A
14 70	26.0	1296.0	4.2	-119.6	6.1	-82.1 *	61.6	493.8 *	2.1	-16.7	108.	82.9	1601.	-1574.3	80.1 **
3 92	21.8	1086.1	5.7	-164.5	2.8	-37.0	64.7	518.5	5.0	-39.9 *	125.	95.6	1409.	-1385.3	73.5
1 22	27.4	1369.4 **	3.7	-107.1	2.0	-26.3	64.4	515.9	2.5	-19.5	102.	78.5	1767.	-1737.8 *	72.9
4 88	24.6	1229.7	5.9	-168.4	2.8	-37.6	64.3	514.9	2.4	-19.1	144.	110.8	1584.	-1557.6	72.6
2 23	26.0	1295.8	3.5	-100.0 **	5.1	-67.9	63.2	505.9	2.3	-18.4	90.	69.3 *	1640.	-1612.8	71.9
8 24	23.6	1176.5	4.8	-139.0	7.2	-96.9	62.1	497.6	2.2	-17.8	114.	87.5	1465.	-1440.0	67.8
10 90	21.3	1063.1	8.8	-251.6 *	1.2	-15.9 **	66.2	530.3	2.5	-20.2	186.	143.0 **	1411.	-1387.0	61.8
6 85	22.3	1111.1	6.2	-176.9	2.8	-38.1	66.3	531.1	2.4	-19.3	137.	105.1	1477.	-1451.7	61.3
11 25	22.5	1122.9	5.0	-143.8	7.5	-100.2 *	62.8	502.8	2.2	-17.8	113.	86.4	1413.	-1389.1	61.2
12 95	21.4	1065.7	6.1	-174.6	2.9	-39.3	67.1	537.2	2.6	-20.3	130.	99.5	1432.	-1408.4	59.7
13 91	21.9	1091.1	5.6	-159.5	2.7	-35.8	67.5	540.5 **	2.4	-19.3	121.	93.1	1476.	-1450.8	59.3
5 96	25.1	1250.3	3.9	-111.0	2.1	-28.3	66.6	533.5	2.4	-18.8	97.	74.2	1669.	-1640.9	59.0
9 89	22.2	1106.2	4.7	-134.9	6.1	-82.1	64.7	518.0	2.3	-18.6	104.	79.8	1434.	-1409.8	58.7
7 94	21.0	1048.1 *	4.7	-133.8	5.8	-77.3	66.5	532.6	2.1	-16.6 **	98.	75.0	1397.	-1373.1 **	54.9 *
AVERAGE	23.3		5.2		4.1		64.9		2.5		121.		1514.		

(header span: -----CROSS PRODUCTS-----)

* MINIMUM VALUE

** MAXIMUM VALUE

286

TABLE 9.7
FITTED VALUES OF THE HEAT OF SOLUTION

NO.	CEMENT	INITIAL	3 DAYS	7 DAYS	28 DAYS	90 DAYS	180 DAYS	365 DAYS	ULTIMATE
1	22	588.4	539.5	530.7	519.0	511.2	507.4	504.0	481.4
2	23	575.6	534.9	526.5	514.9	507.3	503.5	500.2	477.9
3	92	624.1	550.1	541.1	529.1	521.3	517.4	514.0	491.2
4	88	596.7	541.6	532.9	521.1	513.4	509.6	506.2	483.7
5	96	603.4	532.9	525.6	516.0	509.7	506.6	503.8	485.5
6	85	617.9	538.7	531.1	521.1	514.6	511.4	508.5	489.5
7	94	611.4	531.8	525.0	516.0	510.2	507.3	504.7	487.7
8	24	575.4	531.8	523.7	512.8	505.6	502.0	498.9	477.8
9	89	598.1	531.2	524.0	514.4	508.1	505.0	502.3	484.0
10	90	633.0	543.3	535.6	525.5	518.9	515.7	512.8	493.6
11	25	584.0	529.2	521.7	511.8	505.3	502.0	499.2	480.2
12	95	626.1	539.7	532.3	522.6	516.2	513.0	510.2	491.7
13	91	624.0	538.8	531.4	521.8	515.4	512.3	509.5	491.1
14	70	564.2	536.9	528.3	515.9	507.5	503.3	499.6	474.7

influential in determining the effect of magnesia, x_5. The remaining cements have relatively uniform distributions of the other components; silica, ferric oxide, and lime.

In order to determine whether Cements 90 and 92 merely extended the range of the effect of composition, or indeed controlled the form of the equation, two additional runs (not shown) were made, in each of which one of these cements was omitted. The results show that their influence is consistent with that of the remaining cements, and therefore their presence only extends the effective range of applicability.

Fitted values of the initial heat of solution, the heat of solution at specific time periods, and the ultimate heat of solution of each cement are given in Table 9.7.

Estimates of the cumulative heat of hardening are calculated by subtracting the fitted values of the heat of solution at specific time periods from the fitted value of the initial heat of solution. These values and the ultimate heats of hardening are given in Table 9.8.

Table 9.9 displays the residuals from the nonlinear fit of the heat of solution data. Cement 22 at 7 days has the largest residual, 4.9. The fit does not appear to be biased for any of the individual cements. In no case are there fewer than two of the seven residuals of a given cement with the same sign.

TABLE 9.8
FITTED VALUES OF THE CUMULATIVE HEAT OF HARDENING

NO.	CEMENT	3 DAYS	7 DAYS	28 DAYS	90 DAYS	180 DAYS	365 DAYS	ULTIMATE
1	22	49.0	57.7	69.4	77.2	81.0	84.4	107.0
2	23	40.7	49.1	60.6	68.3	72.0	75.4	97.7
3	92	74.1	83.1	95.0	102.9	106.7	110.2	133.0
4	88	55.1	63.9	75.6	83.3	87.1	90.5	113.1
5	96	70.6	77.8	87.5	93.7	96.8	99.6	117.9
6	85	79.2	86.8	96.8	103.3	106.5	109.4	128.4
7	94	79.7	86.5	95.4	101.3	104.1	106.7	123.7
8	24	43.7	51.7	62.6	69.8	73.4	76.6	97.6
9	89	66.9	74.2	83.7	90.0	93.1	95.8	114.1
10	90	89.7	97.4	107.5	114.1	117.3	120.2	139.4
11	25	54.8	62.2	72.2	78.7	81.9	84.8	103.8
12	95	86.4	93.8	103.5	109.9	113.0	115.8	134.4
13	91	85.3	92.6	102.3	108.6	111.7	114.5	132.9
14	70	27.2	35.8	48.2	56.6	60.8	64.6	89.4

Table 9.10 in turn displays the residuals obtained by subtracting the heats of hardening (calculated from the nonlinear equation) from those published by the experimenters. Errors in the published values have been confounded; the errors in the initial and the aged measurements in some cases cancel each other, in other cases are additive. Cement 88 at 90 days now has the largest residual, 5.6, while the residual of Cement 22 at 7 days is reduced to 4.6. Apparently there is more bias—Cement 85 has no negative residuals and Cement 70 only one, while Cements 24 and 91 have only one positive residual.

Comparing these tables of residuals reinforces the lesson that directly measured quantities should be used whenever possible, rather than differences that propagate errors through subsets of data.

The percentage of the ultimate heat of hardening evolved by each cement when observations were made is given in Table 9.11. As previously noted, the majority of the cements had already given off 50% of their potential heat of hardening by the first observation at 3 days. Thus the data provide little information on the early evolution of heat.

TABLE 9.9
RESIDUALS FROM HEAT OF SOLUTION DATA

NO.	CEMENT	0 DAYS	3 DAYS	7 DAYS	28 DAYS	90 DAYS	180 DAYS	365 DAYS
1	22	0.3	-2.0	4.9	-2.8	-3.7	2.8	-0.8
2	23	0.1	0.2	3.5	-0.2	-3.5	-2.1	-0.5
3	92-1	-0.7	-1.3	-2.2	2.3	1.7	1.9	-0.8
3	92-2	0.0						
4	88	-1.2	0.7	-0.5	0.7	4.4	-1.7	-1.3
5	96-1	-0.2						
5	96-2	1.1	-0.3	0.1	-0.3	2.2	2.0	-2.8
6	85-1	0.9	-0.5	-2.6	-3.3	0.5	-1.8	0.5
6	85-2			-0.9	-4.0	-0.3	0.4	0.0
7	94-1	-1.4	0.9	-2.5	1.6	-0.7	0.0	-2.7
7	94-2	0.3						
8	24	-2.4	-0.9	2.5	-1.3	-2.6	-1.5	2.5
9	89	0.9	3.1	-2.9	-1.2	0.8	0.9	-0.3
10	90	0.4	0.6	1.2	-0.9	0.4	1.8	-2.1
11	25	1.9	-2.1	1.4	-0.1	2.5	0.1	3.6
12	95	-0.5	1.1	-1.2	-0.7	1.9	-0.7	-0.0
13	91	-0.0	1.2	1.3	2.3	2.4	2.3	-1.8
14	70	0.5	-0.5	0.1	-1.0			2.5

In the equation for the cumulative heat of hardening,

$$(9.9) \qquad Y = C\left[1 - \left(\frac{F}{t + F}\right)^{B}\right],$$

the proportion of the heat of hardening given off at time t can be calculated from the term

$$1 - \left(\frac{F}{t + F}\right)^{B}.$$

Since B is constant, the *rate* at which heat is evolved is controlled by F, the *amount* by C.

TABLE 9.10

RESIDUALS FROM PUBLISHED HEAT OF HARDENING DATA

NO.	CEMENT	3 DAYS	7 DAYS	28 DAYS	90 DAYS	180 DAYS	3 DAYS
1	22	2.2	-4.6	3.1	4.0	-2.5	1.1
2	23	-0.1	-3.4	0.4	3.6	2.3	0.6
3	92	0.7	1.6	-2.8	-2.3	-2.4	0.2
4	88	-1.9	-0.8	-1.9	-5.6	0.5	0.1
5	96	1.3	1.0	1.3	-1.1	-0.9	3.9
6	85-1	1.4	3.5	4.2	0.4	2.7	0.4
6	85-2		1.8	4.9	1.2	0.5	0.9
7	94	-2.4	1.0	-3.0	-0.8	-1.4	1.3
8	24	-1.6	-4.9	-1.1	0.2	-0.9	-5.0
9	89	-2.2	3.7	2.1	0.1	-0.0	1.2
10	90	-0.2	-0.8	1.3	0.0	-1.4	2.5
11	25	4.0	0.6	2.0	-0.6	1.9	-1.7
12	95	-1.6	0.7	0.2	-2.4	0.3	-0.4
13	91	-1.3	-1.3	-2.4	-2.4	-2.3	1.8
14	70	1.1	0.5	1.6			-2.0

Since $F = (A/C)^{1/B}$, we deduce the following:

1. If two cements have equivalent initial heats of solution and hence equal ultimate heats of hardening, the one with the larger A- or F-value will evolve its heat more slowly.

2. If two cements have equivalent values of A, the one with the smaller ultimate heat of hardening not only will have less heat to evolve but also will do so at a slower rate.

The "half-life" of each cement's heat of hardening, $t_{(1/2)}$, can be calculated from

(9.22)
$$\frac{Y_{0.5}}{C} = \frac{1}{2} = 1 - \left(\frac{F}{t_{(1/2)} + F}\right)^B,$$

or

(9.23)
$$t_{(1/2)} = [(2)^{1/B} - 1]F = 32.0F.$$

TABLE 9.11
PERCENTAGE OF HEAT OF HARDENING GIVEN OFF

NO.	CEMENT	3 DAYS	7 DAYS	28 DAYS	90 DAYS	180 DAYS	365 DAYS
14	70	30.	40.	54.	63.	68.	72.
2	23	42.	50.	62.	70.	74.	77.
8	24	45.	53.	64.	72.	75.	78.
1	22	46.	54.	65.	72.	76.	79.
4	88	49.	56.	67.	74.	77.	80.
11	25	53.	60.	70.	76.	79.	82.
3	92	56.	62.	71.	77.	80.	83.
9	89	59.	65.	73.	79.	82.	84.
5	96	60.	66.	74.	79.	82.	84.
6	85	62.	68.	75.	80.	83.	85.
13	91	64.	70.	77.	82.	84.	86.
12	95	64.	70.	77.	82.	84.	86.
10	90	64.	70.	77.	82.	84.	86.
7	94	64.	70.	77.	82.	84.	86.

The fitted values of A, C, A/C, F, and the "half-life" of each cement are given in Table 9.12. Cement samples 91 and 92 have essentially the same ultimate heat of hardening, C, but with different values of A the half-life of one is about three times as long as the other. The reverse situation can be seen in the comparison of Cements 25 and 90. Although 90 has a slightly higher A-value, the C-value of 25 is sufficiently lower than that of 90 to provide a half-life more than four times as long. In this case a lower C-value is more effective in reducing the heat of hardening than a lower A-value. We see from Table 9.8 that at 3 days the difference in the cumulative heat of hardening between Cements 91 and 92 is only 11 calories per gram, or 15%, whereas the difference between Cements 25 and 90 is 35 calories per gram, or 63%. The values of C and A chosen for specifications will depend on whether a low-early or a low-later heat of hardening is desired. This may determine when refrigeration can be terminated. Cement 85, for example, has a lower heat of hardening value than Cement 94 at 3 days, but from then on 94 evolves less heat than 85.

Figure 9.4 shows, by a single point on an A-C grid, the two basic responses for each cement. From this plot we can see which cements give nearly the same total heats, which have almost the same A-values, and, by their closeness to the lines $A/C =$ constant, which have nearly the same proportional rate constants, F, or half-lives.

From the preceding discussion we realize that not only is considerable

TABLE 9.12
A, C, A/C, F, and Half-Life Values

NO.	CEMENT	A	C	A/C	F, DAYS	HALF—LIFE, DAYS
14	70	80.07	89.4	0.90	0.573	18.34
2	23	71.89	97.7	0.74	0.213	6.81
8	24	67.76	97.6	0.69	0.159	5.08
1	22	72.88	107.0	0.68	0.144	4.61
4	88	72.57	113.1	0.64	0.107	3.42
11	25	61.24	103.8	0.59	0.070	2.23
3	92	73.46	133.0	0.55	0.050	1.60
9	89	58.74	114.1	0.51	0.035	1.12
5	96	59.02	117.9	0.50	0.030	0.97
6	85	61.32	128.4	0.48	0.024	0.77
13	91	59.26	132.9	0.45	0.017	0.54
12	95	59.73	134.4	0.44	0.017	0.54
10	90	61.83	139.4	0.44	0.017	0.53
7	94	54.85	123.7	0.44	0.017	0.53

flexibility available in the combination of C and A to arrive at a given heat of hardening at a specific time, but also there are numerous combinations of composition whereby the specified values of C and A can be attained.

We have analyzed only the heat of solution data from these experiments. The choice of composition for a particular application will depend, of course, on an aggregate of physical properties, as well as on availability and cost.

9.8   Possible Nesting within Cements

Since the data were taken on fourteen cements (with an average of 7.5 observations on each, spread over time), it is natural to inquire whether there may be nesting, as in the data used in Chapter 8. We can imagine that some random impulse affecting a cement might well make all its z-values high (or low) by about the same amount. Such a factor would produce an "among-cement component of variance": we call it σ_1^2. This component is at least conceptually distinct from the component we will call σ_0^2, produced by random

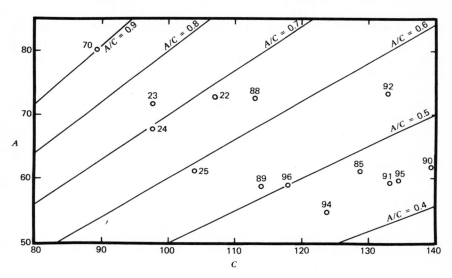

Figure 9.4   Distribution of cements on A versus C grid. Cement number is attached to each point.

variation of measurements taken at different times on the *same* cement.

Statistical habit would lead us to compare the fourteen curves at their *average* values to get an estimate of among-cement random variability with minimal contamination by σ_0^2, but this does not seem appropriate here. The fourteen curves are, more probably, randomly displaced in *two* directions. They may be displaced at their starting points in the z direction, that is, by their D_i, by any random factors that influence initial heats of solution. We are not thinking here of random fluctuations in composition; these will be taken care of (insofar as they appear as changes in oxide concentrations) by the seventeen-coefficient (nonlinear) fit. There may be variations in molecular chemical composition, which are not reflected in the oxide content but nevertheless influence the heat of solution and so move a whole curve up or down by a small amount.

Quite independent of these among-cement variations in heats of solution, there may be random variations in the recorded time of the reactions. There is ample evidence in the uniformity of these data that the experimenters were very careful workers indeed. But it was probably not realized that for five or six of these cements half the total heat of hardening evolves in a day or less and hence that heats of solution at early times should have been measured and time-recorded to the nearest hour. These time errors would produce a cement-to-cement component of variance in F (or equivalently in A).

To return to D, we have a good estimate of σ_0^2, "with 73 degrees of free-dom" in our mean square of 4.0 from the 31-coefficient fit (Pass 4). But we require an estimate of Var. (D_1) based on this value. This appears in the nonlinear fit as a separate value for each D_{1i}. Since all the standard errors estimated for D_{1i} are close to 1.8 and average 1.84, we take the estimated variance of D_i based on within-cement random variation to be 3.4.

We now require estimates of all D_i which (contrary to those found in the 31-coefficient fit) have minimal disturbance from among-cement sources. The best we have are the values from the final 17-coefficient equation (Pass 9). We call these D_{2i} and show them in Table 9.13, beside the corresponding D_{1i} for each cement.

TABLE 9.13

COMPARISON OF FITTED VALUES OF D

Pass 4: Cements Separated and Fitted by Indicator Variables

Pass 9: Cements Fitted by Composition Variables

i	Cement	Pass 4 D_{1i}	Pass 9 D_{2i}	$D_{1i} - D_{2i}$
1	22	588.5	588.4	−0.1
2	23	575.0	575.6	−0.6
3	92	624.1	624.1	0.0
4	88	595.5	596.7	−1.2
5	96	603.9	603.4	0.5
6	85	618.8	617.9	0.9
7	94	610.7	611.4	−0.7
8	24	573.0	575.4	−2.4
9	89	598.9	598.1	0.8
10	90	633.1	633.0	0.1
11	25	586.6	584.0	2.6
12	95	625.6	626.1	−0.5
13	91	624.0	624.0	0.0
14	70	565.0	564.2	0.8

$$SS(D) = 17.42; \quad MS(D) = 17.42/7 = 2.49$$

We see from Table 9.13 that the residual sum of squares for $(D_{1i} - D_{2i})$ is 17.42. We feel that this should reflect about $(14 - 7)$ or 7 degrees of freedom, since we have used 7 composition coefficients in fitting the 14 D_i. The residual mean square is then $17.4/7$ or 2.5. Since this value is *smaller* than the 3.4 we expected if only within-cement random factors were operating, we have no evidence whatever for an among-cement random component. This test must be of low power, both because of the small number of degrees of freedom available, and because of the inexact assumption that mean squares from nonlinear least-squares fits provide the usual unbiased, chi-square distributed estimates of the corresponding variances. All we can say is that we see no sign whatever of any cement-to-cement random variation in the heat of solution, in excess of that expected from the observed within-cement variation.

9.9   Comparison with Previous Linear Least-Square Fits

As the reader will recall, the experimenters used the derived heats of hardening data at specific time periods to fit a series of equations using *linear* least squares. They considered that magnesia was unreactive and then chose the postulated clinker compounds given in Table 9.1 as the independent variables. They used an equation without a b_0 term, namely,

$$(9.24) \qquad Y = b_1 x_1 + b_2 x_2 + b_3 x_3 + b_4 x_4.$$

Finally, they chose the first observation of each sample (except for 85-3, where they used the average) as being "most probable."

The accompanying table shows the results of such fits at the specific time periods (for direct comparison, we have corrected the oxide contents to 100% before converting algebraically to the postulated compounds):

Time Periods, Days	Number of Observations	Number of Coefficients	Degrees of Freedom	RSS	RMS	RRMS
3	14	4	10	155.1	15.5	3.9
7	14	4	10	103.8	10.4	3.2
28	14	4	10	142.9	14.3	3.8
90	13	4	9	113.7	12.6	3.6
180	13	4	9	36.4	4.0	2.0
365	14	4	10	56.5	5.6	2.4
Totals	82	24	58	608.4		

$$\text{Pooled RMS} = \frac{\text{Total RSS}}{\text{Total df}} = \frac{608.4}{58} = 10.5$$

Pooled residual root mean square $= 3.2$

Not only do the residual mean square values vary from time period to time period, but also the pooled RMS of 10.5 is considerably larger than the 3.9 RMS of the nonlinear fit. A larger number of coefficients might be expected to fit the data better (24 vs. 17), but in this case the cross-product terms are needed and the additional coefficients do not include them. There is a further substantial drawback in the inability to interpolate or extrapolate easily to other time periods.

By contrast, a direct fit of all the data (except one observation that is demonstrably defective) has produced equations which provide some aid in understanding the chemical and physical relationships involved. Specifically, we have found that:

1. The ultimate heat of hardening, C, is a linear function of the initial heat of solution, D.

2. The initial heat of solution in turn is a quadratic function of composition (however, see Section 9.10).

3. Both the heat of hardening, Y, and the heat of solution, Z, are nonlinear functions of time.

4. The rate constant F, $(A/C)^{1/B}$, is a complex function of composition, since A and C are quadratic in composition.

A side point of interest in the original analysis was the selection of independent variables. The experimenters calculated the composition of the clinker compounds from the analysis of the oxides, using assumptions which were widely accepted at that time. Since one set of values is derived algebraically from the other, the fit of the equations would have been identical but the coefficients and their interpretation would have been different. No data to challenge these assumptions were taken in this series of experiments. Since then the development of more definitive analytical techniques (see Hansen) has altered these views, and they undoubtedly will continue to change as more is learned about the chemistry of cement. Thus the experimenters and others since may have been misled because the fundamental units of the variables were not used in the early data analysis. This is not a rare occurrence, but the lesson involved is a hard one to learn.

9.10   Fit Using Composition in Mol Percents

The previous fit was made using the weight percents of the clinker oxides. In chemical reactions, it is often more meaningful to chemists and physicists to use mol percents as units of measurement. In order to determine if any different conclusions might be drawn, we repeat the entire sequence of fitting using mol percents.

The number of mols of any component is the weight percent of that component divided by its molecular weight. Its mol fraction, in turn, is the number of mols of that component divided by the sum of the number of mols of all components. Mol percent is the mol fraction multiplied by 100. For example, the mol percent of silica in sample 22 is calculated as follows:

$$\text{Mol}\% \text{ silica} = \frac{\dfrac{W\% \text{ silica}}{MW \text{ silica}}}{\dfrac{W\% \text{ silica}}{MW \text{ silica}} + \dfrac{W\% \text{ alumina}}{MW \text{ alumina}} + \dfrac{W\% \text{ F.O.}}{MW \text{ F.O.}} + \dfrac{W\% \text{ lime}}{MW \text{ lime}} + \dfrac{W\% \text{ magnesia}}{MW \text{ magnesia}}} \times 100$$

$$= \frac{\dfrac{27.68}{60.06}}{\dfrac{27.68}{60.06} + \dfrac{3.76}{101.94} + \dfrac{1.98}{159.68} + \dfrac{64.97}{56.08} + \dfrac{2.48}{40.32}} \times 100$$

$$= 26.64.$$

Table 9.14 gives the analysis of the clinker oxides in mol percents.

TABLE 9.14
CLINKER ANALYSIS, MOL PERCENT

		-------CLINKER ANALYSIS, MOL PERCENT-------				
ANALYSIS	CEMENT	SILICA	ALUMINA	FERRIC OXIDE	LIME	MAGNESIA
1	22	26.64	2.13	0.72	66.96	3.55
2	23	25.70	2.03	1.88	66.96	3.42
3	92	21.13	3.28	1.01	67.32	7.26
4	88	24.26	3.40	1.04	67.78	3.53
5	96	24.33	2.21	0.77	69.27	3.42
6	85	21.92	3.57	1.05	69.91	3.55
7	94	20.95	2.74	2.16	71.05	3.10
8	24	23.81	2.88	2.74	67.19	3.37
9	89	22.17	2.77	2.30	69.27	3.49
10	90	20.97	5.08	0.44	69.79	3.73
11	25	22.77	2.98	2.84	68.03	3.38
12	95	21.00	3.52	1.08	70.64	3.75
13	91	21.43	3.21	0.98	70.83	3.55
14	70	26.00	2.45	2.30	66.10	3.13

When the minimum concentration of each oxide is used, the transformed composition variables of equation 9.13 are:

$$
\begin{aligned}
x_1 &= (\text{mol } \% \text{ silica} & -20.90)/7.43 \\
x_2 &= (\text{mol } \% \text{ alumina} & -2.03)/7.43 \\
x_3 &= (\text{mol } \% \text{ ferric oxide} & -0.44)/7.43 \\
x_4 &= (\text{mol } \% \text{ lime} & -66.10)/7.43 \\
x_5 &= (\text{mol } \% \text{ magnesia} & -3.10)/7.43
\end{aligned}
$$

The C_p searches of the full equations in D and A are made again using these transformed variables. The results (not shown) indicate that in this case D can be estimated quite well ($R_D{}^2 = 0.9938$, RRMS = 2.07) with only the five composition variables while the estimate of $A(R_A{}^2 = 0.9942$, RRMS = 1.24) requires the addition of all the cross product terms.

The resulting 19 variable-equation nonlinear fit, Figures 9.22–9.25, has a residual sum of squares of 339.9, a residual root mean square of 2.0, an $R_z{}^2$ of 0.9972, and a F-value of 1600 with 85 degrees of freedom.

The residuals of the observations at time 0 (the larger fitted Y-values) have a greater dispersion than before (compare Figure 9.25 with Figure 9.21). Hence, since the total residual sum of squares is smaller, the residuals of the observations at the other time periods are slightly smaller than in the fit using weight percents. None of the residuals changed sign.

The untransformed coefficients for equations 9.16 and 9.17 are as follows:

		S. E. Coef.
B = 0.195		0.03
G = 0.72		0.08
H = 315.0		51.0
(9.25)	Initial heat of solution, D =	
	$-0.4 \times (\text{mol } \% \text{ silica})$	0.3
	$+5.9 \times (\text{mol } \% \text{ alumina})$	0.9
	$-7.6 \times (\text{mol } \% \text{ ferric oxide})$	0.7
	$+8.4 \times (\text{mol } \% \text{ lime})$	0.1
	$+7.8 \times (\text{mol } \% \text{ magnesia})$	0.4
(9.26)	Sample constant, A =	
	$+79.0 \times (\text{mol } \% \text{ silica})$	13.0
	$-375.0 \times (\text{mol } \% \text{ alumina})$	107.0
	$+592.0 \times (\text{mol } \% \text{ ferric oxide})$	82.0
	$+11.0 \times (\text{mol } \% \text{ lime})$	2.0
	$-13.0 \times (\text{mol } \% \text{ magnesia})$	7.0
	$+6.0 \times (\text{mol } \% \text{ silica} \times \text{mol } \% \text{ alumina})$	1.0
	$-6.2 \times (\text{mol } \% \text{ silica} \times \text{mol } \% \text{ ferric oxide})$	0.9
	$-1.5 \times (\text{mol } \% \text{ silica} \times \text{mol } \% \text{ lime})$	0.2
	$-14.0 \times (\text{mol } \% \text{ alumina} \times \text{mol } \% \text{ ferric oxide})$	2.0
	$+3.0 \times (\text{mol } \% \text{ alumina} \times \text{mol } \% \text{ lime})$	1.0
	$-6.0 \times (\text{mol } \% \text{ ferric oxide} \times \text{mol } \% \text{ lime})$	1.0

Other than the new coefficients in terms of mol percents (which we will leave for cement chemists to interpret), no additional information was gained. The equation for D, the initial heat of solution, and its correlation with C, the ultimate heat of hardening, is fortunately simpler. The equation for A, however, is more complex. The overall fit using mol percents is as good as that using weight percents. Under these conditions we are more secure about estimating the initial heats of solution and the ultimate heats of hardening, less secure about estimating the time constants F.

9.11 Conclusions

The experimenters collected their data in order to learn more about the effects of chemical composition on the rate and the amount of heat evolved during the setting of Portland cement. *The aims of the data analyst should be the same.*

We have found an equation (9.16) that describes all the data (one point dropped) acceptably. We have shown:

1. How the amount of heat evolved varies with time (equation 9.1).
2. The prediction of the ultimate heat of hardening, C, from the initial acid heat of solution, D (equation 9.20).
3. The dependence of the initial heat on chemical composition (equations 9.18 and 9.25).
4. A table of constants, F, that characterize the rate of evolution of heat for each cement (Table 9.12). F is the time required for 0.128 of the total heat to be evolved; half of the ultimate heat is evolved in time 32.0F.
5. The dependence of F on A (a sample constant) and C, and, in turn, their dependence on chemical composition (equations 9.19, 9.21, and 9.26).

We have tried to meet the requirements given in Chapter 2 for a good method of fitting. All the data have been used (except one point shown to be defective); we have employed 17 coefficients to describe 104 observations with an R_z^2 of 0.9972 and an F-value of 1800; we have estimated and studied the residual error and then have used it to estimate the standard errors of the coefficients in our equation; finally, we have tested the data for nesting and found none.

The experimenters used *linear* least squares to fit a separate equation for each of the six time periods. The single *nonlinear* equation (9.16), with time as a variable, is more compact, more precise, more comprehensive, and hence more informative.

The data have served to introduce nonlinear least squares, to exemplify a massive use of indicator variables, and to permit discovery of an unexpected relation (that between the initial heat of solution and the ultimate heat of hardening). Finally, the example justifies our initial contention that much can be learned from sets of multifactor data taken by careful experimenters, even though the standard statistical requirements of balanced conditions were not met.

APPENDIX 9A

COMPUTER PRINTOUT OF NONLINEAR EXAMPLE

Pass	Figures	Program	Conditions
1	9A.1–9A.4	Nonlinear	Indicator variables, B common
2	9A.5–9A.7	Nonlinear	Indicator variables, 85-3-2 omitted
3	9A.8	Linear	C as a function of D
4	9A.9	Nonlinear	Indicator variables, B, G, and H common
5	9A.10–9A.13	Linear	D as a function of wt.% composition C_p search
6	9A.14–9A.16	Linear	A as a function of wt.% composition, C_p search
7	9A.17	Linear	Selected equation, D vs. wt.% composition
8	9A.18	Linear	Selected equation, A vs. wt.% composition
9	9A.19–9A.21	Nonlinear	Wt.% composition variables, B, G, and H common
10	9A.22–9A.25	Nonlinear	Mol% composition variables, B, G, and H common

NON-LINEAR ESTIMATION   CEMENT HEAT, PASS 1

MODEL NUMBER   2

$Z = D' - C' + (A' / (T + F')EXP\ B)$
Z = ACID HEAT OF SOLUTION
B = EXPONENT COMMON TO ALL CEMENTS
D' = INITIAL ACID HEAT OF SOLUTION = D1V1 + D2V2 + --- + D14V14
A' = CONSTANT = A1V1 + A2V2 + --- + A14V14
C' = ULTIMATE CUMULATIVE HEAT OF HARDENING = C1V1 + C2V2 + --- + C14V14
F' = TIME CORRECTION TERM = (A'/C')EXP(1/B)
T = TIME, MEASURED IN DAYS, AFTER WATER IS ADDED TO THE CEMENT
V1 --- V14 = INDICATOR VARIABLES FOR EACH SAMPLE OF CEMENT
NO OBSERVATIONS OMITTED.

NUMBER OF COEFFICIENTS   43

STARTING LAMBA   0.100

STARTING NU   10.000

MAX NO. OF ITERATIONS   20

DATA FORMAT (6X, A4, 2X, 2F6.1, 6X, 14F1.0)

OBSV. NO.	IDENT.	SEQ.	1-11-21	2-12-22	3-13-23	4-14-24	5-15-25	6-16-26	7-17-27	8-18-28	9-19-29	10-20-30
1	22 0	1	588.700	0.0	1.000	0.0	0.0	0.0	0.0	0.0	0.0	0.0
		2	0.0	0.0	0.0	0.0	0.0	0.0	0.0	0.0	0.0	0.0
2	22 3	1	537.500	0.0	1.000	0.0	0.0	0.0	0.0	0.0	0.0	0.0
		2	0.0	3.000	0.0	0.0	0.0	0.0	0.0	0.0	0.0	0.0
3	22 7	1	535.600	0.0	1.000	0.0	0.0	0.0	0.0	0.0	0.0	0.0
		2	0.0	7.000	0.0	0.0	0.0	0.0	0.0	0.0	0.0	0.0
4	2228	1	516.200	0.0	1.000	0.0	0.0	0.0	0.0	0.0	0.0	0.0
		2	0.0	28.000	0.0	0.0	0.0	0.0	0.0	0.0	0.0	0.0
5	2290	1	507.500	0.0	1.000	0.0	0.0	0.0	0.0	0.0	0.0	0.0
		2	0.0	90.000	0.0	0.0	0.0	0.0	0.0	0.0	0.0	0.0
6	2218	1	510.200	0.0	1.000	0.0	0.0	0.0	0.0	0.0	0.0	0.0
		2	0.0	180.000	0.0	0.0	0.0	0.0	0.0	0.0	0.0	0.0
7	2236	1	503.200	0.0	1.000	0.0	0.0	0.0	0.0	0.0	0.0	0.0
		2	0.0	365.000	0.0	0.0	0.0	0.0	0.0	0.0	0.0	0.0
8	23 0	1	575.700	0.0	0.0	1.000	0.0	0.0	0.0	0.0	0.0	0.0
		2	0.0	0.0	0.0	0.0	0.0	0.0	0.0	0.0	0.0	0.0
9	23 3	1	535.100	0.0	0.0	1.000	0.0	0.0	0.0	0.0	0.0	0.0
		2	0.0	3.000	0.0	0.0	0.0	0.0	0.0	0.0	0.0	0.0
10	23 7	1	530.000	0.0	0.0	1.000	0.0	0.0	0.0	0.0	0.0	0.0
		2	0.0	7.000	0.0	0.0	0.0	0.0	0.0	0.0	0.0	0.0
11	2328	1	514.700	0.0	0.0	1.000	0.0	0.0	0.0	0.0	0.0	0.0
		2	0.0	28.000	0.0	0.0	0.0	0.0	0.0	0.0	0.0	0.0
12	2390	1	503.800	0.0	0.0	1.000	0.0	0.0	0.0	0.0	0.0	0.0
		2	0.0	90.000	0.0	0.0	0.0	0.0	0.0	0.0	0.0	0.0
13	2318	1	501.400	0.0	0.0	1.000	0.0	0.0	0.0	0.0	0.0	0.0
		2	0.0	180.000	0.0	0.0	0.0	0.0	0.0	0.0	0.0	0.0
14	2336	1	499.700	0.0	0.0	1.000	0.0	0.0	0.0	0.0	0.0	0.0
		2	0.0	365.000	0.0	0.0	0.0	0.0	0.0	0.0	0.0	0.0
15	92 0	1	623.600	0.0	0.0	0.0	1.000	0.0	0.0	0.0	0.0	0.0
		2	0.0	0.0	0.0	0.0	0.0	0.0	0.0	0.0	0.0	0.0
16	92 0	1	624.100	0.0	0.0	0.0	1.000	0.0	0.0	0.0	0.0	0.0
		2	0.0	0.0	0.0	0.0	0.0	0.0	0.0	0.0	0.0	0.0
17	92 3	1	548.800	0.0	0.0	0.0	1.000	0.0	0.0	0.0	0.0	0.0
		2	0.0	3.000	0.0	0.0	0.0	0.0	0.0	0.0	0.0	0.0
18	92 7	1	538.900	0.0	0.0	0.0	1.000	0.0	0.0	0.0	0.0	0.0
		2	0.0	7.000	0.0	0.0	0.0	0.0	0.0	0.0	0.0	0.0
19	9228	1	531.400	0.0	0.0	0.0	1.000	0.0	0.0	0.0	0.0	0.0
		2	0.0	28.000	0.0	0.0	0.0	0.0	0.0	0.0	0.0	0.0

Figure 9A.1

Row	ID	k	Val1	Val2	B1	B2	B3	B4	B5	B6	B7	B8
20	9290	1	523.000	90.000	0.0	0.0	1.000	0.0	0.0	0.0	0.0	0.0
21	9218	2	519.300	180.000	0.0	0.0	1.000	0.0	0.0	0.0	0.0	0.0
22	9236	1	513.200	365.000	0.0	0.0	1.000	0.0	0.0	0.0	0.0	0.0
23	88 0	2	595.500	0.0	0.0	0.0	1.000	0.0	0.0	0.0	0.0	0.0
24	88 3	1	542.300	3.000	0.0	0.0	0.0	1.000	0.0	0.0	0.0	0.0
25	88 7	2	532.400	7.000	0.0	0.0	0.0	1.000	0.0	0.0	0.0	0.0
26	8828	1	521.800	28.000	0.0	0.0	0.0	1.000	0.0	0.0	0.0	0.0
27	8890	2	517.800	90.000	0.0	0.0	0.0	1.000	0.0	0.0	0.0	0.0
28	8818	1	507.900	180.000	0.0	0.0	0.0	1.000	0.0	0.0	0.0	0.0
29	8836	2	504.900	365.000	0.0	0.0	0.0	1.000	0.0	0.0	0.0	0.0
30	96 0	1	604.500	0.0	0.0	0.0	0.0	0.0	1.000	0.0	0.0	0.0
31	96 0	2	603.200	0.0	0.0	0.0	0.0	0.0	1.000	0.0	0.0	0.0
32	96 3	1	532.600	3.000	0.0	0.0	0.0	0.0	1.000	0.0	0.0	0.0
33	96 7	2	525.700	7.000	0.0	0.0	0.0	0.0	1.000	0.0	0.0	0.0
34	9628	1	515.700	28.000	0.0	0.0	0.0	0.0	1.000	0.0	0.0	0.0
35	9690	2	511.900	90.000	0.0	0.0	0.0	0.0	1.000	0.0	0.0	0.0
36	9618	1	508.600	180.000	0.0	0.0	0.0	0.0	1.000	0.0	0.0	0.0
37	9636	2	501.000	365.000	0.0	0.0	0.0	0.0	1.000	0.0	0.0	0.0
38	85 0	1	618.800	0.0	0.0	0.0	0.0	0.0	0.0	1.000	0.0	0.0
39	85 3	2	538.200	3.000	0.0	0.0	0.0	0.0	0.0	1.000	0.0	0.0
40	85 3	1	544.800	3.000	0.0	0.0	0.0	0.0	0.0	1.000	0.0	0.0
41	85 7	2	528.500	7.000	0.0	0.0	0.0	0.0	0.0	1.000	0.0	0.0
42	85 7	1	530.200	7.000	0.0	0.0	0.0	0.0	0.0	1.000	0.0	0.0
43	8528	2	517.800	28.000	0.0	0.0	0.0	0.0	0.0	1.000	0.0	0.0
44	8528	1	517.100	28.000	0.0	0.0	0.0	0.0	0.0	1.000	0.0	0.0
45	8590	2	515.100	90.000	0.0	0.0	0.0	0.0	0.0	1.000	0.0	0.0
46	8590	1	514.300	90.000	0.0	0.0	0.0	0.0	0.0	1.000	0.0	0.0
47	8518	2	509.600	180.000	0.0	0.0	0.0	0.0	0.0	1.000	0.0	0.0
48	8518	1	511.800	180.000	0.0	0.0	0.0	0.0	0.0	1.000	0.0	0.0
49	8536	2	509.000	365.000	0.0	0.0	0.0	0.0	0.0	1.000	0.0	0.0
50	8536	1	508.500	365.000	0.0	0.0	0.0	0.0	0.0	1.000	0.0	0.0
51	94 0	2	610.000	0.0	0.0	0.0	0.0	0.0	0.0	0.0	1.000	0.0
52	94 0	1	611.700	0.0	0.0	0.0	0.0	0.0	0.0	0.0	1.000	0.0

Figure 9A.1 (*continued*)

Obs	ID	N	X1	X2	D1	D2	D3	D4	D5	D6	D7	D8	R1	R2	R3	R4
54	94 7	2	522.500	0.0	0.0	0.0	0.0	0.0	0.0	0.0	0.0	0.0	0.0	1.000	0.0	0.0
55	9428	1	517.600	7.000	0.0	0.0	0.0	0.0	0.0	0.0	0.0	0.0	0.0	1.000	0.0	0.0
56	9490	2	509.500	28.000	0.0	0.0	0.0	0.0	0.0	0.0	0.0	0.0	0.0	1.000	0.0	0.0
57	9418	1	507.300	90.000	0.0	0.0	0.0	0.0	0.0	0.0	0.0	0.0	0.0	1.000	0.0	0.0
58	9436	2	502.000	180.000	0.0	0.0	0.0	0.0	0.0	0.0	0.0	0.0	0.0	1.000	0.0	0.0
59	24 0	1	573.000	365.000	0.0	0.0	0.0	0.0	0.0	0.0	0.0	0.0	1.000	0.0	0.0	0.0
60	24 3	2	530.900	0.0	0.0	0.0	0.0	0.0	0.0	0.0	0.0	0.0	1.000	0.0	0.0	0.0
61	24 7	1	526.200	3.000	0.0	0.0	0.0	0.0	0.0	0.0	0.0	0.0	1.000	0.0	0.0	0.0
62	2428	2	511.500	7.000	0.0	0.0	0.0	0.0	0.0	0.0	0.0	0.0	1.000	0.0	0.0	0.0
63	2490	1	503.000	28.000	0.0	0.0	0.0	0.0	0.0	0.0	0.0	0.0	1.000	0.0	0.0	0.0
64	2418	2	500.500	90.000	0.0	0.0	0.0	0.0	0.0	0.0	0.0	0.0	1.000	0.0	0.0	0.0
65	2436	1	501.400	180.000	0.0	0.0	0.0	0.0	0.0	0.0	0.0	0.0	0.0	0.0	0.0	0.0
66	89 0	2	599.000	365.000	0.0	0.0	0.0	0.0	0.0	0.0	0.0	0.0	0.0	0.0	0.0	0.0
67	89 3	1	534.300	0.0	0.0	0.0	0.0	0.0	0.0	0.0	0.0	0.0	0.0	0.0	0.0	0.0
68	89 7	2	521.100	3.000	0.0	0.0	0.0	0.0	0.0	0.0	0.0	0.0	0.0	0.0	0.0	0.0
69	8928	1	513.200	7.000	0.0	0.0	0.0	0.0	0.0	0.0	0.0	0.0	0.0	0.0	0.0	0.0
70	8990	2	508.900	28.000	0.0	0.0	0.0	0.0	0.0	0.0	0.0	0.0	0.0	0.0	0.0	0.0
71	8918	1	505.900	90.000	0.0	0.0	0.0	0.0	0.0	0.0	0.0	0.0	0.0	0.0	0.0	0.0
72	8936	2	502.000	180.000	0.0	0.0	0.0	0.0	0.0	0.0	0.0	0.0	0.0	0.0	0.0	0.0
73	90 0	1	633.400	365.000	0.0	0.0	0.0	0.0	0.0	0.0	0.0	0.0	0.0	0.0	0.0	0.0
74	90 3	2	543.900	0.0	0.0	0.0	0.0	0.0	0.0	0.0	0.0	0.0	0.0	0.0	0.0	0.0
75	90 7	1	536.800	3.000	0.0	0.0	0.0	0.0	0.0	0.0	0.0	0.0	0.0	0.0	0.0	0.0
76	9028	2	524.600	7.000	0.0	0.0	0.0	0.0	0.0	0.0	0.0	0.0	0.0	0.0	0.0	0.0
77	9090	1	519.300	28.000	1.000	0.0	0.0	0.0	0.0	0.0	0.0	0.0	0.0	0.0	0.0	0.0
78	9018	2	517.500	90.000	1.000	0.0	0.0	0.0	0.0	0.0	0.0	0.0	0.0	0.0	0.0	0.0
79	9036	1	510.700	180.000	1.000	0.0	0.0	0.0	0.0	0.0	0.0	0.0	0.0	0.0	0.0	0.0
80	25 0	2	585.900	365.000	1.000	0.0	0.0	0.0	0.0	0.0	0.0	0.0	0.0	0.0	0.0	0.0
81	25 3	1	527.100	0.0	1.000	0.0	0.0	0.0	0.0	0.0	0.0	0.0	0.0	0.0	0.0	0.0
82	25 7	2	523.100	3.000	1.000	0.0	0.0	0.0	0.0	0.0	0.0	0.0	0.0	0.0	0.0	0.0
83	2528	1	511.700	7.000	1.000	0.0	0.0	0.0	0.0	0.0	0.0	0.0	0.0	0.0	0.0	0.0
84	2590	1	507.800	28.000	1.000	0.0	0.0	0.0	0.0	0.0	0.0	0.0	0.0	0.0	0.0	0.0
85	2518	2	502.100	90.000	1.000	0.0	0.0	0.0	0.0	0.0	0.0	0.0	0.0	0.0	0.0	0.0

Figure 9A.1 (*continued*)

303

#	ID	k													
86	2536	1	502.800	365.000	0.0	0.0	0.0	0.0	0.0	0.0	0.0	0.0	0.0	0.0	0.0
		2	0.0	0.0	1.000	0.0	0.0	0.0	0.0	0.0	0.0	0.0	0.0	0.0	0.0
87	95 0	2	625.600	0.0	0.0	1.000	0.0	0.0	0.0	0.0	0.0	0.0	0.0	0.0	0.0
88	95 3	1	540.800	0.0	1.000	0.0	0.0	0.0	0.0	0.0	0.0	0.0	0.0	0.0	0.0
		2	0.0	3.000	1.000	0.0	0.0	0.0	0.0	0.0	0.0	0.0	0.0	0.0	0.0
89	95 7	1	531.100	0.0	1.000	0.0	0.0	0.0	0.0	0.0	0.0	0.0	0.0	0.0	0.0
		2	0.0	7.000	1.000	0.0	0.0	0.0	0.0	0.0	0.0	0.0	0.0	0.0	0.0
90	9528	1	521.900	0.0	1.000	0.0	0.0	0.0	0.0	0.0	0.0	0.0	0.0	0.0	0.0
		2	0.0	28.000	1.000	0.3	0.0	0.0	0.0	0.0	0.0	0.0	0.0	0.0	0.0
91	9590	1	518.100	0.0	1.000	0.0	0.0	0.0	0.0	0.0	0.0	0.0	0.0	0.0	0.0
		2	0.0	90.000	1.000	0.0	0.0	0.0	0.0	0.0	0.0	0.0	0.0	0.0	0.0
92	9518	1	512.300	0.0	1.000	0.0	0.0	0.0	0.0	0.0	0.0	0.0	0.0	0.0	0.0
		2	0.0	180.000	1.000	0.0	0.0	0.0	0.0	0.0	0.0	0.0	0.0	0.0	0.0
93	9536	1	510.200	0.0	1.000	0.0	0.0	0.0	0.0	0.0	0.0	0.0	0.0	0.0	0.0
		2	0.0	365.000	0.0	0.0	0.0	0.0	0.0	0.0	0.0	0.0	0.0	0.0	0.0
94	91 0	1	624.000	0.0	0.0	0.0	1.000	0.0	0.0	0.0	0.0	0.0	0.0	0.0	0.0
		2	0.0	0.0	0.0	1.000	0.0	0.0	0.0	0.0	0.0	0.0	0.0	0.0	0.0
95	91 3	1	540.000	3.000	0.0	1.000	0.0	0.0	0.0	0.0	0.0	0.0	0.0	0.0	0.0
		2	0.0	0.0	0.0	1.000	0.0	0.0	0.0	0.0	0.0	0.0	0.0	0.0	0.0
96	91 7	1	532.700	7.000	0.0	1.000	0.0	0.0	0.0	0.0	0.0	0.0	0.0	0.0	1.000
		2	0.0	0.0	0.0	1.000	0.0	0.0	0.0	0.0	0.0	0.0	0.0	0.0	0.0
97	9128	1	524.100	28.000	0.0	1.000	0.0	0.0	0.0	0.0	0.0	0.0	0.0	0.0	1.000
		2	0.0	0.0	0.0	1.000	0.0	0.0	0.0	0.0	0.0	0.0	0.0	0.0	0.0
98	9190	1	517.800	90.000	0.0	1.000	0.0	0.0	0.0	0.0	0.0	0.0	0.0	0.0	1.000
		2	0.0	0.0	0.0	1.000	0.0	0.0	0.0	0.0	0.0	0.0	0.0	0.0	0.0
99	9118	1	514.600	180.000	0.0	1.000	0.0	0.0	0.0	0.0	0.0	0.0	0.0	0.0	1.000
		2	0.0	0.0	0.0	1.000	0.0	0.0	0.0	0.0	0.0	0.0	0.0	0.0	0.0
100	9136	1	507.700	365.000	0.0	1.000	0.0	0.0	0.0	0.0	0.0	0.0	0.0	0.0	1.000
		2	0.0	0.0	0.0	0.0	0.0	0.0	0.0	0.0	0.0	0.0	0.0	0.0	0.0
101	70 0	1	564.700	0.0	0.0	0.0	1.000	0.0	0.0	0.0	0.0	0.0	0.0	0.0	0.0
		2	0.0	0.0	0.0	0.0	1.000	0.0	0.0	0.0	0.0	0.0	0.0	0.0	0.0
102	70 3	1	536.400	3.000	0.0	0.0	1.000	0.0	0.0	0.0	0.0	0.0	0.0	0.0	0.0
		2	0.0	0.0	0.0	0.0	1.000	0.0	0.0	0.0	0.0	0.0	0.0	0.0	0.0
103	70 7	1	528.400	7.000	0.0	0.0	1.000	0.0	0.0	0.0	0.0	0.0	0.0	0.0	0.0
		2	0.0	0.0	0.0	0.0	1.000	0.0	0.0	0.0	0.0	0.0	0.0	0.0	0.0
104	7028	1	514.900	28.000	0.0	0.0	1.000	0.0	0.0	0.0	0.0	0.0	0.0	0.0	0.0
		2	0.0	0.0	0.0	0.0	0.0	0.0	0.0	0.0	0.0	0.0	0.0	0.0	0.0
105	7036	1	502.100	365.000	0.0	0.0	1.000	0.0	0.0	0.0	0.0	0.0	0.0	0.0	0.0
		2	0.0	0.0	0.0	0.0	0.0	0.0	0.0	0.0	0.0	0.0	0.0	0.0	0.0
106		1	0.0	999.000	0.0	0.0	0.0	0.0	0.0	0.0	0.0	0.0	0.0	0.0	0.0
		2	0.0	0.0	0.0	0.0	0.0	0.0	0.0	0.0	0.0	0.0	0.0	0.0	0.0

Figure 9A.1 (*continued*)

DEP.VAR: MIN Y = 4.997D+02   MAX Y = 6.340D+02   RANGE Y = 1.337D+02

$Z = D - C + (A / (T + F)EXP\ B)$
A = ACID HEAT OF SOLUTION
B = CONSTANT = B(16)*X(2) + B(17)*X(3)+----+B(29)*X(15)
B = EXPONENT OR RATE OF HARDENING = B(1)
C = ULTIMATE CUMULATIVE HEAT OF HARDENING = B(J0)*X(2)+----+B(43)*X(15)
D = INITIAL ACID HEAT OF SOLUTION = B(2)*X(2)+----+B(15)*X(15)
F = TIME CORRECTION FACTOR = (A/C)EXP(1/B)
T = TIME IN DAYS = X(1)

IND.VAR(I)	NAME	COEF.B(I)	S.E. COEF	T-VALUE	95% CONFIDENCE LIMITS LOWER	UPPER
1	EXP B	2.20000D-01	3.660-02	6.0	1.470D-01	2.930D-01
2	D 22	5.886750+02	2.200+00	267.9	5.840+02	5.930+02
3	D 23	5.756540+02	2.200+00	262.0	5.710+02	5.800+02
4	D 92	6.237480+02	1.550+00	401.4	6.210+02	6.270+02
5	D 88	5.954900+02	2.200+00	271.0	5.910+02	6.000+02
6	D 96	6.038490+02	1.550+00	388.6	6.010+02	6.070+02
7	D 85	6.188130+02	2.200+00	281.6	6.140+02	6.240+02
8	D 94	6.108050+02	1.550+00	393.1	6.080+02	6.140+02
9	D 24	5.729930+02	2.200+00	260.7	5.690+02	5.770+02
10	D 89	5.990080+02	2.200+00	272.5	5.950+02	6.030+02
11	D 90	6.333990+02	2.200+00	288.2	6.290+02	6.380+02
12	D 25	5.858970+02	2.200+00	266.6	5.820+02	5.900+02
13	D 95	6.256010+02	2.200+00	284.6	6.210+02	6.300+02
14	D 91	6.239980+02	2.200+00	283.9	6.200+02	6.280+02
15	D 70	5.647290+02	6.250+00	257.1	5.600+02	5.690+02
16	A 22	7.254320+01	6.250+00	11.6	0.000+01	8.500+01
17	A 23	7.808640+01	6.500+00	12.0	6.510+01	9.110+01
18	A 92	6.547990+01	6.000+00	10.9	5.330+01	7.750+01
19	A 88	7.125840+01	6.200+00	11.5	5.890+01	8.370+01
20	A 96	5.672470+01	5.760+00	9.9	4.520+01	6.820+01
21	A 85	6.201690+01	4.740+00	13.1	5.260+01	7.150+01
22	A 94	5.537010+01	5.710+00	9.7	4.390+01	6.680+01
23	A 24	6.663080+01	6.140+00	10.8	5.430+01	7.890+01
24	A 89	5.806910+01	5.800+00	10.0	4.650+01	7.000+01
25	A 90	6.209420+01	5.890+00	10.5	5.030+01	7.390+01
26	A 25	5.183940+01	5.650+00	9.2	4.050+01	6.310+01
27	A 95	5.785360+01	5.780+00	10.0	4.630+01	6.940+01
28	A 91	5.895200+01	5.810+00	10.1	4.730+01	7.060+01
29	A 70	7.147530+01	7.380+00	9.7	5.670+01	8.620+01
30	C 22	1.049570+02	6.470+00	16.2	9.200+01	1.180+02
31	C 23	9.867220+01	6.760+00	14.6	8.510+01	1.120+02
32	C 92	1.260410+02	5.880+00	21.4	1.140+02	1.380+02
33	C 88	1.083790+02	6.400+00	16.9	9.560+01	1.210+02
34	C 96	1.151390+02	5.310+00	21.7	1.050+02	1.260+02
35	C 85	1.281550+02	5.580+00	22.9	1.170+02	1.390+02
36	C 94	1.221510+02	5.230+00	23.4	1.120+02	1.330+02
37	C 24	9.253720+01	6.080+00	15.2	8.040+01	1.050+02
38	C 89	1.126220+02	5.610+00	20.1	1.010+02	1.240+02
39	C 90	1.377270+02	5.880+00	23.4	1.260+02	1.490+02
40	C 25	9.831720+01	5.230+00	18.8	8.790+01	1.090+02
41	C 95	1.309630+02	5.610+00	23.3	1.200+02	1.420+02
42	C 91	1.294660+02	5.680+00	22.8	1.180+02	1.410+02
43	C 70	8.273260+01	6.640+00	12.5	6.950+01	9.600+01

NO. OF OBSERVATIONS   105
NO. OF COEFFICIENTS   43
RESIDUAL DEGREES OF FREEDOM   62
RESIDUAL ROOT MEAN SQUARE   2.19780161
RESIDUAL MEAN SQUARE   4.83033190
RESIDUAL SUM OF SQUARES   299.48057768

Figure 9A.2

ORDERED BY OBS. Y — COMPUTER INPUT

OBS. NO.	OBS. Y	FITTED Y	RESIDUAL
22 0	588.700	588.675	0.025
22 3	537.600	539.935	-2.435
22 7	535.600	530.725	4.875
2228	516.200	518.519	-2.319
2290	507.500	510.663	-3.163
2218	510.200	506.857	3.343
2236	503.200	503.526	-0.326
23 0	575.700	575.654	0.046
23 3	535.100	536.851	-1.751
23 7	530.000	527.339	2.661
2328	514.700	514.396	0.304
2390	503.800	505.974	-2.174
2318	501.400	501.884	-0.484
2336	499.700	498.302	1.398
92 0	624.400	623.748	-0.348
92 0	624.100	623.748	0.352
92 3	548.800	548.932	-0.132
92 7	538.900	540.310	-1.410
9228	531.400	529.149	2.251
9290	523.000	522.033	0.967
9218	519.300	518.594	0.706
9236	513.200	515.586	-2.386
88 3	595.500	595.490	0.010
88 3	542.300	542.477	-0.177
8828	532.400	533.339	-0.939
8828	521.800	521.305	0.495
8890	517.800	513.580	4.220
8818	507.900	509.841	-1.941
8836	504.000	506.568	-1.668
96 0	604.500	603.849	0.651
96 0	603.200	603.849	-0.649
96 3	532.600	533.125	-0.525
96 7	525.700	525.633	0.067
9628	515.700	515.953	-0.253
9690	511.900	509.786	2.114
9618	508.600	506.806	1.794
9636	501.000	504.200	-3.200
85 0	618.800	618.813	-0.013
85 3	544.800	539.229	5.571
85 3	538.200	539.229	-1.029
85 7	524.500	531.031	-2.531
85 7	530.200	531.031	-0.831
8528	517.800	520.444	-2.644
8528	517.100	520.444	-3.344
8590	515.100	513.701	1.399
8590	514.300	513.701	0.599
8518	509.600	510.443	-0.843
8518	511.800	510.443	1.357
8536	509.000	507.594	1.406
8536	508.500	507.594	0.936

ORDERED BY RESIDUALS

OBS. NO.	OBS. Y	FITTED Y	ORDERED RESID.	SEQ.
85 3	544.800	539.229	5.571	1
22 7	535.600	530.725	4.875	2
8890	517.800	513.580	4.220	3
2218	510.200	506.857	3.343	4
2436	501.400	498.649	2.751	5
23 7	530.000	527.339	2.661	6
24 7	526.200	523.581	2.619	7
89 3	534.300	531.825	2.475	8
9428	517.600	515.295	2.305	9
9228	531.400	529.149	2.251	10
9690	511.900	509.786	2.114	11
9018	508.600	506.806	2.018	12
9590	518.100	516.135	1.965	13
9618	508.600	506.806	1.794	14
25 7	523.100	521.308	1.792	15
8536	509.000	507.594	1.406	16
8590	515.100	513.701	1.399	17
2336	499.700	498.302	1.398	18
9190	517.800	516.436	1.364	19
8518	511.800	510.443	1.357	20
9118	514.600	513.339	1.261	21
9128	524.100	522.847	1.253	22
2536	502.800	501.736	1.064	23
8918	505.900	504.912	0.988	24
9290	523.000	522.033	0.967	25
2590	507.800	506.840	0.960	26
8990	508.900	507.962	0.938	27
9418	507.300	506.364	0.936	28
8536	508.500	507.594	0.906	29
94 0	611.700	610.850	0.850	30
95 3	540.800	539.990	0.810	31
9218	519.300	518.594	0.706	32
90 7	536.200	536.108	0.092	33
96 0	604.500	603.849	0.651	34
94 3	532.700	532.094	0.606	35
8590	514.300	513.701	0.599	36
7036	502.100	501.509	0.591	37
9090	519.300	519.745	0.555	38
70 7	528.400	527.859	0.541	39
8828	521.800	521.305	0.495	40
92 0	624.100	623.748	0.352	41
2328	514.700	514.396	0.304	42
9490	—	—	0.227	43
70 3	530.430	530.205	0.195	44
96 7	525.700	525.633	0.067	45
23 0	575.700	575.654	0.046	46
22 0	588.700	588.675	0.025	47
86 0	—	—	0.010	48
24 0	—	—	0.007	49
25 0	—	—	0.003	50

Figure 9A.2 (*continued*)

306

Table (rotated on page; two side-by-side data blocks):

idx	col2	col3	col4	col5
51	0.002	623.998	624.400	91 0
52	-0.001	633.399	633.400	90 0
53	-0.001	625.601	625.600	95 0
54	-0.008	599.008	599.000	85 0
55	-0.013	618.813	618.800	70 0
56	-0.029	604.729	604.700	92 3
57	-0.132	548.922	548.800	88 3
58	-0.177	542.477	542.300	91 7
59	-0.219	532.919	532.700	9536
60	-0.237	510.437	510.200	8936
61	-0.244	502.244	502.000	9628
62	-0.253	503.526	503.200	2236
63	-0.326	515.953	515.700	92 0
64	-0.348	623.748	623.200	90 3
65	-0.440	544.340	543.900	2318
66	-0.484	501.884	501.400	96 3
67	-0.525	533.125	532.600	9528
68	-0.527	522.427	521.900	96 0
69	-0.049	603.049	603.200	91 3
70	-0.732	540.732	540.900	2528
71	-0.774	512.474	511.700	9518
72	-0.795	513.095	512.300	85 7
73	-0.851	531.031	531.200	8518
74	-0.843	510.443	509.600	94 0
75	-0.850	610.850	610.000	9028
76	-0.527	525.498	524.600	2428
77	-0.049	512.411	511.500	88 7
78	-0.939	533.339	533.200	25 3
79	-1.028	528.120	527.100	85 3
80	-1.029	539.229	538.200	24 3
81	-1.050	531.950	530.900	8928
82	-1.074	514.274	513.200	2418
83	-1.208	532.413	532.400	2418
84	-1.415	513.415	513.200	95 7
85	-1.498	516.198	514.900	92 7
86	-1.410	540.310	538.900	8836
87	-1.008	500.368	504.900	9236
88	-1.751	539.851	535.100	22 3
89	-1.820	503.820	502.000	85 7
90	-1.929	512.662	510.700	8528
91	-1.941	509.841	507.700	9130
92	-2.017	504.117	502.100	89 7
93	-2.174	505.974	505.000	2290
94	-2.202	502.202	522.300	9636
95	-2.655	524.723	516.200	94 7
96	-2.319	516.519	510.200	9228
97	-2.386	515.586	537.500	22 3
98	-2.435	524.935	517.800	85 7
99	-2.531	531.031	520.100	8528
100	-2.444	520.444	521.100	9130
101	-2.930	510.630	524.100	89 7
102	-3.075	524.175	507.500	2290
103	-3.163	510.603	501.000	9636
104	-3.200	510.600	511.100	8528
105	-3.384	520.444		

col6	col7	col8	idx2
-0.850	610.850	610.000	94 0
0.850	610.850	611.700	94 0
0.606	532.094	532.700	94 3
-2.255	524.755	522.500	94 7
2.305	515.295	517.600	9428
0.227	509.273	509.500	9490
0.936	506.364	507.300	9418
-1.820	503.820	502.000	9436
0.007	572.993	573.000	24 0
-1.056	531.956	530.900	24 3
2.619	523.581	526.200	24 7
-0.911	512.411	511.500	2428
-2.202	505.202	503.000	2490
-1.208	501.708	500.500	2418
2.751	498.649	501.400	2436
-0.008	599.008	599.000	89 0
2.475	531.825	534.300	89 3
-3.075	524.175	521.100	89 7
-1.074	514.274	513.200	8928
0.938	507.962	508.900	8990
0.988	504.912	505.900	8918
-0.244	502.244	502.000	8936
-0.001	633.399	633.400	90 0
-0.440	544.340	543.900	90 3
0.692	536.108	536.800	90 7
-0.898	525.498	524.600	9028
0.555	518.745	519.300	9090
2.018	515.482	517.500	9018
-1.929	512.629	510.700	25 0
0.003	585.897	585.900	25 3
-1.028	528.128	527.100	25 7
1.792	521.308	523.100	2528
-0.774	511.474	511.700	2590
-0.960	506.840	507.800	2536
-2.017	504.117	504.200	2518
1.064	501.736	502.100	95 0
-0.001	625.601	502.800	95 3
0.810	539.990	540.800	95 7
-1.215	532.315	531.100	9528
-0.527	522.427	521.900	9590
1.965	516.135	518.100	9518
-0.795	513.095	512.300	9536
-0.237	510.437	510.200	91 0
0.002	623.998	624.000	91 3
-0.732	540.732	540.000	91 7
-0.219	522.847	532.700	9128
1.253	524.149	524.100	9190
1.364	516.436	517.800	9118
1.261	513.339	514.600	9136
-2.930	510.630	507.700	70 0
-0.029	564.729	564.700	70 3
0.195	536.205	536.400	70 7
0.541	527.859	528.400	7028
-1.298	516.198	514.900	7036
0.591	501.509	502.100	

Figure 9A.2 (*continued*)

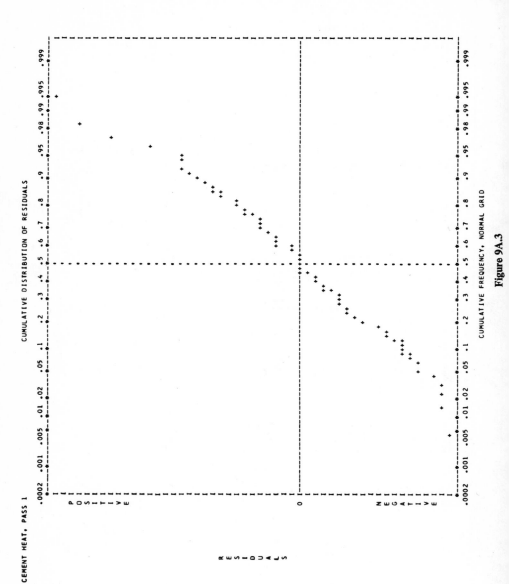

Figure 9A.3

CEMENT HEAT, PASS 1

RESIDUAL VS. FITTED Z

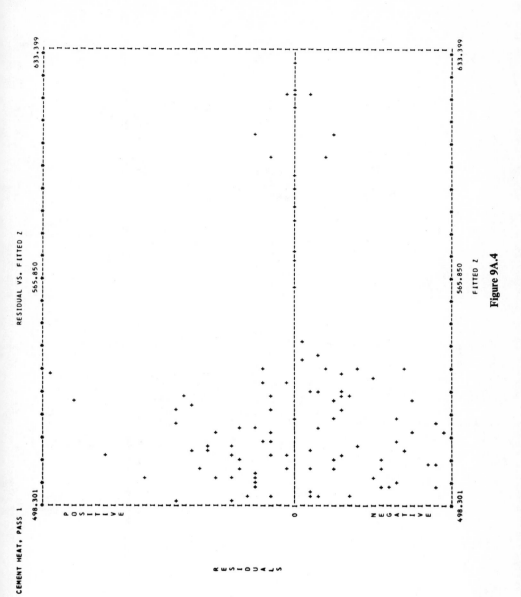

FITTED Z

Figure 9A.4

309

NON-LINEAR LEAST-SQUARES CURVE FITTING PROGRAM

CEMENT HEAT, PASS 2

Z = D' - C' + (A' / (T + F')EXP B)
Z = ACID HEAT OF SOLUTION
B = EXPONENT COMMON TO ALL CEMENTS
D' = INITIAL ACID HEAT OF SOLUTION = D1V1 + D2V2 + --- + D14V14
A' = CONSTANT = A1V1 + A2V2 + --- + A14V14
C' = ULTIMATE CUMULATIVE HEAT OF HARDENING = C1V1 + C2V2 + --- + C14V14
T = TIME CORRECTION TERM = (A'/C')EXP(1/B)
V1 --- V14 = INDICATOR VARIABLES FOR EACH SAMPLE OF CEMENT
T, TIME, MEASURED IN DAYS, AFTER WATFR IS ADDED TO THE CEMENT
OBSERVATION 85-3-2 OMITTED.

IND.VAR(I)	NAME		COEF-B(I)	S.E. COEF.	T-VALUE	95% CONFIDENCE LIMITS	
						LOWER	UPPER
1	B	22	1.99732E-01	3.40E-02	5.9	1.32E-01	2.68E-01
2	D1	22	5.88682E 02	2.04E 00	289.1	5.85E 02	5.93E 02
3	D2	23	5.75667E 02	2.04E 00	282.7	5.72E 02	5.80E 02
4	D3	92	6.23849E 02	1.44E 00	433.1	6.21E 02	6.27E 02
5	D4	88	5.95493E 02	2.04E 00	292.4	5.91E 02	6.00E 02
6	D5	96	6.03849E 02	1.44E 00	419.2	6.01E 02	6.07E 02
7	D6	85	6.18803E 02	2.04E 00	303.8	6.15E 02	6.23E 02
8	D7	94	6.10850E 02	1.44E 00	424.1	6.08E 02	6.14E 02
9	D8	24	5.72999E 02	2.04E 00	281.3	5.69E 02	5.77E 02
10	D9	89	5.99007E 02	2.04E 00	294.1	5.95E 02	6.03E 02
11	D10	90	6.33400E 02	2.04E 00	311.0	6.29E 02	6.37E 02
12	D11	25	5.85898E 02	2.04E 00	287.6	5.82E 02	5.90E 02
13	D12	95	6.25601E 02	2.04E 00	307.1	6.22E 02	6.30E 02
14	D13	91	6.23999E 02	2.04E 00	306.3	6.20E 02	6.28E 02
15	D14	70	5.64736E 02	2.04E 00	277.3	5.61E 02	5.69E 02
16	A1	22	7.47356E 01	6.63E 00	11.3	6.15E 01	8.80E 01
17	A2	23	8.02845E 01	6.87E 00	11.7	6.65E 01	9.40E 01
18	A3	92	6.77074E 01	6.34E 00	10.7	5.50E 01	8.04E 01
19	A4	88	7.35522E 01	6.58E 00	11.2	6.04E 01	8.67E 01
20	A5	96	5.86780E 01	5.99E 00	9.8	4.67E 01	7.07E 01
21	A6	85	5.75986E 01	5.10E 00	11.3	4.74E 01	6.78E 01
22	A7	94	5.72813E 01	5.94E 00	9.6	4.54E 01	6.92E 01
23	A8	24	6.86008E 01	6.42E 00	10.7	5.58E 01	8.14E 01
24	A9	89	6.42191E 01	6.04E 00	9.9	5.18E 01	7.20E 01
25	A10	90	5.99506E 01	6.21E 00	9.9	4.79E 01	7.20E 01
26	A11	25	5.35725E 01	5.81E 00	9.2	4.20E 01	6.52E 01
27	A12	95	5.98080E 01	6.04E 00	9.9	4.77E 01	7.19E 01
28	A13	91	6.10152E 01	6.09E 00	10.0	4.88E 01	7.32E 01
29	A14	70	7.32536E 01	7.43E 00	9.9	5.84E 01	8.81E 01
30	C1	22	1.08332E 02	7.07E 00	15.3	9.42E 01	1.22E 02
31	C2	23	1.02250E 02	7.41E 00	13.8	8.74E 01	1.17E 02
32	C3	92	1.29293E 02	6.46E 00	20.0	1.16E 02	1.42E 02
33	C4	88	1.17759E 02	7.00E 00	16.0	9.78E 01	1.26E 02
34	C5	96	1.17876E 02	5.79E 00	20.4	1.06E 02	1.29E 02
35	C6	85	1.28488E 02	5.57E 00	23.1	1.17E 02	1.40E 02
36	C7	94	1.24822E 02	5.69E 00	21.9	1.13E 02	1.36E 02
37	C8	24	9.56193E 01	6.60E 00	14.5	8.24E 01	1.09E 02
38	C9	89	1.15363E 02	6.05E 00	19.1	1.03E 02	1.27E 02
39	C10	90	1.40714E 02	6.37E 00	22.1	1.28E 02	1.53E 02
40	C11	25	1.00795E 02	5.60E 00	18.0	8.96E 01	1.12E 02
41	C12	95	1.33731E 02	6.05E 00	22.1	1.22E 02	1.46E 02
42	C13	91	1.32326E 02	6.14E 00	21.5	1.20E 02	1.45E 02
43	C14	70	8.58738E 01	7.14E 00	12.0	7.16E 01	1.00E 02

NO. OF OBSERVATIONS   104
NO. OF COEFFICIENTS   43
RESIDUAL DEGREES OF FREEDOM   61
RESIDUAL ROOT MEAN SQUARE   2.03694439
RESIDUAL MEAN SQUARE   4.14914322
RESIDUAL SUM OF SQUARES   253.09777832

Figure 9A.5

CEMENT HEAT, PASS 2

	----ORDERED BY COMPUTER INPUT----				----ORDERED BY RESIDUALS----			
OBS. NO.	OBS. Z	FITTED Z	RESIDUAL	OBS. NO.	OBS. Z	FITTED Z	ORDERED RESID.	SEQ.
22 2	588.700	588.681	0.019	22 7	535.600	530.795	4.805	1
22 3	537.500	539.756	-2.256	8890	517.800	513.669	4.131	2
22 7	535.600	530.795	4.805	2218	510.200	506.835	3.365	3
2228	516.200	518.720	-2.521	2436	501.400	498.490	2.909	4
2290	507.500	510.762	-3.262	89 3	534.300	531.663	2.637	5
2218	510.200	506.835	3.365	23 7	530.000	527.396	2.604	6
2236	503.200	503.349	-0.149	24 7	526.200	523.641	2.559	7
23 0	575.700	575.667	0.033	9428	517.600	515.466	2.134	8
23 3	535.100	536.676	-1.576	9228	531.400	529.347	2.052	9
23 7	530.000	527.396	2.604	9018	517.500	515.448	2.052	10
2328	503.800	506.077	0.104	9690	511.900	509.857	2.042	11
2318	501.800	501.864	-0.464	9590	518.100	516.214	1.885	12
2336	499.700	498.122	1.578	9618	508.600	506.770	1.830	13
92 0	623.600	623.849	-0.249	25 7	523.100	521.380	1.720	14
92 3	624.100	623.849	0.251	85 3	538.200	536.510	1.690	15
92 0	548.800	548.783	0.017	2336	499.700	498.122	1.578	16
92 7	538.900	540.408	-1.508	8590	515.100	513.761	1.339	17
9228	531.400	529.347	2.052	9118	514.600	513.299	1.301	18
9290	523.000	522.116	0.884	9190	517.800	516.509	1.291	19
9218	519.300	518.554	0.746	2536	502.800	501.591	1.209	20
9236	513.200	515.394	-2.194	9128	524.100	523.029	1.071	21
88 0	595.500	595.493	0.007	8518	511.800	510.730	1.070	22
88 7	542.390	542.325	-0.075	9418	505.900	504.893	1.007	23
8828	532.400	533.429	-1.029	95 3	507.300	506.330	0.969	24
8890	521.800	521.509	0.291	8536	540.800	539.837	0.963	25
8818	517.800	513.669	4.131	2590	509.000	508.042	0.958	26
8836	507.900	509.802	-1.903	9290	507.800	506.909	0.891	27
96 0	504.900	506.371	-1.471	85 7	523.000	522.116	0.884	28
96 0	604.500	603.849	0.651	8990	530.200	529.344	0.855	29
96 3	603.200	603.849	-0.649	94 0	508.900	508.047	0.853	30
96 7	532.600	532.995	-0.395	9218	611.700	610.850	0.850	31
9628	525.700	525.720	-0.020	94 3	519.300	518.554	0.746	32
9690	515.700	516.126	-0.426	7036	532.700	531.962	0.738	33
9618	511.900	509.857	2.042	96 0	502.100	501.402	0.698	34
9636	508.600	506.770	1.830	90 7	604.500	603.849	0.651	35
85 0	501.000	504.032	-3.032	8590	536.800	536.199	0.600	36
85 3	618.800	618.803	-0.003	70 7	514.300	513.761	0.539	37
85 7	538.200	536.510	1.690	70 7	528.400	527.909	0.491	38
8528	528.500	529.344	-0.844	9090	519.300	518.827	0.473	39
8528	530.200	529.344	0.855	8536	508.500	508.042	0.458	40
8590	517.100	519.916	-2.816	70 3	536.400	536.060	0.340	41
8590	515.100	513.761	1.339	8828	521.800	521.509	0.291	42
8518	514.300	513.761	0.539	92 0	624.100	623.849	0.251	43
8518	509.600	510.730	-1.130	9490	509.500	509.344	0.156	44
8536	511.800	510.730	1.070	8590	514.700	514.596	0.104	45
8536	509.000	508.042	0.958	23 0	575.700	575.667	0.033	46
	508.500	508.042	0.458	22 0	588.700	588.681	0.019	47
				92 3	548.800	548.783	0.017	48
				88 0	595.500	595.493	0.007	49

Figure 9A.5 (*continued*)

Data table (Figure 9A.5, continued).

Label			
94 0	610.000	610.850	-0.850
94 0	611.700	610.850	0.850
94 3	532.700	531.962	0.738
94 7	522.500	524.840	-2.340
9428	517.600	515.466	2.134
9490	507.300	509.344	-2.134
9418	507.300	506.330	0.969
9436	507.300	503.557	-1.657
24 3	573.000	572.999	0.001
24 7	530.200	531.794	-0.894
24 7	526.200	523.641	-2.559
2428	511.500	512.592	-1.092
2490	503.000	505.293	-2.293
2418	500.500	501.689	-1.189
2436	501.400	498.490	-2.909
89 0	599.000	599.007	-0.007
89 3	534.300	531.663	-2.637
89 7	521.100	524.245	-3.145
8928	513.200	514.450	-1.250
8990	508.900	508.047	0.853
8918	505.900	504.893	1.007
8936	502.000	502.095	-0.095
90 3	633.400	633.399	0.000
90 7	543.900	544.185	-0.285
90 7	536.800	536.199	0.600
9028	524.600	525.689	-1.089
9090	519.300	518.827	0.473
9018	517.500	515.448	2.052
9036	510.700	512.450	-1.750
25 3	585.900	585.898	0.002
25 3	527.100	528.000	-0.901
25 7	523.100	521.380	1.720
2528	511.700	512.630	-0.930
2590	507.800	506.909	0.891
2518	502.100	501.591	-1.991
2536	502.800	501.591	-1.209
95 0	625.600	625.601	-0.001
95 3	540.800	539.837	0.963
95 3	531.100	532.396	-1.296
9528	521.900	522.606	-0.706
9590	518.100	516.214	1.885
9518	510.200	510.276	-0.076
9536	624.000	623.998	0.002
91 0	540.000	540.599	-0.599
91 3	532.700	533.014	-0.314
91 7	524.100	523.029	1.071
9128	517.800	516.509	1.291
9190	514.600	513.299	1.301
9118	507.700	510.451	-2.751
9136	564.700	564.736	-0.036
70 0	536.400	536.060	0.340
70 3	528.400	527.909	0.491
70 7	514.900	516.394	-1.494
7028	502.100	501.402	0.698

				Index
0.002	585.898	585.900	25 0	50
0.002	623.998	624.000	91 0	51
0.001	572.999	633.400	24 0	52
0.000	633.399	633.400	90 0	53
-0.001	625.601	625.600	95 0	54
-0.003	618.803	618.800	85 0	55
-0.007	599.007	599.000	96 7	56
-0.020	525.720	525.700	89 0	57
-0.025	542.325	542.300	88 7	58
-0.036	564.736	564.700	70 0	59
-0.076	510.276	510.200	9536	60
-0.095	502.095	502.000	8936	61
-0.149	503.349	503.200	2236	62
-0.249	623.849	623.600	92 0	63
-0.285	544.185	543.900	90 3	64
-0.314	533.014	532.700	91 7	65
-0.395	532.995	532.600	96 3	66
-0.426	516.126	515.700	9628	67
-0.464	501.864	501.400	2318	68
-0.599	540.599	540.000	91 3	69
-0.649	603.849	603.200	96 0	70
-0.706	522.606	521.900	9528	71
-0.768	513.067	512.300	9518	72
-0.844	529.344	528.500	85 7	73
-0.850	610.850	610.000	94 0	74
-0.894	531.794	530.900	24 3	75
-0.901	528.000	527.100	25 3	76
-0.930	512.630	511.700	2528	77
-1.029	533.429	532.400	88 7	78
-1.089	525.689	524.600	9028	79
-1.092	512.592	511.500	2428	80
-1.130	510.730	509.600	8518	81
-1.189	501.689	500.500	2418	82
-1.250	514.450	513.200	8928	83
-1.296	532.396	531.100	95 7	84
-1.471	516.394	514.900	8836	85
-1.494	506.371	504.900	7028	86
-1.508	540.408	538.900	92 7	87
-1.576	536.676	535.100	23 3	88
-1.657	503.657	502.000	9436	89
-1.750	512.450	510.700	9036	90
-1.903	509.802	507.900	8818	91
-1.991	504.091	502.100	2518	92
-2.116	519.916	517.800	8528	93
-2.194	515.394	513.200	9236	94
-2.256	539.756	537.500	22 3	95
-2.278	506.077	503.800	2390	96
-2.293	505.293	503.000	2490	97
-2.340	524.840	522.500	94 7	98
-2.521	518.720	516.200	2228	99
-2.751	510.451	507.700	9136	100
-2.751	510.451	517.100	70 0	101
-2.816	519.916	517.100	8528	102
-3.032	504.032	501.000	9636	103
-3.145	524.245	521.100	89 7	104
-3.262	510.762	507.500	2290	

Figure 9A.5 (*continued*)

312

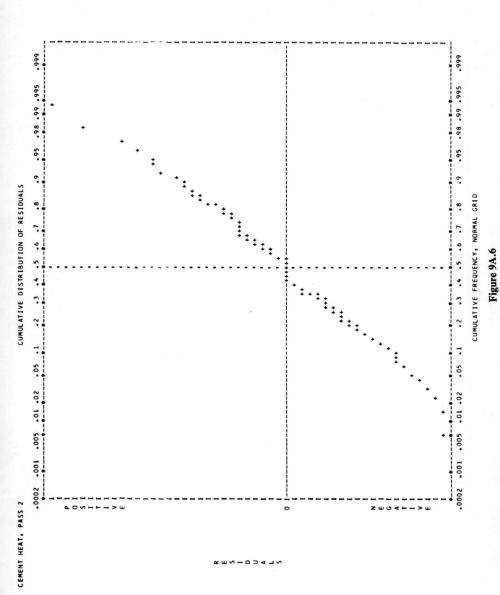

Figure 9A.6

313

CEMENT HEAT, PASS 2

RESIDUAL VS. FITTED Z

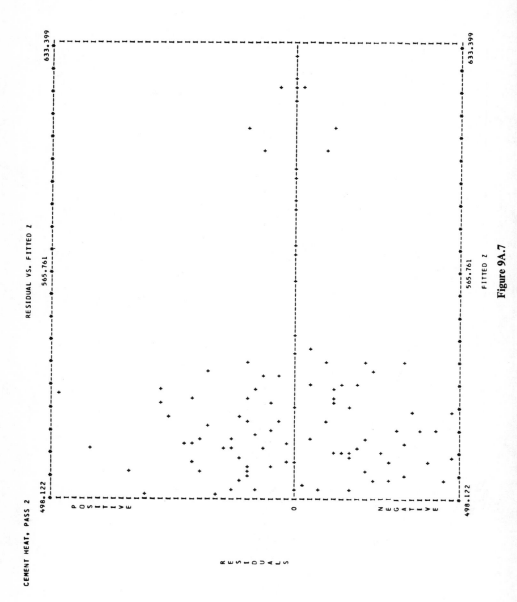

Figure 9A.7

314

LINEAR LEAST-SQUARES CURVE FITTING PROGRAM

CEMENT HEAT PASS 3   DEP VAR 1: C

MIN Y = 8.587D 01   MAX Y = 1.407D 02   RANGE Y = 5.484D 01

$Y = B(0) + B(1)X(1);$ $C = H + GD;$ $C = - 324. + 0.73D$
B(0) = C, THE ULTIMATE CUMULATIVE HEAT OF HARDENING, CALORIES/GRAM
B(0) = H, A CONSTANT
X = D, THE INITIAL ACID HEAT OF SOLUTION, CALORIES/GRAM
C AND D VALUES FROM NON-LINEAR FIT WITH B COMMON, PASS 2.

IND.VAR(I)	NAME	COEF.B(I)	S.E. COEF.	T-VALUE	R(I)SQRD	MIN X(I)	MAX X(I)	RANGE X(I)	REL.INF.X(I)
0		-3.23563D 02							
1	G	7.31004D-01	3.07D-02	23.8	0.0	5.647D 02	6.334D 02	6.866D 01	0.92

NO. OF OBSERVATIONS	14
NO. OF IND. VARIABLES	1
RESIDUAL DEGREES OF FREEDOM	12
F-VALUE	567.9
RESIDUAL ROOT MEAN SQUARE	2.43035632
RESIDUAL MEAN SQUARE	5.90663185
RESIDUAL SUM OF SQUARES	70.87958217
TOTAL SUM OF SQUARES	3425.35599575
MULT. CORREL. COEF. SQUARED	.9793

-------ORDERED BY COMPUTER INPUT-------

IDENT.	OBSV.	MS DISTANCE	OBS. Y	FITTED Y	RESIDUAL
D1	22	15.	108.332	106.766	1.566
D2	23	61.	102.250	97.252	4.998
D3	92	45.	129.293	132.473	-3.180
D4	88	3.	111.759	111.745	0.013
D5	96	0.	117.876	117.853	0.023
D6	85	27.	128.488	128.784	-0.296
D7	94	8.	124.822	122.971	1.851
D8	24	74.	95.619	95.301	0.318
D9	89	1.	115.363	114.314	1.049
D10	90	91.	140.714	139.455	1.259
D11	25	22.	100.795	104.731	-3.936
D12	95	52.	133.731	133.754	-0.022
D13	91	45.	132.326	132.582	-0.257
D14	70	123.	85.874	89.261	-3.387

-------ORDERED BY RESIDUALS-------

OBSV.	OBS. Y	FITTED Y	ORDERED RESID.	SEQ
23	102.250	97.252	4.998	1
94	124.822	122.971	1.851	2
22	108.332	106.766	1.566	3
90	140.714	139.455	1.259	4
89	115.363	114.314	1.049	5
24	95.619	95.301	0.318	6
96	117.876	117.853	0.023	7
88	111.759	111.745	0.013	8
95	133.731	133.754	-0.022	9
91	132.326	132.582	-0.257	10
85	128.488	128.784	-0.296	11
92	129.293	132.473	-3.180	12
70	85.874	89.261	-3.387	13
25	100.795	104.731	-3.936	14

Figure 9A.8

CEMENT HEAT, PASS 4

NON-LINEAR LEAST-SQUARES CURVE FITTING PROGRAM

Z = H + (G - 1)D' + (A' / (T + F')EXP B)
Z = ACID HEAT OF SOLUTION
B, G AND H ARE COMMON TO ALL CEMENTS
D' = INITIAL ACID HEAT OF SOLUTION = D1V1 + D2V2 + --- + D14V14
A' = CONSTANT = A1V1 + A2V2 + --- + A14V14
F' = TIME CORRECTION TERM = (A' / (GD' - H))EXP(1/B)
T = TIME, MEASURED IN DAYS, AFTER WATER IS ADDED TO THE CEMENT
V1 --- V14 = INDICATOR VARIABLES FOR EACH SAMPLE OF CEMENT
OBSERVATION 85-3-2 OMITTED.

IND.VAR(I)	NAME	COEF.B(I)	S.E. COEF.	T-VALUE	95% CONFIDENCE LIMITS LOWER	UPPER
1	B	1.92961E-01	2.78E-02	6.9	1.37E-01	2.49E-01
2	D1 22	5.88475E 02	1.96E 00	300.0	5.85E 02	5.92E 02
3	D2 23	5.75046E 02	1.97E 00	292.1	5.71E 02	5.79E 02
4	D3 92	6.24083E 02	1.40E 00	446.0	6.21E 02	6.27E 02
5	D4 88	5.95496E 02	1.96E 00	303.7	5.92E 02	5.99E 02
6	D5 96	6.03860E 02	1.40E 00	432.6	6.01E 02	6.07E 02
7	D6 85	6.18885E 02	1.94E 00	319.8	6.15E 02	6.23E 02
8	D7 94	6.10704E 02	1.40E 00	437.2	6.08E 02	6.13E 02
9	D8 24	5.73016E 02	1.97E 00	291.2	5.69E 02	5.77E 02
10	D9 89	5.98872E 02	1.96E 00	306.3	5.95E 02	6.03E 02
11	D10 90	6.33143E 02	1.96E 00	323.1	6.29E 02	6.37E 02
12	D11 25	5.86587E 02	1.96E 00	299.5	5.83E 02	5.91E 02
13	D12 95	6.25581E 02	1.96E 00	319.7	6.22E 02	6.29E 02
14	D13 91	6.24014E 02	1.96E 00	319.0	6.20E 02	6.28E 02
15	D14 70	5.64969E 02	1.98E 00	284.7	5.61E 02	5.69E 02
16	A1 22	7.31728E 01	3.89E 00	18.8	6.54E 01	8.09E 01
17	A2 23	7.29480E 01	4.27E 00	17.1	6.44E 01	8.15E 01
18	A3 92	7.35550E 01	3.42E 00	21.5	6.67E 01	8.04E 01
19	A4 88	7.44962E 01	3.66E 00	20.4	6.72E 01	8.18E 01
20	A5 96	5.96849E 01	4.36E 00	13.7	5.10E 01	6.84E 01
21	A6 85	5.88155E 01	4.16E 00	14.1	5.05E 01	6.71E 01
22	A7 94	5.50717E 01	4.55E 00	12.1	4.60E 01	6.42E 01
23	A8 24	6.95947E 01	4.54E 00	15.3	6.05E 01	7.87E 01
24	A9 89	5.93339E 01	4.50E 00	13.2	5.03E 01	6.83E 01
25	A10 90	6.24585E 01	4.10E 00	15.2	5.43E 01	7.07E 01
26	A11 25	6.16220E 01	4.64E 00	13.3	5.23E 01	7.09E 01
27	A12 95	6.03975E 01	4.19E 00	14.4	5.20E 01	6.88E 01
28	A13 91	6.20098E 01	4.10E 00	15.1	5.38E 01	7.02E 01
29	A14 70	8.03994E 01	4.10E 00	19.6	7.22E 01	8.86E 01
30	G	7.17966E-01	3.46E-02	20.8	6.49E-01	7.87E-01
31	H	3.14493E 02	2.29E 01	13.7	2.69E 02	3.60E 02

NO. OF OBSERVATIONS   104
NO. OF COEFFICIENTS   31
RESIDUAL DEGREES OF FREEDOM   73
RESIDUAL ROOT MEAN SQUARE   1.99544048
RESIDUAL MEAN SQUARE   3.98178577
RESIDUAL SUM OF SQUARES   290.67041016

Figure 9A.9

NON-LINEAR LEAST-SQUARES CURVE FITTING PROGRAM

CEMENT HEAT, PASS 4

——ORDERED BY COMPUTER INPUT——				——ORDERED BY RESIDUALS——				
OBS. NO.	OBS. Z	FITTED Z	RESIDUAL	OBS. NO.	OBS. Z	FITTED Z	ORDERED RESID.	SEQ.
22 0	588.700	588.475	0.225	22 7	535.600	530.548	5.052	1
22 3	537.500	539.165	-1.665	8890	517.800	513.699	4.100	2
22 7	535.600	530.548	5.052	23 7	530.000	526.499	3.501	3
2228	516.200	518.897	-2.697	2536	502.800	499.669	3.131	4
2290	507.500	511.163	-3.663	2436	501.400	498.394	3.006	5
2218	510.200	507.323	2.876	89 3	534.300	531.298	3.002	6
2236	503.200	503.900	-0.700	2218	510.200	507.323	2.876	7
23 0	575.700	575.044	0.656	24 7	526.200	523.678	2.521	8
23 3	535.100	534.915	0.185	7036	502.100	499.580	2.520	9
23 7	530.000	526.499	3.501	9228	531.400	529.163	2.237	10
2328	514.700	514.970	-0.270	9690	511.900	509.849	2.051	11
2390	503.800	507.275	-3.476	2590	507.800	505.789	2.011	12
2318	501.400	503.451	-2.051	9428	517.600	515.682	1.918	13
2336	499.700	500.039	-0.339	9618	508.600	506.714	1.886	14
92 0	623.600	624.083	-0.483	9590	518.100	516.274	1.825	15
92 0	624.100	624.083	0.017	9218	519.300	517.509	1.791	16
92 3	548.800	549.838	-1.038	85 3	538.200	536.570	1.630	17
92 7	538.900	540.972	-2.072	9290	523.000	521.372	1.628	18
9228	531.400	529.163	2.237	9018	517.500	515.991	1.509	19
9290	523.000	521.372	1.628	94 3	532.700	531.241	1.458	20
9218	519.300	517.509	1.791	8590	515.100	513.722	1.378	21
9236	513.200	514.066	-0.866	9118	514.600	513.252	1.348	22
88 0	595.500	595.496	0.004	9190	517.800	516.509	1.291	23
88 3	542.300	542.272	0.027	8518	511.800	510.633	1.167	24
88 7	532.400	533.458	-1.058	8536	509.000	507.879	1.121	25
8828	521.800	521.577	0.223	95 3	540.800	539.739	1.061	26
8890	517.800	513.699	4.100	9128	524.100	523.082	1.018	27
8818	507.900	509.790	-1.890	94 0	611.700	610.704	0.996	28
8836	504.900	506.304	-1.404	25 7	523.100	522.195	0.905	29
96 0	604.500	603.860	0.640	90 7	536.800	535.949	0.850	30
96 0	603.200	603.860	-0.660	85 7	530.200	529.425	0.775	31
96 3	532.600	533.000	-0.400	8918	505.900	505.178	0.722	32
96 7	525.700	525.772	-0.072	23 0	575.700	575.044	0.656	33
9628	515.700	516.174	-0.474	96 0	604.500	603.860	0.640	34
9690	511.900	509.849	2.051	8536	508.500	507.879	0.621	35
9618	508.600	506.714	1.886	8990	508.900	508.294	0.605	36
9636	501.000	503.920	-2.920	8590	514.300	513.722	0.578	37
85 0	618.800	618.885	-0.085	90 3	543.900	543.539	0.361	38
85 3	538.200	536.570	1.630	9418	507.300	506.951	0.349	39
85 7	528.500	529.425	-0.925	90 0	633.400	633.143	0.257	40
85 7	530.200	529.425	0.775	22 0	588.700	588.475	0.225	41
8528	517.800	519.957	-2.157	8828	521.800	521.577	0.223	42
8528	517.100	519.957	-2.857	23 3	535.100	534.915	0.185	43
8590	515.100	513.722	1.378	89 0	599.000	598.872	0.128	44
8590	514.300	513.722	0.578	70 7	528.400	528.303	0.097	45
8518	509.600	510.633	-1.033	9090	519.300	519.272	0.028	46
8518	511.800	510.633	1.167	88 3	542.300	542.272	0.027	47
8536	509.000	507.879	1.121	95 0	625.600	625.581	0.019	48
8536	508.500	507.879	0.621	92 0	624.100	624.083	0.017	49

Figure 9A.9 (continued)

317

Table (left block):

94 0	610.000	610.704	-0.704
94 0	611.700	610.704	0.996
94 7	532.700	531.241	1.458
94 7	522.500	524.549	-2.049
9428	517.600	515.682	1.918
9490	509.500	509.844	-0.344
9418	507.300	506.951	0.349
9436	502.000	502.000	0.000
24 0	573.000	573.015	-0.015
24 3	530.900	531.775	-0.875
24 7	526.200	523.678	2.521
2428	511.500	512.646	-1.146
2490	503.000	505.299	-2.299
2418	500.500	501.649	-1.149
2436	501.400	498.394	3.006
89 0	599.000	598.872	0.128
89 3	534.300	531.298	3.002
89 7	521.100	524.120	-3.020
8928	513.200	514.582	-1.382
8990	508.900	508.294	0.605
8918	505.900	505.178	0.722
8936	502.000	502.401	-0.401
90 0	633.000	633.143	0.257
90 3	543.900	543.539	0.361
90 7	536.800	535.949	-0.850
9028	524.600	525.893	-1.293
9090	519.300	519.272	0.028
9018	517.500	515.991	1.509
9036	510.700	513.067	-2.367
25 0	585.900	586.587	-0.687
25 3	527.100	529.597	-2.497
25 7	523.100	522.195	0.905
2528	511.700	512.314	-0.614
2590	507.800	505.789	2.011
2518	502.100	502.553	-0.453
2536	502.800	499.669	3.131
95 0	625.600	625.581	0.019
95 3	540.800	539.739	1.061
95 7	531.100	532.401	-1.301
9528	521.900	522.677	-0.777
9590	518.100	516.274	1.825
9518	512.300	513.102	-0.802
9536	510.200	510.274	-0.074
91 0	624.000	624.014	-0.014
91 3	540.000	540.591	-0.591
91 7	532.700	533.082	-0.363
9128	524.100	523.082	1.018
9190	517.800	516.509	1.291
9118	514.600	513.252	1.348
9136	507.700	510.349	-2.649
70 0	564.700	564.968	-0.268
70 3	536.400	536.891	-0.491
70 7	528.400	528.303	0.097
7028	514.900	515.951	-1.051
7036	502.100	499.580	2.520

Table (right block):

88 0	595.500	595.496	-0.004	50
91 0	624.000	624.014	-0.014	51
24 0	573.000	573.015	-0.015	52
96 7	525.700	525.772	-0.072	53
9536	510.200	510.274	-0.074	54
85 0	618.800	618.885	-0.085	55
70 0	564.700	564.968	-0.268	56
2328	514.700	514.970	-0.270	57
2336	499.700	500.039	-0.339	58
9490	509.500	509.844	-0.344	59
91 7	532.700	533.063	-0.363	60
96 3	532.600	533.000	-0.400	61
8936	502.100	502.553	-0.453	62
2518	515.700	516.174	-0.474	63
9628	623.600	624.083	-0.483	64
92 0	536.400	536.891	-0.491	65
70 3	540.000	540.591	-0.591	66
91 3	511.700	512.314	-0.614	67
2528	603.200	603.860	-0.660	68
96 0	585.900	586.587	-0.687	69
25 0	503.200	503.900	-0.700	70
2236	610.000	610.704	-0.704	71
94 0	521.900	522.677	-0.777	72
9528	512.300	513.102	-0.802	73
9518	513.200	514.066	-0.866	74
9236	530.900	531.775	-0.875	75
24 3	528.500	529.425	-0.925	76
85 7	509.600	510.633	-1.033	77
8518	548.800	549.838	-1.038	78
92 3	514.900	515.951	-1.051	79
7028	532.400	533.458	-1.058	80
88 7	511.500	512.646	-1.146	81
2428	500.500	501.649	-1.149	82
2418	524.600	525.893	-1.293	83
9028	531.100	532.401	-1.301	84
95 7	513.200	514.582	-1.382	85
8928	504.900	506.304	-1.404	86
22 3	537.500	539.165	-1.665	87
8836	507.900	509.790	-1.890	88
94 7	522.500	524.549	-2.049	89
2318	501.400	503.451	-2.051	90
92 7	538.900	540.972	-2.072	91
8528	517.800	519.957	-2.157	92
2490	503.000	505.299	-2.299	93
9036	510.700	513.067	-2.367	94
25 3	502.000	504.373	-2.373	95
9436	527.100	529.597	-2.497	96
9136	507.700	510.349	-2.649	97
8528	516.200	518.897	-2.697	98
2228	517.100	519.957	-2.857	99
9636	501.000	503.920	-2.920	100
89 7	521.100	524.120	-3.020	101
2390	503.800	507.275	-3.476	102
2290	507.500	511.163	-3.663	103

Figure 9A.9 (continued)

318

LINEAR LEAST-SQUARES CURVE-FITTING PROGRAM

CEMENT HEAT PASS 5

DATA INPUT   36 INDEPENDENT VARIABLES   2 DEPENDENT VARIABLE(S)

OBSV.	SEQ.	1-11-21	2-12-22	3-13-23	4-14-24	5-15-25	6-16-26	7-17-27	8-18-28	9-19-29	10-20-30
22	0	27.680	3.760	1.980	2.480	64.970	0.0	0.0	0.0	0.0	0.0
	1	0.0	0.0	0.0	0.0	0.0	0.0	588.475	73.173	0.0	0.0
	2	0.0	0.0	0.0	0.0	0.0	0.0	0.0	0.0	0.718	0.0
	3	0.0	0.0	0.0	0.0	0.0	0.0	0.0	0.0	0.0	0.0

DATA TRANSFORMATIONS

POSITION	CODE	OPERATION	CONSTANT	LOCATION	OMIT	VARIABLE	NAME
1		NONE		1	1		
2		NONE		2	1		
3		NONE		3	1		
4		NONE		4	1		
5		NONE		5	1		
6		NONE		6	1		
7	11	ADD POSITIONS, 1+ 2		7	1		
8	11	ADD POSITIONS, 3+ 7		8	1		
9	11	ADD POSITIONS, 4+ 8		9	1		
10	11	ADD POSITIONS, 5+ 9		11	1		
11	7	DIVIDE BY CONSTANT	100.	11	1		
12	10	DIVIDE POSITIONS, 1/11		20	1		
13	10	DIVIDE POSITIONS, 2/11		21	1		
14	10	DIVIDE POSITIONS, 3/11		22	1		
15	10	DIVIDE POSITIONS, 4/11		23	1		
16	10	DIVIDE POSITIONS, 5/11		24	1		
17		NONE		37	1		
18		NONE		38	1		
19		NONE		39	1		
20	8	ADD CONSTANT	-21.0	20	1		
21	8	ADD CONSTANT	-3.48	21	1		
22	8	ADD CONSTANT	-1.18	22	1		
23	8	ADD CONSTANT	-2.08	23	1		
24	8	ADD CONSTANT	-61.6	24	1		
25	8	ADD CONSTANT	10.66	25	1		
26	10	DIVIDE POSITIONS, 20/25		26	0	1	X1
27	10	DIVIDE POSITIONS, 21/25		27	0	2	X2
28	10	DIVIDE POSITIONS, 22/25		28	0	3	X3
29	10	DIVIDE POSITIONS, 23/25		29	0	4	X4
30	10	DIVIDE POSITIONS, 24/25		30	0	5	X5
31	9	MULTIPLY POSITIONS, 26X27		31	0	6	X1X2
32	9	MULTIPLY POSITIONS, 26X28		32	0	7	X1X3
33	9	MULTIPLY POSITIONS, 26X29		33	0	8	X1X4
34	9	MULTIPLY POSITIONS, 27X28		34	0	9	X2X3
35	9	MULTIPLY POSITIONS, 27X29		35	0	10	X2X4
36	9	MULTIPLY POSITICNS, 28X29		36	0	11	X3X4
37		NONE		37	0	12	D
38		NONE		38	0	13	A

Figure 9A.10

319

DATA AFTER TRANSFORMATIONS   THE FITTED EQUATION HAS 11 INDEPENDENT VARIABLES,   2 DEPENDENT VARIABLES

OBSV.	X1 / X3X4 (1-11-21)	X2 / D (2-12-22)	A (3-13-23)	X3 (4-14-24)	X4 (5-15-25)	X5 (6-16-26)	X1X2 (7-17-27)	X1X3 (8-18-28)	X1X4 (9-19-29)	X2X3 (10-20-30)	X2X4
22	6.042460D-01	7.344490D-02	2.632680D-01	3.551690D-02	1.403310D-02	4.437880D-02	1.592600D-01	1.705700D-03			6.121160D-03
	1.935770D-02	5.884750D 02	7.311280D 01								
	5.359160D-02										
23	4.660220D-01	3.641200D-01	1.471810D-01	2.257940D-02	4.565410D-05	1.696880D-01	6.858960D-01	3.567120D-05			1.441870D-05
	9.796560D-05	5.750440D 02	7.294800D 01								
	5.750440D 02										
92	7.169210D-02	1.480170D-01	2.941060D-01	2.756020D-01	1.509710D-02	1.061170D-02	2.108500D-02	3.116990D-02			6.193350D-02
	2.105830D-01	6.240830D 02	7.355500D 01								
88	3.416410D-01	1.524170D-01	2.522790D-01	3.040220D-02	7.627510D-02	5.207200D-02	8.618870D-02	3.402880D-02			5.632400D-02
	2.232610D-01	5.954960D 02	7.449620D 01								
	3.845160D-02										
96	3.803970D-01	8.736090D-02	4.699780D-01	2.639930D-02	1.364280D-02	3.323180D-02	1.787780D-01	3.133170D-03			1.685560D-02
	3.586470D-02	8.736090D 02	5.968490D 01								
	4.105770D-02										
85	1.188140D-01	5.560200D-01	4.419190D-01	3.228720D-02	2.981760D-02	1.853740D-02	5.250630D-02	3.915480D-02			1.109040D-01
	2.509590D-01	5.881550D 02	5.968490D 01								
	6.894830D-02										
94	3.294830D-04	1.296590D-01	4.589800D-01	6.853250D-04	3.635740D-05	1.415650D-04	1.512260D-04	4.741150D-02			5.064700D-02
	1.103470D-01	1.034700D 02	5.507170D 01								
	1.972050D-01										
24	2.415910D-01	1.273220D-01	4.909920D-02	1.532510D-02	3.076000D-02	1.369050D-01	1.185710D-02	7.215130D-02			6.248880D-03
	2.781230D-01	5.730160D 02	6.954470D 01								
	1.096040D-01										
89	1.337390D-02	1.138950D-01	2.887400D-01	2.457930D-02	1.248340D-02	5.076660D-02	3.164720D-02	5.275410D-02			3.288620D-02
	1.138950D-01	4.631810D 02	5.933390D 01								
	2.848560D-02										
90	9.881260D-05	2.281830D-04	4.330410D-01	4.350100D-02	1.409310D-02	6.499940D-06	1.233540D-02	1.128920D-04			2.142450D-01
	1.409480D-01	5.947440D 02	6.245850D 01								
	6.544350D-02										
25	1.430590D-01	5.898190D-01	1.109550D-01	1.521970D-02	2.016380D-02	8.313360D-02	1.563890D-02	8.437870D-02			1.587310D-02
	5.898190D-01	5.865870D 02	6.162200D 01								
	6.255810D 02										
95	3.329720D-02	2.433630D-02	5.137860D-01	4.519640D-02	8.103320D-03	5.472650D-03	1.710760D-02	3.999860D-02			1.250370D-01
	2.433630D-02	2.558100D 02	6.039750D 01								
	8.444460D-02										
91	1.942900D-01	1.398610D-01	5.516480D-01	3.299550D-02	1.577750D-02	1.135760D-02	4.479720D-02	2.717350D-02			1.071790D-01
	1.398610D-01	1.942900D 02	6.200980D 01								
	7.715410D-02										
70	4.664100D-02	6.382320D-02	4.501450D-03	1.911220D-03	2.976780D-03	2.161130D-01	2.099520D-03	2.957280D-02			2.872970D-04
	6.382320D-02	4.633540D 02	8.039940D 01								
	2.085770D-03										

MEANS OF VARIABLES

2.203350D-01	1.596310D-01	2.711230D-01	3.056970D-01	4.301430D-02	2.000690D-02	5.945830D-02	5.014590D-02	3.305580D-02	5.746820D-02
6.092300D-02	6.016230D 02	6.596860D 01							

ROOT MEAN SQUARES OF VARIABLES

3.011850D-01	2.087770D-01	3.446870D-01	3.639560D-01	8.156650D-02	2.804650D-02	9.204130D-02	7.671260D-02	4.318530D-02	8.619720D-02
8.226610D-02	6.247200D 02	6.889770D 01							

Figure 9A.10 (*continued*)

LINEAR LEAST-SQUARES CURVE FITTING PROGRAM

CEMENT HEAT PASS 5 DEP VAR 1: D

MIN Y = 5.650D+02 MAX Y = 6.331D+02 RANGE Y = 6.817D+01

FULL EQUATION - - - SEARCH FCF INFLUENTIAL VARIABLES
$D = B(1)X1 + B(2)X2 + B(3)X3 + B(4)X4 + E(5)X5 + B(6)X1X2 + B(7)X1X3 + B(8)X1X4 + B(9)X2X3 + B(10)X2X4 + B(11)X3X4$

D = INITIAL ACID HEAT OF SOLUTION, CALORIES/GRAM
X1 = (PERCENT SILICA IN THE CLINKER - 21.0) / 10.66
X2 = (PERCENT ALUMINA - 3.48) / 10.66
X3 = (PERCENT FERRIC OXIDE - 1.18) / 10.66
X4 = (PERCENT LIME - (1.6) / 10.66
X5 = (PERCENT MAGNESIA - 2.08) / 10.66
D VALUES FROM NON-LINEAR FIT WITH B, G AND H COMMON.

B(0) = 0

INC.VAR(I)	NAME	COEF.B(I)	S.E. COEF.	T-VALUE	R(I)SQRD	MIN X(I)	MAX X(I)	RANGE X(I)	REL.INF.X(I)
1	X1	5.65902D+02	1.26D+01	44.4	0.9741	3.295D-04	6.042D-01	6.039D-01	5.01
2	X2	5.71201D+02	1.07D+02	5.3	0.9992	9.757D-05	4.947D-01	4.946D-01	4.14
3	X3	5.23790D+02	6.05D+01	9.6	0.9991	2.282D-04	5.898D-01	5.896D-01	5.04
4	X4	6.27214D+02	2.17D+01	28.9	0.9939	4.501D-03	5.516D-01	5.471D-01	5.03
5	X5	6.61219D+02	2.89D+01	22.9	0.9311	6.853D-04	2.756D-01	2.749D-01	2.67
6	X1X2	-1.25580D+01	2.02D+02	0.1	0.9881	3.636D-05	7.626D-02	7.624D-02	0.01
7	X1X3	-5.19942D+01	1.11D+02	0.5	0.9963	6.500D-06	2.161D-01	2.161D-01	0.16
8	X1X4	2.11143D+01	5.27D+01	0.4	0.9766	1.512D-04	1.788D-01	1.786D-01	0.06
9	X2X3	2.06661D+01	1.83D+02	0.1	0.9939	3.567D-05	8.434D-02	8.434D-02	0.03
10	X2X4	1.57850D+02	2.58D+02	0.6	0.9992	1.442D-05	2.142D-01	2.142D-01	0.50
11	X3X4	7.43544D-01	6.18D+01	0.0	0.9916	9.881D-05	1.972D-01	1.971D-01	0.00

NO. OF OBSERVATIONS   14
NO. OF IND. VARIABLES   11
RESIDUAL DEGREES OF FREEDOM   3
RESIDUAL ROOT MEAN SQUARE   2.22854155
RESIDUAL MEAN SQUARE   4.96639746
RESIDUAL SUM OF SQUARES   14.89519238
TOTAL SUM OF SQUARES   6271.14395981
MULT. CORREL. COEF. SQUARED   .9976

ORDERED BY COMPUTER INPUT

IDENT.	OBSV.	WSS DISTANCE	OBS.Y	FITTED Y	RESIDUAL
1	22	13560.	588.475	588.677	-0.202
2	23	8203.	575.044	575.746	-0.702
3	92	7406.	624.083	623.959	0.124
4	88	2421.	595.496	595.787	-0.291
5	96	3616.	603.860	603.499	0.361
6	85	7014.	618.885	618.167	0.718
7	94	11356.	610.704	610.955	-0.252
8	24	3498.	573.016	575.729	-2.713
9	89	16180.	598.872	598.510	0.362
10	90	10433.	633.143	633.129	0.014
11	25	6953.	586.587	584.803	1.785
12	95	7321.	625.581	626.366	-0.785
13	51	7321.	624.014	623.936	0.078
14	70	14391.	564.969	563.466	1.502

ORDERED BY RESIDUALS

OBSV.	OBS. Y	FITTED Y	ORDERED RESID.	STUD.RESID.	SEQ
25	586.587	584.803	1.785	1.1	1
70	564.969	563.466	1.502	1.2	2
85	618.885	618.167	0.718	0.4	3
89	598.872	598.510	0.362	0.2	4
96	603.860	603.499	0.361	0.5	5
92	624.083	623.959	0.124	0.2	6
51	624.014	623.936	0.078	0.1	7
90	633.143	633.129	0.014	0.0	8
22	588.475	588.677	-0.202	-0.3	9
94	610.704	610.955	-0.252	-0.3	10
88	595.496	595.787	-0.291	-0.3	11
23	575.044	575.746	-0.702	-0.6	12
95	625.581	626.366	-0.785	-0.5	13
24	573.016	575.729	-2.713	-1.5	14

Figure 9A.11

321

CP VALUES FOR THE SELECTION OF VARIABLES

CEMENT HEAT PASS 5   DEP VAR 1: D

NUMBER OF OBSERVATIONS   14
NUMBER OF VARIABLES IN FULL EQUATION   11
NUMBER OF VARIABLES IN BASIC EQUATION   5
REMAINDER OF VARIABLES TO BE CONSIDERED   6

P	CP	VARIABLES IN EQUATION										
7	3.4	X1	X2	X3	X4	X5		X1X3			X2X4	
7	4.2	X1	X2	X3	X4	X5	X1X2	X1X3				
5	4.3	X1	X2	X3	X4	X5						
6	4.4	X1	X2	X3	X4	X5	X1X2					
7	4.8	X1	X2	X3	X4	X5	X1X2			X2X3		
8	5.1	X1	X2	X3	X4	X5		X1X3	X1X4		X2X4	
6	5.1	X1	X2	X3	X4	X5		X1X3				
8	5.2	X1	X2	X3	X4	X5		X1X3			X2X4	X3X4
8	5.3	X1	X2	X3	X4	X5	X1X2	X1X3			X2X4	
8	5.4	X1	X2	X3	X4	X5		X1X3		X2X3	X2X4	
8	5.4	X1	X2	X3	X4	X5	X1X2			X2X3		X3X4
6	5.8	X1	X2	X3	X4	X5					X2X4	
7	5.9	X1	X2	X3	X4	X5	X1X2					X3X4
8	5.9	X1	X2	X3	X4	X5	X1X2	X1X3		X2X3		
7	6.0	X1	X2	X3	X4	X5	X1X2		X1X4			
6	6.1	X1	X2	X3	X4	X5						X3X4
8	6.1	X1	X2	X3	X4	X5	X1X2	X1X3	X1X4			
7	6.2	X1	X2	X3	X4	X5	X1X2				X2X4	
9	7.0	X1	X2	X3	X4	X5		X1X3	X1X4	X2X3	X2X4	
10	9.0	X1	X2	X3	X4	X5		X1X3	X1X4	X2X3	X2X4	X3X4
11	11.0	X1	X2	X3	X4	X5	X1X2	X1X3	X1X4	X2X3	X2X4	X3X4

Figure 9A.12

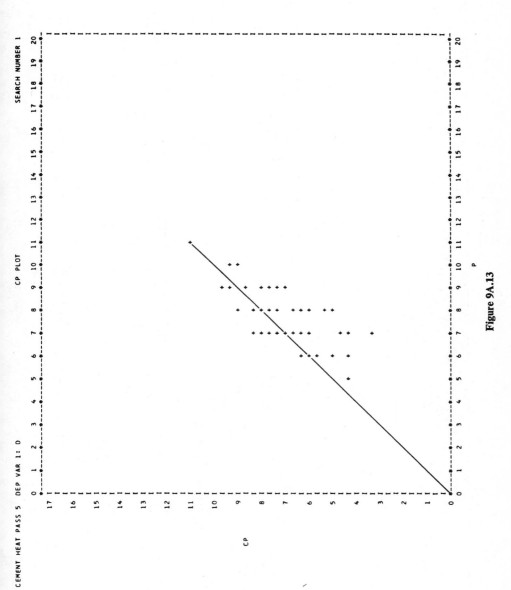

Figure 9A.13

LINEAR LEAST-SQUARES CURVE FITTING PROGRAM

MIN Y = 5.507D+01   MAX Y = 8.040D+01   RANGE Y = 2.533D+01

```
FULL EQUATION - - - SEARCH FOR INFLUENTIAL VARIABLES
A = B(1)X1 + B(2)X2 + B(3)X3 + E(4)X4 + E(5)X5 + B(6)X1X2 + B(7)X1X3
    B(8)X1X4 + B(9)X2X3 + B(10)X2X4 + B(11)X3X4

A  = COEFFICIENT IN NCN-LINEAR EQUATION
X1 = (PERCENT SILICA IN THE CLINKER - 21.0) / 10.66
X2 = (PERCENT ALUMINA - 3.48) / 10.66
X3 = (PERCENT FERRIC OXIDE - 1.18) / 10.66
X4 = (PERCENT LIME - 61.6) / 10.66
X5 = (PERCENT MAGNESIA - 2.08) / 10.66
A VALUES FROM NCN-LINEAR FIT WITH B, G AND H COMMON.
```

INC.VAR(I)	NAME	COEF.B(II)	S.E. COEF.	T-VALUE	R(II)SQRD	MIN X(I)	MAX X(I)	RANGE X(I)	REL.INF.X(I)
0		B(0) = 0							
1	X1	1.10375D+02	9.44D+00	11.7	0.9741	3.295D-04	6.042D-01	6.035D-01	2.63
2	X2	-3.83723D+01	7.96D+00	0.5	0.9992	9.797D-05	4.947D-01	4.946D-01	0.75
3	X3	1.22400D+02	4.48D+01	2.7	0.9991	2.228D-04	5.898D-01	5.896D-01	2.85
4	X4	6.2042D+01	1.61D+01	3.9	0.9939	4.501D-03	5.516D-01	5.471D-01	1.34
5	X5	1.31025D+02	2.14D+01	6.1	0.9311	6.883D-04	2.756D-01	2.749D-01	1.42
6	X1X2	3.20575D+02	1.49D+02	2.1	0.9881	3.636D-05	7.628D-02	7.624D-02	0.97
7	X1X3	-1.26387D+02	8.19D+01	1.5	0.9963	6.500D-06	2.161D-01	2.161D-01	1.08
8	X1X4	-1.26724D+02	3.90D+01	3.3	0.9766	1.512D-04	1.78D-01	1.786D-01	0.89
9	X2X3	-2.212E6D+02	1.35D+02	1.7	0.9939	3.567D-05	8.438D-02	8.434D-02	0.77
10	X2X4	1.99748D+02	1.91D+02	1.0	0.9992	1.442D-05	2.142D-01	2.142D-01	1.69
11	X3X4	-1.05438D+02	6.06D+02	1.7	0.9916	9.881D-05	1.972D-01	1.971D-01	0.82

```
NO. OF CBSERVATIONS             14
NO. OF INC. VARIABLES           11
RESIDUAL DEGREES OF FREEDOM      3
RESIDUAL ROOT MEAN SQUARE       1.6501427S
RESIDUAL MEAN SQUARE            2.72297122
RESIDUAL SUM OF SQUARES         8.16691367
TOTAL SUM OF SQUARES          783.62088279
MULT. CORREL. COEF. SQUARED      .9896
```

	ORDERED BY COMPUTER INPUT						ORDERED BY RESIDUALS				
IDENT.	OBSV.	WSS DISTANCE	OBS. Y	FITTEC Y	RESIDUAL	OBSV.	OBS. Y	FITTED Y	ORDERED RESID.	STUD.RESID.	SEQ
1	22	1026.	73.173	73.298	-0.126	95	60.398	58.815	1.583	1.5	1
2	23	528.	72.948	72.312	0.636	88	74.496	73.738	0.758	1.1	2
3	92	547.	73.555	73.712	-0.157	24	69.595	68.890	0.704	0.5	3
4	88	281.	74.496	73.738	0.755	23	72.948	72.312	0.636	0.7	4
5	96	496.	59.685	59.757	-0.072	90	62.458	62.472	-0.014	-0.0	5
6	85	207.	58.815	60.630	-1.815	89	59.334	59.402	-0.068	-0.1	6
7	94	531.	55.072	55.291	-0.220	96	59.685	59.757	-0.072	-0.1	7
8	24	702.	69.595	68.890	0.704	91	62.010	62.117	-0.107	-0.1	8
9	89	304.	59.334	59.402	-0.068	22	73.173	73.298	-0.126	-0.3	9
10	90	1080.	62.458	62.472	-0.014	92	73.555	73.712	-0.157	-0.3	10
11	25	732.	61.622	61.878	-0.256	94	55.072	55.291	-0.220	-0.3	11
12	95	383.	60.398	58.815	1.583	25	61.622	61.878	-0.256	-0.2	12
13	91	321.	62.010	62.117	-0.107	70	80.399	81.248	-0.848	-0.9	13
14	70	841.	80.399	81.248	-0.848	85	58.815	60.630	-1.815	-1.5	14

Figure 9A.14

CP VALUES FOR THE SELECTION OF VARIABLES

CEMENT HEAT PASS 6   DEP VAR 2: A

NUMBER OF OBSERVATIONS   14
NUMBER OF VARIABLES IN FULL EQUATION   11
NUMBER OF VARIABLES IN BASIC EQUATION   5
REMAINDER OF VARIABLES TO BE CONSIDERED   6

P   CP   VARIABLES IN EQUATION

 7   6.4   X1   X2   X3   X4   X5   X1X2   X1X4
 8   8.0   X1   X2   X3   X4   X5   X1X2   X1X4   X3X4
 8   8.1   X1   X2   X3   X4   X5   X1X2   X1X4 X2X3
 8   8.3   X1   X2   X3   X4   X5   X1X2 X1X3 X1X4
 8   8.4   X1   X2   X3   X4   X5   X1X2   X1X4   X2X4
 9   9.5   X1   X2   X3   X4   X5   X1X2   X1X4 X2X3   X3X4
 9   9.9   X1   X2   X3   X4   X5   X1X2   X1X4   X2X4 X3X4
10 10.1   X1   X2   X3   X4   X5   X1X2 X1X3 X1X4 X2X3   X3X4
11 11.0   X1   X2   X3   X4   X5   X1X2 X1X3 X1X4 X2X3 X2X4 X3X4

Figure 9A.15

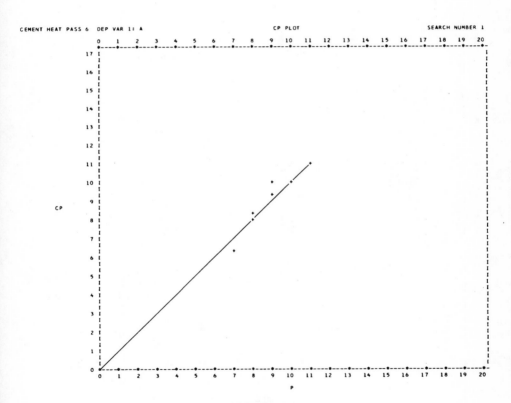

Figure 9A.16

325

CEMENT HEAT PASS 7   DEP VAR 1: D

LINEAR LEAST-SQUARES CURVE FITTING PROGRAM

MIN Y = 5.650D+02   MAX Y = 6.331D+02   RANGE Y = 6.817D+01

SELECTED EQUATION

D = B(1)X1 + B(2)X2 + B(3)X3 + E(4)X4 + E(5)X5 + B(6)X1X3 + E(7)X2X4

C = INITIAL ACID HEAT OF SOLUTION, CALORIES/GRAM
X1 = (PERCENT SILICA IN THE CLINKER - 21.0) / 10.66
X2 = (PERCENT ALUMINA - 3.4E) / 10.66
X3 = (PERCENT FERRIC CXIDE - 1.18) / 10.66
X4 = (PERCENT LIME - 61.6) / 10.66
X5 = (PERCENT MAGNESIA - 2.CE) / 10.66
D VALUES FROM NCN-LINEAR FIT WITH B, C AND H COMMON.

IND.VAR(I)	NAME	COEF.B(I)	S.E. COEF.	T-VALUE	R(I)SQRD	MIN X(I)	MAX X(I)	RANGE X(I)	REL.INF.X(I)
0		E(0) = 0							
1	X1	5.7C172D+02	5.33D+00	107.0	0.9278	3.295D-04	6.042D-01	6.039D-01	5.05
2	X2	5.77082D+02	1.70D+01	34.0	0.9852	9.797D-05	4.947D-01	4.946D-01	4.19
3	X3	5.83677D+02	5.51D+00	105.9	0.9485	2.262D-04	5.898D-01	5.896D-01	5.05
4	X4	6.557C3D+02	7.70D+00	85.7	0.5289	6.853D-04	2.756D-01	2.749D-01	2.66
5	X5	6.31448D+02	6.33D+C0	99.7	0.9650	4.5C1D-03	5.516D-01	5.471D-01	5.07
6	X1X3	-6.2C689D+01	2.07D+01	3.0	0.9487	6.500D-06	2.161D-01	2.161D-01	0.20
7	X2X4	1.35560D+02	4.94D+01	2.7	0.9897	1.442D-05	2.142D-01	2.142D-01	0.43

```
NO. OF CBSERVATIONS              14
NO. OF IND. VARIABLES             7
RESIDUAL DEGREES CF FREEDCM       7
RESIDUAL ROOT MEAN SQUARE        1.55473912
RESIDUAL MEAN SQUARE             2.41721372
RESIDUAL SUM CF SQUARES         16.92049603
TOTAL SUM OF SQUARES          6271.14395981
MULT. CORREL. COEF. SQUARED      .9973
```

----ORDERED BY COMPUTER INPUT----

IDENT.	OBSV.	WSS DISTANCE	OBS. Y	FITTED Y	RESIDUAL
1	22	28228.	588.475	588.730	-0.255
2	23	17103.	575.044	575.600	-0.556
3	92	15238.	624.083	624.061	0.022
4	88	5029.	595.496	596.358	-0.862
5	56	14841.	6C3.860	602.984	0.875
6	85	7515.	618.835	617.669	1.016
7	94	14582.	610.704	611.781	-1.077
8	24	23530.	572.016	575.434	-2.418
9	89	7238.	598.872	598.415	0.456
10	90	33647.	633.143	633.071	0.072
11	25	21591.	586.587	584.280	2.3C8
12	95	14466.	625.581	626.215	-0.635
13	91	15223.	624.014	623.587	0.027
14	70	29938.	564.969	563.543	1.C26

----ORDERED BY RESIDUALS----

OBSV.	OBS. Y	FITTED Y	ORDERED RESID.	STUD.RESID.	SEQ
25	586.587	584.280	2.308	1.8	1
70	564.969	563.943	1.026	0.9	2
85	618.885	617.869	1.016	0.7	3
96	6C3.860	602.984	0.875	0.9	4
89	598.872	598.415	0.456	0.3	5
90	633.143	633.071	0.072	0.1	6
91	624.014	623.987	0.027	0.0	7
92	624.083	624.061	0.022	0.0	8
22	588.475	588.730	-0.255	-0.4	9
23	575.044	575.600	-0.556	-0.5	10
95	625.581	625.215	-0.635	-0.5	11
88	595.496	596.358	-0.862	-0.8	12
94	610.704	611.781	-1.077	-0.9	13
24	573.016	575.434	-2.418	-1.8	14

Figure 9A.17

LINEAR LEAST-SQUARES CURVE FITTING PROGRAM

CEMENT HEAT PASS 8   DEP VAR 1: A

MIN Y = 5.507D+01   MAX Y = 8.040D+01   RANGE Y = 2.533D+01

SELECTED EQUATION

$A = B(1)X1 + B(2)X2 + B(3)X3 + B(4)X4 + B(5)X5 + B(6)X1X2 + B(7)X1X4$

A = COEFFICIENT IN NON-LINEAR EQUATION
X1 = (PERCENT SILICA IN THE CLINKER - 21.0) / 10.66
X2 = (PERCENT ALUMINA - 3.48) / 10.66
X3 = (PERCENT FERRIC OXIDE - 1.18) / 10.66
X4 = (PERCENT LIME - 61.6) / 10.66
X5 = (PERCENT MAGNESIA - 2.08) / 10.66
A VALUES FROM NON-LINEAR FIT WITH B, G AND H COMMON.

IND.VAR(I)	NAME	COEF.=B(I)	S.E. COEF.	T-VALUE	R(I)SQRD	MIN X(I)	MAX X(I)	RANGE X(I)	REL.INF.x(I)
0		B(0) = 0							
1	X1	1.08936D+02	4.14D+00	26.3	0.8776	3.295D-04	6.042D-01	6.039D-01	2.60
2	X2	5.7310D+01	5.56D+00	10.3	0.8585	9.757D-05	4.947D-01	4.946D-01	1.12
3	X3	5.0504D+01	2.35D+00	21.5	0.7100	2.282D-04	5.898D-01	5.896D-01	1.18
4	X4	5.5687D+01	3.75D+00	15.9	0.8979	4.501D-03	5.516D-01	5.471D-01	1.29
5	X5	1.07567D+02	6.50D+00	16.6	0.3206	6.853D-04	2.756D-01	2.749D-01	1.17
6	X1X2	1.00855D+02	2.88D+01	3.5	0.7072	3.636D-05	7.628D-02	7.624D-02	0.30
7	X1X4	-1.15227D+02	1.83D+01	6.3	0.9031	1.512D-04	1.788D-01	1.786D-01	0.81

NO. OF OBSERVATIONS   14
NO. OF IND. VARIABLES   7
RESIDUAL DEGREES OF FREEDOM   7
RESIDUAL ROOT MEAN SQUARE   1.57455660
RESIDUAL MEAN SQUARE   2.47922846
RESIDUAL SUM OF SQUARES   17.35459938
TOTAL SUM OF SQUARES   783.62888279
MULT. CORREL. COEF. SQUARED   .9779

	----ORDERED BY COMPUTER INPUT----					----ORDERED BY RESIDUALS----					
IDENT.	OBSV.	WSS DISTANCE	OBS. Y	FITTED Y	RESIDUAL	OBSV.	OBS. Y	FITTED Y	ORDERED RESID.	STUD.RESID.	SEQ
1	22	837.	73.173	73.481	-0.309	91	62.010	59.951	2.059	1.5	1
2	23	373.	72.948	72.478	0.470	24	69.595	68.552	1.042	0.8	2
3	92	382.	73.555	73.648	-0.092	88	74.496	73.802	0.694	0.8	3
4	88	115.	74.496	73.802	0.694	23	72.948	72.478	0.470	0.4	4
5	96	307.	59.685	59.574	0.110	90	62.458	61.995	0.463	0.7	5
6	85	102.	58.815	62.014	-3.199	95	60.398	60.250	0.147	0.1	6
7	94	318.	55.072	55.517	-0.445	96	59.685	59.574	0.110	0.1	7
8	24	200.	69.595	68.552	1.042	89	59.334	59.353	-0.019	-0.0	8
9	89	103.	59.334	59.353	-0.019	92	73.555	73.648	-0.093	-0.2	9
10	90	432.	62.458	61.995	0.463	25	61.622	61.836	-0.214	-0.2	10
11	25	199.	61.622	61.836	-0.214	22	73.173	73.481	-0.309	-0.3	11
12	95	257.	60.398	60.250	0.147	94	55.072	55.517	-0.445	-0.4	12
13	91	200.	62.010	59.951	2.059	70	80.399	81.106	-0.706	-0.7	13
14	70	491.	80.399	81.106	-0.706	85	58.815	62.014	-3.199	-2.1	14

Figure 9A.18

NON-LINEAR LEAST-SQUARES CURVE FITTING PROGRAM

CEMENT HEAT, PASS 9

$Z = H + (G - 11D' + (A' / (T + F')EXP B)$
$Z = $ ACID HEAT OF SOLUTION AT TIME "T"
B, G AND H ARE COMMON TO ALL CEMENTS
$D' = D1X1 + D2X2 + D3X3 + D4X4 + D5X5 - D13X1X3 + D24X2X4$
$A' = A1X1 + A2X2 + A3X3 + A4X4 + A5X5 + A12X1X2 - A14X1X4$
$F' = $ TIME CORRECTION TERM $= (A' / (GD' - H))EXP(1/B)$

$X1 = $ (PERCENT SILICA IN THE CLINKER - 21.0) / 10.66
$X2 = $ (PERCENT ALUMINA - 3.48) / 10.66
$X3 = $ (PERCENT FERRIC OXIDE - 1.18) / 10.66
$X4 = $ (PERCENT LIME - 61.6) / 10.66
$X5 = $ (PERCENT MAGNESIA - 2.08) / 10.66

OBSERVATION 85-3-2 OMITTED.

IND.VAR(I)	NAME	COEF.B(I)	S.E. COEF.	T-VALUE	95% CONFIDENCE LIMITS LOWER	UPPER
1	B	1.98204E-01	2.80E-02	7.1	1.42E-01	2.54E-01
2	D1 S	5.68108E 02	5.84E 00	97.3	5.56E 02	5.80E 02
3	D2 A	5.84257E 02	1.90E 01	30.7	5.46E 02	6.22E 02
4	D3 F	5.80589E 02	5.58E 00	104.0	5.69E 02	5.92E 02
5	D4 L	6.34258E 02	6.40E 00	99.1	6.21E 02	6.47E 02
6	D5 M	6.58238E 02	7.66E 00	85.9	6.43E 02	6.74E 02
7	D13 SF	5.21298E 01	2.14E 01	2.4	9.36E 00	9.49E 01
8	D24 AL	1.13566E 02	5.34E 01	2.1	6.73E 00	2.20E 02
9	A1 S	1.07869E 02	5.30E 00	20.4	9.73E 01	1.18E 02
10	A2 A	5.84145E 01	6.92E 00	8.4	4.46E 01	7.23E 01
11	A3 F	5.00791E 01	5.98E 00	8.4	3.81E 01	6.20E 01
12	A4 L	5.83754E 01	5.59E 00	10.4	4.72E 01	6.95E 01
13	A5 M	1.08437E 02	7.37E 00	14.7	9.37E 01	1.23E 02
14	A12 SA	8.71551E 01	3.00E 01	2.9	2.71E 01	1.47E 02
15	A14 SL	1.11729E 02	2.22E 01	5.0	6.73E 01	1.56E 02
16	G	7.25998E-01	3.25E-02	22.3	6.61E-01	7.91E-01
17	H	3.20155E 02	2.14E 01	15.0	2.77E 02	3.63E 02

NO. OF OBSERVATIONS 104
NO. OF COEFFICIENTS 17
RESIDUAL DEGREES OF FREEDOM 87
RESIDUAL ROOT MEAN SQUARE 1.97992229
RESIDUAL MEAN SQUARE 3.92009544
RESIDUAL SUM OF SQUARES 341.04833984

Figure 9A.19

328

NON-LINEAR LEAST-SQUARES CURVE FITTING PROGRAM

CEMENT HEAT, PASS 9

	ORDERED BY COMPUTER INPUT				ORDERED BY RESIDUALS			
OBS. NO.	OBS. Z	FITTED Z	RESIDUAL	OBS. NO.	OBS. Z	FITTED Z	ORDERED RESID.	SEQ.
22 0	588.700	588.418	0.281	22 7	535.600	530.741	4.859	1
22 3	537.500	539.461	-1.961	8890	517.800	513.397	4.403	2
22 7	535.600	530.741	4.859	2536	502.800	499.178	3.622	3
2228	516.200	518.996	-2.796	23 7	530.000	526.460	3.540	4
2290	507.500	511.246	-3.746	89 3	534.300	531.186	3.114	5
2218	510.200	507.417	2.783	2218	510.200	507.417	2.783	6
2236	503.200	504.015	-0.815	2590	507.800	505.257	2.543	7
23 0	575.700	575.582	0.118	2436	501.400	498.867	2.533	8
23 3	535.100	534.909	0.190	7036	502.100	499.595	2.505	9
23 7	530.000	526.460	3.540	24 7	526.200	523.697	2.503	10
2328	514.700	514.949	-0.250	9190	517.800	515.432	2.367	11
2390	503.800	507.318	-3.519	9128	524.100	521.755	2.345	12
2318	501.400	503.543	-2.144	9228	531.400	529.106	2.294	13
2336	499.700	500.189	-0.489	9118	514.600	512.315	2.285	14
92 0	623.600	624.133	-0.533	9690	511.900	509.683	2.217	15
92 0	624.100	624.133	-0.033	9618	508.600	506.579	2.021	16
92 3	546.800	550.061	-1.261	25 0	585.900	583.960	1.939	17
92 7	538.900	541.050	-2.150	9590	518.100	516.178	1.922	18
9228	531.400	529.106	2.294	9218	519.300	517.412	1.887	19
9290	523.000	521.276	1.724	9018	517.500	515.687	1.813	20
9218	519.300	517.412	1.887	9290	523.000	521.276	1.724	21
9236	513.200	513.982	-0.782	9428	517.000	516.020	1.580	22
88 0	595.500	596.726	-1.226	25 7	523.100	521.719	1.381	23
88 3	542.300	541.625	0.675	91 3	532.700	531.421	1.279	24
88 7	532.400	532.857	-0.457	91 7	540.000	538.756	1.244	25
8828	521.800	521.121	0.679	90 7	536.800	535.622	1.178	26
8890	517.800	513.397	4.403	95 3	540.800	539.685	1.115	27
8818	507.900	509.583	-1.683	96 0	604.500	603.428	1.072	28
8836	504.900	506.195	-1.295	94 3	532.700	531.757	0.943	29
96 0	604.500	603.428	1.072	85 0	618.800	617.880	0.920	30
96 0	603.200	603.428	-0.228	8918	505.900	505.031	0.869	31
96 3	532.600	532.868	-0.268	89 0	599.000	598.133	0.867	32
96 7	525.600	525.590	-0.110	8990	508.900	508.121	0.779	33
9628	515.700	515.977	-0.277	8828	521.800	521.121	0.679	34
9690	511.900	509.683	2.217	88 3	542.300	541.625	0.675	35
9618	508.600	506.579	2.021	90 3	543.900	543.275	0.625	36
9636	501.000	503.823	-2.823	70 0	564.700	564.159	0.541	37
85 0	618.800	617.880	0.920	8590	515.100	514.588	0.512	38
85 3	536.200	536.698	-0.498	8536	509.000	508.498	0.502	39
85 7	528.500	531.123	-2.623	8518	511.800	511.362	0.438	40
85 7	530.200	531.123	-0.923	90 0	633.400	632.997	0.403	41
8528	517.800	521.128	-3.328	9090	519.300	518.940	0.360	42
8528	517.100	521.128	-4.028	22 0	588.700	588.418	0.281	43
8590	515.100	514.588	0.512	94 0	611.700	611.421	0.279	44
8590	514.300	514.588	-0.288	23 3	535.100	534.909	0.190	45
8518	509.600	511.362	-1.762	23 0	575.700	575.582	0.118	46
8518	511.800	511.362	0.438	96 7	525.700	525.590	0.110	47
8536	509.000	508.498	0.502	2518	502.100	502.037	0.063	48
8536	508.500	508.498	0.002	70 7	528.400	528.342	0.058	49

Figure 9A.19 (continued)

329

94 0	610.000	611.421	-1.421	9418	507.300	507.282	0.018	50
94 0	611.700	611.421	0.279	8536	508.500	508.498	0.002	51
94 3	532.700	531.757	0.943	92 0	624.100	624.133	-0.033	52
94 7	522.500	524.967	-2.467	91 0	624.000	624.037	-0.037	53
9428	517.600	516.020	1.580	9536	510.200	510.247	-0.047	54
9490	509.500	510.168	-0.668	2528	511.700	511.781	-0.081	55
9418	507.300	507.282	0.018	96 0	603.200	603.428	-0.228	56
9436	502.000	504.720	-2.720	2328	514.700	514.949	-0.250	57
24 0	573.000	575.438	-2.437	96 3	532.600	532.868	-0.268	58
24 3	530.900	531.773	-0.873	9628	515.700	515.977	-0.277	59
24 7	526.200	523.697	2.503	8590	514.300	514.588	-0.288	60
2428	511.500	512.792	-1.292	8936	502.000	502.288	-0.288	61
2490	503.000	505.590	-2.590	88 7	532.400	532.857	-0.457	62
2418	500.500	502.030	-1.530	95 0	625.600	626.066	-0.466	63
2436	501.400	498.867	2.533	2336	499.700	500.189	-0.489	64
89 0	599.000	598.133	0.867	85 3	538.200	538.698	-0.498	65
89 3	534.300	531.186	3.114	92 0	623.600	624.133	-0.533	66
89 7	521.100	523.950	-2.850	70 3	536.400	536.948	-0.548	67
8928	513.200	514.385	-1.185	9528	521.900	522.550	-0.650	68
8990	508.900	508.121	0.779	9490	509.500	510.168	-0.668	69
8918	505.900	505.031	0.869	9518	512.300	513.036	-0.737	70
8936	502.000	502.288	-0.288	9236	513.200	513.982	-0.782	71
90 0	633.400	632.997	0.403	2236	503.200	504.015	-0.815	72
90 3	543.900	543.275	0.625	24 3	530.900	531.773	-0.873	73
90 7	536.800	535.622	1.178	85 7	530.200	531.123	-0.923	74
9028	524.600	525.536	-0.937	9028	524.600	525.536	-0.937	75
9090	519.300	518.940	0.360	7028	514.900	515.937	-1.037	76
9018	517.500	515.687	1.813	8928	513.200	514.385	-1.185	77
9036	510.700	512.799	-2.099	95 7	531.100	532.292	-1.192	78
25 0	585.900	583.960	1.939	88 0	595.500	596.726	-1.226	79
25 3	527.100	529.191	-2.091	92 3	548.800	550.061	-1.261	80
25 7	523.100	521.719	1.381	2428	511.500	512.792	-1.292	81
2528	511.700	511.781	-0.081	8836	504.900	506.195	-1.295	82
2590	507.800	505.257	2.543	94 0	610.000	611.421	-1.421	83
2518	502.100	502.037	0.063	2418	500.500	502.030	-1.530	84
2536	502.800	499.178	3.622	8818	507.900	509.583	-1.683	85
95 3	625.600	626.066	-0.466	8518	509.600	511.362	-1.762	86
95 7	540.800	539.685	-1.115	9136	507.700	509.547	-1.847	87
9528	531.100	532.292	-1.192	22 3	537.500	539.461	-1.961	88
9518	521.900	522.550	-0.650	25 3	527.100	529.191	-2.091	89
9536	518.100	516.178	1.922	9036	510.700	512.799	-2.099	90
91 0	512.300	513.036	-0.737	2318	501.400	503.543	-2.144	91
91 3	510.200	510.247	-0.047	92 7	538.900	541.050	-2.150	92
91 7	624.000	624.037	-0.037	24 0	573.000	575.438	-2.437	93
9128	532.700	538.756	1.244	94 7	522.500	524.967	-2.467	94
9190	524.100	531.421	1.279	2490	503.000	505.590	-2.590	95
9118	517.800	521.755	2.345	85 7	528.500	531.123	-2.623	96
9136	514.600	515.432	2.367	9436	502.000	504.720	-2.720	97
70 0	507.700	512.315	2.285	2228	516.200	518.996	-2.796	98
70 3	564.700	509.547	-1.847	9636	501.000	503.823	-2.823	99
70 7	536.400	564.159	0.541	89 7	521.100	523.950	-2.850	100
7028	528.400	536.948	-0.548	8528	517.800	521.128	-3.328	101
7035	514.900	528.342	0.058	2390	503.800	507.318	-3.519	102
	502.100	515.937	-1.037	2290	507.500	511.246	-3.746	103
		499.595	2.505	8528	517.100	521.128	-4.028	104

Figure 9A.19 (continued)

330

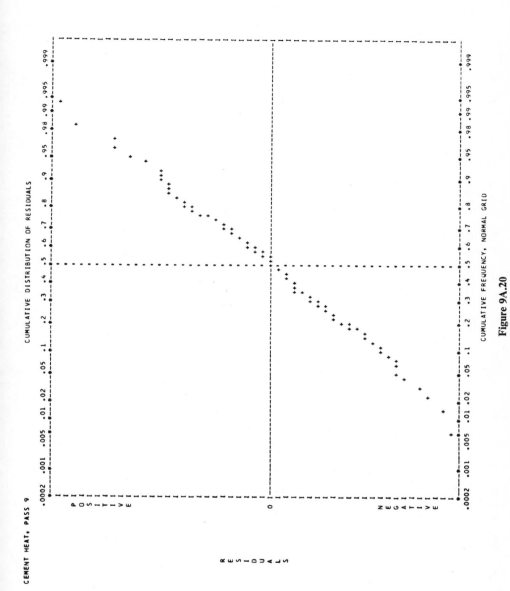

Figure 9A.20

331

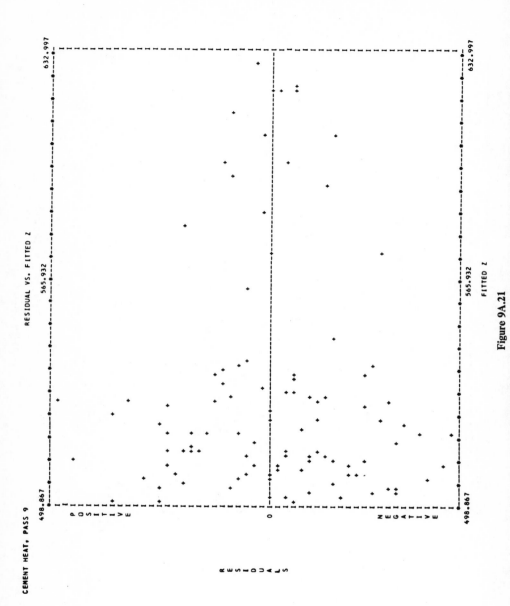

Figure 9A.21

NONLINEAR LEAST-SQUARES CURVE-FITTING PROGRAM

CEMENT HEAT, PASS 10   DEP.VAR.: MIN Y = 4.997D+02   MAX Y = 6.334D+02   RANGE Y = 1.337D+02

$$Z = H + (G - 1)D' + (A' / (T + F')^{EXP\ B})$$

Z = ACID HEAT OF SOLUTION AT TIME "T"
B, G AND H ARE COMMON TO ALL CEMENTS

$$A' = A1X1 - A2X2 + A3X3 + A4X4 + A5X5 + A12X1X2 - A13X1X3 - A14X1X4 - A23X2X3 + A24X2X4 - A34X3X4$$

$$D' = D1X1 + D2X2 + D3X3 + D4X4 + D5X5$$

F' = TIME CORRECTION TERM = $(A' / (GD' - H))^{EXP(1/B)}$

X1 = (MOL PERCENT SILICA IN THE CLINKER - 20.9) / 7.43
X2 = (MOL PERCENT ALUMINA - 2.03) / 7.43
X3 = (MOL PERCENT FERRIC OXIDE - 0.44) / 7.43
X4 = (MOL PERCENT LIME - 66.1) / 7.43
X5 = (MOL PERCENT MAGNESIA - 3.10) / 7.43 ; OBSV. 85-3-2 OMITTED.

IND.VAR(I)	NAME	COEF.B(I)	S.E. COEF	T-VALUE	95% CONFIDENCE LIMITS LOWER	UPPER
1	3	1.95889D-01	2.81D-02	7.0	1.40D-01	2.52D-01
2	G	7.20147D-01	3.39D-02	21.2	6.52D-01	7.88D-01
3	H	3.16318D+02	2.23D+01	14.2	2.72D+02	3.61D+02
4	A1	8.49908D+01	4.80D+00	17.7	7.54D+01	9.46D+01
5	A2	4.65436D+01	6.42D+01	0.7	-8.19D+01	1.75D+02
6	A3	3.26844D+02	1.26D+02	2.6	7.50D+01	5.79D+02
7	A4	6.53313D+01	1.10D+01	5.9	4.33D+01	8.73D+01
8	A5	9.22943D+01	1.03D+01	9.0	7.17D+01	1.13D+02
9	A12	3.33207D+02	1.23D+02	2.7	8.77D+01	5.79D+02
10	A13	3.46716D+02	1.71D+02	2.0	5.37D+00	6.88D+02
11	A14	8.54590D+01	2.07D+01	4.1	4.41D+01	1.27D+02
12	A23	8.11138D+02	3.52D+02	2.3	1.08D+02	1.51D+03
13	A24	1.94884D+02	1.31D+02	1.5	-0.73D+01	4.57D+02
14	A34	3.52489D+02	1.62D+02	2.2	2.86D+01	6.76D+02
15	D1	5.75543D+02	1.62D+00	356.4	5.72D+02	5.79D+02
16	D2	6.22704D+02	5.47D+00	113.8	6.12D+02	6.34D+02
17	D3	5.21679D+02	4.22D+00	123.6	5.13D+02	5.30D+02
18	D4	6.40787D+02	2.08D+00	308.8	6.37D+02	6.45D+02
19	D5	6.36472D+02	2.94D+00	216.3	6.31D+02	6.42D+02

NO. OF OBSERVATIONS   104
NO. OF COEFFICIENTS   19
RESIDUAL DEGREES OF FREEDOM   85
RESIDUAL ROOT MEAN SQUARE   1.99988029
RESIDUAL MEAN SQUARE   3.99952118
RESIDUAL SUM OF SQUARES   339.95930046

Figure 9A.22

333

| ORDERED BY COMPUTER INPUT | | | | | ORDERED BY RESIDUALS | | | |
OBS. NO.	OBS. Y	FITTED Y	RESIDUAL		OBS. NO.	OBS. Y	FITTED Y	URDERED RESID.	SEQ.
22 0	588.700	585.463	3.237		22 7	535.600	530.920	4.680	1
22 3	537.500	539.714	-2.214		8890	517.800	513.630	4.170	2
22 7	535.600	530.920	4.680		2536	502.800	499.023	3.777	3
2228	516.200	518.988	-2.788		23 7	530.000	526.293	3.707	4
2290	507.500	511.075	-3.575		22 0	588.700	585.463	3.237	5
2218	510.200	507.155	3.045		2218	510.200	507.155	3.045	6
2236	503.200	503.666	-0.466		89 3	534.300	531.466	2.834	7
23 0	575.700	575.275	0.425		2590	507.800	505.154	2.646	8
23 3	535.100	534.707	0.393		7036	502.100	499.554	2.546	9
23 7	530.000	526.293	3.707		25 0	535.900	533.411	2.489	10
2328	514.700	514.806	-0.106		9228	531.400	529.151	2.249	11
2390	503.800	507.170	-3.370		2436	501.400	499.160	2.240	12
2318	501.400	503.384	-1.984		9590	518.100	515.864	2.236	13
2336	499.700	500.015	-0.315		9690	511.900	509.853	2.047	14
92 0	623.600	624.187	-0.587		24 7	526.200	524.247	1.953	15
92 0	624.100	624.187	-0.087		9618	508.000	506.769	1.831	16
92 3	548.800	549.931	-1.131		95 3	540.800	538.998	1.802	17
92 7	538.900	541.006	-2.106		9218	519.300	517.504	1.796	18
9228	531.400	529.151	2.249		9428	517.600	515.811	1.789	19
9290	523.000	521.358	1.642		9290	523.000	521.358	1.642	20
9218	519.300	517.504	1.796		9018	517.500	515.926	1.574	21
9236	513.200	514.077	-0.877		25 7	523.100	521.675	1.425	22
88 0	595.500	598.129	-2.629		94 3	532.700	531.417	1.283	23
88 3	542.300	541.621	0.679		9118	514.600	513.326	1.274	24
88 7	532.400	532.936	-0.536		9190	517.800	516.568	1.232	25
8828	521.800	521.303	0.497		89 0	599.000	597.782	1.218	26
8890	517.800	513.630	4.170		8590	515.100	513.930	1.170	27
8818	507.900	509.833	-1.933		8518	511.800	510.800	1.000	28
8836	504.900	506.456	-1.556		8536	509.000	508.016	0.984	29
96 0	604.500	604.740	-0.240		85 3	538.200	537.221	0.979	30
96 0	603.200	604.740	-1.540		9128	524.100	523.130	0.970	31
96 3	532.600	532.782	-0.182		96 3	633.400	632.467	0.933	32
96 7	525.700	525.600	0.100		90 7	536.800	536.008	0.792	33
9628	515.700	516.094	-0.394		8918	505.900	505.118	0.782	34
9690	511.900	509.853	2.047		85 0	618.800	618.019	0.781	35
9618	508.600	506.769	1.831		94 0	611.700	610.995	0.705	36
9636	501.000	504.026	-3.026		88 3	542.300	541.621	0.679	37
85 0	618.800	618.019	0.781		8990	508.900	508.245	0.655	38
85 3	538.200	537.221	0.979		85 3	521.800	521.303	0.497	39
85 7	528.500	529.917	-1.417		8828	508.500	508.016	0.484	40
85 7	530.200	529.917	0.283		23 0	575.700	575.275	0.425	41
8528	517.800	520.264	-2.464		23 3	535.100	534.707	0.393	42
8528	517.100	520.264	-3.164		8590	514.300	513.930	0.370	43
8590	515.100	513.930	1.170		85 7	530.200	529.917	0.283	44
8590	514.300	513.930	0.370		8590	528.400	528.175	0.225	45
8518	509.600	510.800	-1.200		8518	543.900	543.684	0.216	46
8518	511.800	510.800	1.000		9536	510.200	509.993	0.207	47
8536	509.000	508.016	0.984		9418	507.300	507.106	0.194	48
8536	508.500	508.016	0.484		2518	502.100	501.909	0.191	49

Figure 9A.23

94 0	610.000	610.995	-0.995	525.700	96 7	525.600	0.100	50
94 3	611.700	610.995	0.705	519.300	9090	519.213	0.087	51
94 7	532.700	531.417	1.283	625.000	95 0	625.576	0.024	52
9428	522.500	524.693	-2.193	624.100	2528	624.187	-0.013	53
9490	517.600	515.811	1.789	624.700	92 0	624.806	-0.087	54
9418	509.500	509.985	-0.485	514.700	2328	514.806	-0.106	55
9436	507.300	507.106	0.194	532.600	96 3	532.782	-0.182	56
24 0	502.000	504.546	-2.546	604.500	96 0	604.740	-0.240	57
24 3	573.000	576.096	-3.096	521.900	9528	522.153	-0.253	58
24 7	530.900	532.367	-1.467	624.000	91 0	624.281	-0.281	59
2428	526.200	524.247	1.953	499.700	2336	500.015	-0.315	60
2490	511.500	513.256	-1.756	504.700	70 0	565.030	-0.330	61
2418	503.000	505.975	-2.975	502.000	8936	502.336	-0.336	62
2436	500.500	502.369	-1.869	536.400	70 3	535.785	-0.385	63
89 0	501.400	499.160	2.240	515.700	9628	516.094	-0.394	64
89 3	599.000	597.782	1.218	532.700	917	533.130	-0.430	65
89 7	534.300	531.466	2.834	512.300	9518	512.757	-0.457	66
8928	521.100	524.201	-3.101	509.200	9490	503.666	-0.466	67
8990	513.200	514.571	-1.371	532.400	88 7	509.985	-0.485	68
8918	508.900	508.245	0.655	623.000	92 0	532.936	-0.536	69
8936	505.900	505.118	0.782	531.100	95 7	624.187	-0.587	70
90 0	502.000	502.336	-0.336	540.000	91 3	531.739	-0.639	71
90 3	633.400	632.467	0.933	515.200	9236	540.696	-0.696	72
90 7	543.300	543.684	0.216	514.900	7028	540.696	-0.677	73
9028	536.800	536.008	0.792	610.000	94 0	515.629	-0.929	74
9090	524.600	525.866	-1.266	548.800	92 3	549.931	-0.995	75
9018	519.200	519.213	0.087	509.600	8518	510.800	-1.131	76
9036	517.500	515.926	1.574	524.000	9028	525.860	-1.200	77
25 0	510.700	513.002	-2.302	513.200	8928	514.571	-1.266	78
25 3	585.900	583.411	2.489	528.500	85 7	529.917	-1.371	79
25 7	527.100	529.146	-2.046	530.900	24 3	532.367	-1.417	80
2528	523.100	521.675	1.425	603.200	96 0	604.740	-1.467	81
2590	511.700	511.713	-0.013	504.900	8836	506.456	-1.540	82
2518	507.800	505.154	0.191	511.800	2428	513.256	-1.556	83
2536	502.000	501.909	0.191	500.500	8818	511.256	-1.756	84
95 0	502.800	499.023	3.777	507.900	2318	507.930	-1.869	85
95 3	625.576	625.576	0.024	501.400	25 3	501.400	-1.933	86
95 7	540.800	538.998	1.802	527.100	2490	527.100	-1.984	87
9528	531.100	531.739	-0.639	538.900	9636	541.006	-2.046	88
9590	521.900	522.153	-0.253	510.700	24 0	524.693	-2.106	89
9518	518.100	512.757	-0.457	544.500	89 7	539.714	-2.193	90
9536	512.300	509.993	-0.207	537.500	94 7	513.002	-2.214	91
91 0	510.200	624.000	-0.281	510.700	22 3	520.264	-2.302	92
91 3	624.000	540.000	-0.696	517.800	9036	504.546	-2.464	93
91 7	540.000	533.130	-0.430	502.000	8528	502.540	-2.540	94
9128	532.700	523.130	0.970	595.500	9436	598.129	-2.629	95
9190	524.100	516.568	1.232	507.700	88 0	510.442	-2.742	96
9118	517.600	513.326	1.274	516.200	9136	516.988	-2.788	97
9136	514.600	510.442	-2.742	503.000	2228	503.975	-2.975	98
70 0	507.700	565.030	-0.330	501.000	2490	501.026	-3.026	99
70 3	564.700	536.785	-0.385	573.000	9636	576.096	-3.096	100
70 7	536.785	528.175	0.225	521.100	24 0	524.201	-3.101	101
7028	528.400	515.829	-0.929	517.100	8528	520.264	-3.164	102
7036	514.900	499.554	2.546	503.800	2390	507.170	-3.370	103
	502.100			507.500	2290	511.075	-3.575	104

Figure 9A.23 (*continued*)

335

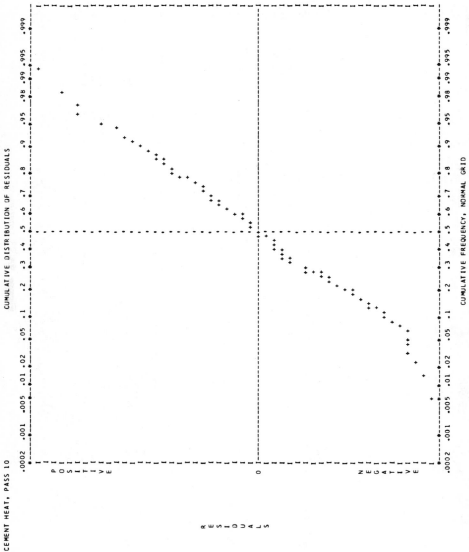

CEMENT HEAT, PASS 10   CUMULATIVE DISTRIBUTION OF RESIDUALS

CUMULATIVE FREQUENCY, NORMAL GRID

Figure 9A.24

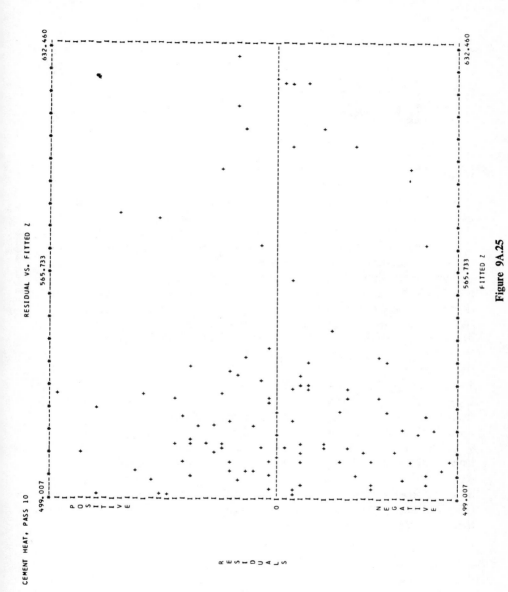

CEMENT HEAT, PASS 10

RESIDUAL VS. FITTED Z

Figure 9A.25

337

Glossary

Conventions

A clear distinction is drawn between parameters and estimates, that is, between quantities which characterize the "true" situation and estimates of these quantities calculated from observations. When possible, a Greek letter is used to represent the parameter and the corresponding Roman letter to represent the estimate. Thus σ represents the standard deviation of the total population of values, and s the estimate of σ. Likewise, $\eta = \beta_0 + \beta_1 x_1$ represents the true equation, and $Y = b_0 + b_1 x_1$ its estimate. Notice that Y is used for the estimated or fitted value of the dependent variable and y for the observed value. Since computers to date print only capital letters, Y is listed on our printout as FITTED VALUE and y as OBS. VALUE.

Independent variables and their coefficients are indexed by the subscript i from 1 through K, that is, $b_1, b_2, \ldots, b_i, \ldots, b_K$. Dependent variables are indexed by the subscript m from 1 through Q: $y_1, y_2, \ldots, y_m, \ldots, y_Q$. Observations are indexed by the second subscript, j, from 1 through N; that is, the ith independent variable's observations are $x_{i1}, x_{i2}, \ldots, x_{ij}, \ldots, x_{iN}$, and the mth dependent variable's observations are $y_{m1}, y_{m2}, \ldots, y_{mj}, \ldots, y_{mN}$. When there is only one dependent variable, however, the subscript m is usually dropped and only the observation subscript j is used; $y_1, y_2, \ldots, y_j, \ldots, y_N$ instead of $y_{11}, y_{12}, \ldots, y_{1j}, \ldots, y_{1N}$. The computer printout lists variables numerically under IND. VAR. (I) and coefficients under COEF. B(I). Observations are listed in the order in which they are given to the computer.

Symbols

$\beta_1, \beta_2, \beta_3$	"Beta (1), beta (2), beta (3)," true values of the coefficients of three independent variables.
b_1, b_2, b_3	Estimates of the coefficients of three independent variables.

338

$c_{ii'}$ Elements of the inverse of the matrix of simple correlation coefficients among the independent variables, i indexes rows; i', columns.

C_p An estimate of the standardized total squared error.

C_{ij} A component of the full fitting equation:

$$C_{ij} = b_i(x_{ij} - \bar{x}_i).$$

df   Degrees of freedom.

e_j Random error associated with the jth observation.

$E\{\ \}$ The expected value of the quantity in braces.

Est. Var. ( )   Estimated variance of the quantity in parentheses.

F F-value, Fisher's variance ratio: the ratio of two independent estimates of variance:

$$F = \frac{s_1^2}{s_2^2}.$$

Γ_p "Gamma sub p," the standardized total squared error:

$$\Gamma_p = E\{C_p\}.$$

i Subscript i, index of the independent variable x, its parameter β, and its coefficient b.

(I)   Computer index of the independent variable X.

j Subscript j, index of the observation.

K The total number of independent variables, the terminal value of index i.

m Subscript m, index for the dependent variables, y_m: $m = 1, \ldots, m, \ldots, Q$.

N The total number of observations, the terminal value of index j.

η "Eta," the "true" value of a response, or dependent variable.

p The number of constants to be estimated. Used in plots of p and C_p; $p_{\max} = k + 1$ with b_0 present, $p_{\max} = k$ without b_0.

Q The total number of dependent variables, the terminal value of index m.

r_{12}^2 The degree of orthogonality between x_1 and x_2.

R_i^2 When the x_i are correlated, the overall linear dependence of x_i on all the $x_{i'}$ ($i' \neq i$) may be expressed by $X_i = \sum_{i'} a_{i'} x_{i'}$.

$$R_i^2 = \frac{[X_i X_i]}{[x_i x_i]}.$$

R_y^2 The multiple correlation coefficient squared (sometimes called the coefficient of determination): the fraction of the total variation accounted for by the fitted equation:

$$R_y^2 = \frac{[YY]}{[yy]}.$$

RMS   The residual mean square, the residual sum of squares divided by $(N - p)$ degrees of freedom:

$$RMS = \frac{RSS}{N - p} \equiv \frac{[rr]}{N - p}.$$

RSS   The residual sum of squares, that part of the total sum of squares not accounted for by the fitted equation:

$$RSS \equiv [rr].$$

$\sigma(y)$ "Sigma," the standard deviation of y.

$\sigma^2(y)$ "Sigma squared," the variance of y.

Σ The summation of the values which follow; for example

$$\sum_{j=1}^{N} x_{ij}$$

is the summation of the values of the x_i variable from 1 through N observations.

$s(b_i)$ The estimated standard error of the coefficient, b_i.

$s(x_i)$ The root-mean-square deviation of x_i.

$s(y)$ The estimate of the standard deviation of y, $\sigma(y)$.

$s^2(y)$ The estimate of the variance of y, $\sigma^2(y)$.

SSB   The sum of squares due to bias; one of the components of C_p.

SSFE	The sum of squares accounted for by the fitted equation: $\text{SSFE} \equiv [YY]$.
TSS	The total sum of squares: $\text{TSS} \equiv [yy]$.
v_j	A dimensionless distance from the centroid of factor space:

$$v_j = \frac{1}{N} + \left[\sum_{i=1}^{K} \sum_{i'=1}^{K} \frac{(x_{ij} - \bar{x}_i)(x_{i'j} - \bar{x}_{i'})(c_{ii'})}{(s(x_i))(s(x_{i'}))(df)} \right].$$

Var. ( )	The variance of the quantity in parentheses.
$V(R)$	Computer designation for variance of residual:

$$\text{variance (residual of fitted } Y_j) = (1 - v_j)s_{(y)}^2$$
$$= (1 - v_j)\,\text{RMS}.$$

$$\text{variance (residual of new observation}_j) = (1 + v_j)s_{(y)}^2$$
$$= (1 + v_j)\,\text{RMS}.$$

$V(Y)$	Computer designation for variance of fitted Y:

$$\text{variance (fitted } Y_j) = v_j s_{(y)}^2 = v_j\,\text{RMS}.$$

w_i	The range of observations, $x_{i,\max} - x_{i,\min}$.
w_j	In weighted least squares, $w_j = $ weight given, otherwise $w_j = 1$.
WSSD	The weighted squared standardized distance for point j from the centroid of all the observations

$$\text{WSSD}_j = \sum_{i=1}^{K} \left[\frac{b_i(x_{ij} - \bar{x}_i)}{s_y} \right]^2$$

or between any two points j and j':

$$\text{WSSD}_{jj'} = \sum_{i=1}^{K} \left[\frac{b_i(x_{ij} - x_{ij'})}{s_y} \right]^2.$$

x_{ij}	The independent variable indexed by i from 1 to K with observations indexed by j from 1 to N.
$\bar{x}_i$	The arithmetic mean of N observations of the ith independent variable:

$$\bar{x}_i \equiv \frac{\sum_{j=1}^{N} x_{ij}}{N}.$$

y_{mj} The dependent variable indexed by m from 1 to Q with observations indexed by j from 1 to N.

$\bar{y}_m$ The arithmetic mean of N observations of the mth dependent variable:

$$\bar{y}_m \equiv \frac{\displaystyle\sum_{j=1}^{N} y_{mj}}{N}.$$

Y_{mj} The estimated or fitted value of the dependent variable indexed by the subscript m from 1 to Q with observations indexed by j from 1 to N.

z A random normal deviate with a standard deviation of 1 and a mean of 0.

[ ] Gaussian brackets, indicating summation of products obtained by multiplying deviations from means of factors shown inside the brackets.

$[rr]$ The residual sum of squares:

$$[rr] \equiv \sum_{j=1}^{N} (y_j - Y_j)^2.$$

$[yy]$ The total sum of squares:

$$[yy] \equiv \sum_{j=1}^{N} (y_j - \bar{y})^2.$$

$[YY]$ The sum of squares accounted for by the fitted equation:

$$[YY] \equiv \sum_{j=1}^{N} (Y_j - \bar{y})^2.$$

$[1\,y]$ The corrected (for the means) sum of the products of the independent and dependent variables, x_{1j} and y_j:

$$[1\,y] \equiv \sum_{j=1}^{N} (x_{1j} - \bar{x}_1)(y_j - \bar{y}).$$

$[1\,2]$ The corrected (for the means) sum of the products of two independent variables, x_{1j} and x_{2j}:

$$[1\,2] \equiv \sum_{j=1}^{N} (x_{1j} - \bar{x}_1)(x_{2j} - \bar{x}_2).$$

Computer Terms

This portion of the glossary is a connecting link between the computer Linear and Nonlinear Least-Squares Curve-Fitting Programs and the text. Because of the limitations of both space and computer printing symbols, many of the computer listings are necessarily abbreviations and are restricted to capital letters. In the text, on the other hand, it was necessary to use mathematical abbreviations to avoid cumbersome rewriting of certain summations. Thus we have the residual sum of squares in four forms:

$$\sum_{j=1}^{N}(y_j - Y_j)^2 \equiv [rr] \equiv \text{RESIDUAL SUM OF SQUARES} \equiv \text{RSS}$$

(Normal form)   (Gaussian bracket abbreviation)   (Computer listing)   (Abbreviation)

The Glossary of Computer Terms is arranged in the alphabetical order of the computer listings. Here, along with the definition of terms, we have included enough of the mathematics of the curve-fitting programs so that the reader can understand what the computer is doing. Those wishing a more complete exposition of the mathematics may refer to Brownlee, Hald, Bennett and Franklin, or Anderson and Bancroft.

B ZERO = CALCULATED VALUE. This statement on the computer printout indicates that the b_0 term in the fitted equation is a calculated value and has not been set equal to zero.

B ZERO = ZERO. This statement on the computer printout indicates that the b_0 term in the fitted equation has been set, a priori, equal to zero. (See the User's Manual for instructions on control card options.)

C(I, I PRIME). Computer designation for $c_{ii'}$.

COEF. B(I). Under this heading, the coefficients of the variables are in-dexed by i and the values calculated for the fitted equation are tabulated starting with b_0. (See FITTED EQUATION.)

COMPONENT. A component of the equation of fitted Y. The component effect of each observation (indexed by j) on each variable (indexed by i) is $b_i(x_{ij} - \bar{x}_i)$. This has the units of Y and is used in both the component-effect table and in the component and component-plus-residual plots versus x_i.

COMPONENT-PLUS-RESIDUAL.   Adding the residual of each observation to the component effect of each observation allows us to obtain some insight into the patterns of the residuals, and their dependence on each component of the fitted equation.

CUMULATIVE DISTRIBUTION OF RESIDUALS.   The program plots the ranked residuals, OBSERVED Y − FITTED Y, on a CUMULATIVE FREQUENCY NORMAL PROBABILITY GRID. If the residuals are normally distributed, the points should fall approximately on a straight line with mean zero.

On a normal probability grid, the abscissa is specially ruled so that the cumulative distribution function of a normal distribution can be plotted as a straight line against the residuals as ordinate. Plotting points for the abscissa, P, are calculated from the formula:

$$P = \frac{i - \frac{1}{2}}{N},$$

where i = rank of observation, and N = total number of observations. Examples of plots for samples ranging in size from 8 to 384, drawn from a population of known normal distribution with mean 0 and variance 1, are shown in Appendix 3A (lines on the graphs represent the true population line). In the long run, the larger the sample the better is the approximation to a straight line. (See Blom, pages 71–72, and Hald, Sections 6.6 and 6.8, for additional examples.)

DATA INPUT.   This is a listing of the data read into the computer from the data input cards. The format card determines which columns are used for each variable. Instructions for submitting problems to the computer are given in the User's Manual.

DATA TRANSFORMATIONS.   This is a listing of the transformed values of each variable as used in the fitting equation. The program can convert the DATA INPUT to several forms such as sums, differences, products, ratios, and logarithms. Also, the input and transformed variables can be put in or left out in any desired pattern. (See the User's Manual for specific instructions.)

DEP. VAR. (M).   The dependent variables are indexed by m from 1 to Q.

FITTED EQUATION

FIRST-ORDER EQUATION. This is an equation with K independent variables, indexed by i. It is generally of the form:

$$Y = b_0 + b_1x_1 + b_2x_2 + b_3x_3 + b_ix_i + \cdots + b_Kx_K,$$

where $Y =$ FITTED Y, the dependent variable; $x_1, x_2, x_3, \ldots, x_K =$ IND. VAR. (I), the independent variables indexed by i from 1 to K; $b_0 =$ is listed under COEF. B(I) when I is 0; and $b_1, b_2, \ldots, b_i, \ldots, b_K$ are listed under COEF. B(I) when $I = 1, 2, \ldots, i, \ldots, K$.

SECOND-ORDER EQUATION. This is an equation in K variables which includes K linear terms, the $K(K-1)/2$ cross-product (or interaction) terms, and the K square terms:

$$Y = b_0 + b_1x_1 + \cdots + b_Kx_K + b_{12}x_1x_2 + \cdots + b_{K-1,K}x_{K-1}x_K$$
$$+ b_{11}x_1^2 + \cdots + b_{KK}x_K^2.$$

FITTED Y. OBSERVED Y is the observed response, and FITTED Y is the corresponding value calculated from the fitted equation. Both are indexed by the OBSERVATION NUMBER, j from 1 to N.

F-VALUE. The F-VALUE can be compared with tabled values to give a joint test of the hypothesis that all the coefficients are zero against the alternative that the equation as a whole produces a significant reduction in the total sum of squares. The F-value is calculated from the multiple correlation coefficient squared, the residual degrees of freedom, and the degrees of freedom associated with the fitted equation:

$$F\text{-value} = \frac{\text{SSFE}/K}{\text{RSS}/(N-K-1)} = \left(\frac{N-K-1}{K}\right)\left(\frac{R^2}{1-R^2}\right).$$

The ratio can then be compared to the corresponding value from an F-table with K and $(N-K-1)$ degrees of freedom.

IND. VAR. (I). The independent variables, indexed by i, are tabulated under IND. VAR. (I), starting with 0 (to identify the b_0 term) and proceeding through the K variables, $0, 1, 2, \ldots, i, \ldots, K$. (See FITTED EQUATION.)

INVERSE, C(I, I PRIME). Elements of the inverse matrix are listed in the following form:

$$
\begin{matrix}
C_{11} & C_{12} & C_{13} & \cdots & C_{1i'} & \cdots & C_{1K} \\
C_{21} & C_{22} & C_{23} & \cdots & C_{2i'} & \cdots & C_{2K} \\
C_{31} & C_{32} & C_{33} & \cdots & C_{3i'} & \cdots & C_{3K} \\
\cdot & \cdot & \cdot & \cdots & \cdot & \cdots & \cdot \\
\cdot & \cdot & \cdot & \cdots & \cdot & \cdots & \cdot \\
\cdot & \cdot & \cdot & \cdots & \cdot & \cdots & \cdot \\
C_{i1} & C_{i2} & C_{i3} & \cdots & C_{ii'} & \cdots & C_{iK} \\
\cdot & \cdot & \cdot & \cdots & \cdot & \cdots & \cdot \\
\cdot & \cdot & \cdot & \cdots & \cdot & \cdots & \cdot \\
\cdot & \cdot & \cdot & \cdots & \cdot & \cdots & \cdot \\
C_{K1} & C_{K2} & C_{K3} & \cdots & C_{Ki'} & \cdots & C_{KK}
\end{matrix}
$$

(See Appendix 2B for details on the inverse matrix as used in the computer program.)

LAMBDA. In the nonlinear program, LAMBDA is a multiplier to scale the space or size of steps taken.

LINEAR LEAST-SQUARES ESTIMATES. See Appendix 2B for a detailed description of linear least-squares and computer computations.

MAX Y(M). The maximum value of each dependent variable is given, indexed by m.

MAX X(I). The largest value of each independent variable is tabulated under this heading.

MEANS OF VARIABLES. The arithmetic means of N observations of each independent and dependent variable are listed under this heading in the same order that they appear on the data cards or as dictated by transformation instructions.

MIN Y(M). The minimum value of each dependent variable is given, indexed by m.

MIN X(I). The minimum value of each independent variable is tabulated under this heading.

MULT. CORREL. COEF. SQUARED. The multiple correlation co-efficient squared (sometimes called the coefficient of determination) is defined as follows:

MULT. CORREL. COEF. SQUARED $\equiv$

$$\frac{\text{Sum of squares due to the fitted equation}}{\text{Total sum of squares}}.$$

(See "Partitioning Sums of Squares" in Appendix 2B.)

MULTIPLIER. In the nonlinear program, MULTIPLIER is used to change the coefficients from one iteration to the next.

NU. In the nonlinear program, NU is a divider or multiplier to change the size of lambda depending on whether the sum of squares of that iteration is relatively close to or far from the value of the previous iteration.

OBS. NO. The observation number is read from the computer card, and the numbers are listed in the order in which the cards are given to the computer. If the observations are in the same order in which they were taken, their time independence can be verified by a "run-of-signs" test on the residuals.

OBS. Y. The observed value of the dependent variable, y, is indexed by j, the observation number from 1 to N.

ORDERED RESID. Under this heading the residuals are ranked in order from the largest positive to the largest negative value. The observation numbers are listed with the ordered residuals so that they can be easily identified.

RANGE Y(M). The range of values of each dependent variable is given, indexed by m.

RANGE X(I). The range of each independent variable, w_i, is determined by subtracting its minimum value from its maximum value.

RAW SUMS OF SQUARES + CROSS PRODUCTS. The raw sum of squares for the first variable over the N observations is the quantity $\sum_{j=1}^{N} x_{1j}^{2}$.

The raw sum of cross products for the first and second variables over the N observations is the quantity $\sum_{j=1}^{N} x_{1j}x_{2j}$.

The computer printout lists the raw sums of squares + cross products for all the independent and dependent variables in the following form:

	Col. 1	Col. 2	Col. 3	$\cdots$	Col. $(K + Q)$
Row 1	$\sum_{j=1}^{N} x_{1j}^{2}$				
Row 2	$\Sigma x_{2j}x_{1j}$				
Row 3	$\Sigma x_{3j}x_{1j}$	$\Sigma x_{3j}x_{2j}$	Σx_{3j}^{2}		
	.	.	.		
	.	.	.		
	.	.	.		
Row $(K + Q)$	$\Sigma x_{1j}x_{(K+Q)j}$	$\Sigma x_{2j}x_{(K+Q)j}$		$\cdots$	$\Sigma x_{(K+Q)}^{2}$

where K = number of independent variables, and Q = number of dependent variables.

REL. INF. X(I). The relative influence of x_i describes the fraction of the total change in Y that can be accounted for by the accompanying total change in the ith variable:

$$\text{Relative influence } x_i = \frac{b_i w_i}{w_y},$$

where b_i is the coefficient, w_i is the range of the ith independent variable, and w_y is the range of the observed dependent variable.

REQUIRED X(I) PRECISION. Since random variation in the x_i violates the standard least-squares assumption that all x_i are exactly known, it is useful to be able to put some limit on the acceptable precision of each x_i. The index of precision, h_i, specifies the number of decimal places required for each x_i so that the corresponding b_i is not biased by more than its estimated standard error. The number of digits to the right of the decimal is $h_i/2$ when h_i is positive. The number of digits to the left of the decimal is $(h_i/2) - 1$ when h_i is negative.

$$h_i = \log_{10}[N \times K(\text{Est. Var. } b_i)/(12 \text{ RMS})]$$

The h_i value is used to indicate how well the x_i data must be known to cause the resulting coefficient to be unbiased. Some data are known to fewer places than are recorded; that is, the observed number of unemployed may be recorded, but it may be impossible to measure the actual number of unemployed closer than the nearest 10,000. If the index requires the number to be known to the nearest 1000, the corresponding coefficient is biased by more than its standard error. If the index requires the number to be known to the nearest 100,000, the corresponding coefficient is acceptable.

An example of the index is shown in Figure 26, page 410. The value of -1 is interpreted to mean that the x_1 data must be known to one place left of the decimal. Since this requirement is met by the x_1 data in Table 2, page 408, we know that the bias of b_1 is less than its estimated standard error.

RESIDUAL. The residual is the difference between the observed value of the dependent variable, y_j, and its fitted value, Y_j:

$$\text{Residual} = y_j - Y_j.$$

The residuals of the fitted equation are listed twice, first in the order of the observation number, and then by rank, from the largest positive to the largest negative value.

RESIDUAL DEGREES OF FREEDOM. The residual degrees of freedom of a set of observations is the number of observations minus the number of constants estimated from the same set of observations.

RESIDUAL MEAN SQUARE. The residual mean square is calculated by dividing the residual sum of squares by the residual degrees of freedom ($N - K - 1$):

$$\text{RMS} = \frac{[rr]}{N - K - 1} = \frac{\text{RSS}}{N - K - 1}.$$

RESIDUAL ROOT MEAN SQUARE. The residual root mean square is the square root of the residual mean square:

$$\text{RRMS} = (\text{RMS})^{1/2}$$

RESIDUAL SUM OF SQUARES. The residual sum of squares is a measure of the squared scatter of the observed values around those calculated by the fitted equation:

$$\text{RSS} = \sum_{j=1}^{N}(y_j - Y_j)^2 \equiv [rr] = [yy] - [YY].$$

(See "Partitioning Sums of Squares" in Appendix 2B.)

RESIDUAL SUMS OF SQUARES AND CROSS PRODUCTS. The residual sum of squares for the first variable about its mean is the quantity

$$\sum_{j=1}^{N}(x_{1j} - \bar{x}_1)^2 \equiv [11].$$

The residual sum of cross products for the first and the second variables about their means is the quantity

$$\sum_{j=1}^{N}(x_{1j} - \bar{x}_1)(x_{2j} - \bar{x}_2) \equiv [12] \equiv [21].$$

The computer printout lists the residual sums for all independent and dependent variables in the following form:

[11]
[21]   [22]
[31]   [32]   [33]
 .   .   .
 .   .   .
 .   .   .
[(K + Q)1]   [(K + Q)2] $\cdots$ [(K + Q)(K + Q)]

where $K = $ number of independent variables, and $Q = $ number of dependent variables.

RESIDUAL VS FITTED Y.   An option of the computer program is a plot of each residual versus its corresponding fitted Y-value.

R(I)SQRD.   When one x_i is correlated with the other $x_{i'}$ $(i' \neq i)$, we can express this as a linear least-squares equation, showing X_i as a function of the other $x_{i'}$:

$$X_i = \sum_{i'} a_{i'} x_{i'}.$$

The squared multiple correlation coefficient, R_i^2, shows how well this equation fits and hence measures the degree of linear dependence of x_i on the other $x_{i'}$. It is then a measure of the nonorthogonality of the ith independent variable. The simple(!) correlation coefficient between X_{ij} and x_{ij} is R_i.

ROOT MEAN SQUARES OF VARIABLES. The computer lists $s(x_i)$ for each independent variable and $s(y_m)$ for each dependent variable:

$$s(x_i) = \left[\frac{\sum\limits_{j=1}^{N} (x_{ij} - \bar{x}_i)^2}{N - 1} \right]^{1/2}$$

$$s(y_m) = \left[\frac{\sum\limits_{j=1}^{N} (y_{mj} - \bar{y}_m)^2}{N - 1} \right]^{1/2}.$$

S. E. COEF. Under this heading is listed the estimated standard error of the coefficient, $\sqrt{\text{Var.}\ (b_i)}$. (See Appendix 2B.)

SIMPLE CORRELATION COEFFICIENT, R(I, I PRIME). The simple correlation coefficient is a measure of the linear interdependence between two variables. It is a pure number varying between -1 and $+1$; a value of 0 indicates no linear correlation. The limiting values of -1 and $+1$ indicate perfect negative or positive linear correlation. The correlation coefficient for the first and second variables is:

$$r_{12} = \frac{\sum\limits_{j=1}^{N} (x_{1j} - \bar{x}_1)(x_{2j} - \bar{x}_2)}{\sqrt{\left[\sum\limits_{j=1}^{N}(x_{1j} - \bar{x}_1)^2 \right]\left[\sum\limits_{j=1}^{N}(x_{2y} - \bar{x}_2)^2 \right]}} \equiv \frac{[12]}{\sqrt{[11][22]}}.$$

The computer printout lists the simple correlation coefficients between all independent and dependent variables in the following form:

$$
\begin{array}{llll}
r_{11} & & & \\
r_{21} & r_{22} & & \\
r_{31} & r_{32} & r_{33} & \\
\ \cdot & \ \cdot & \ \cdot & \\
\ \cdot & \ \cdot & \ \cdot & \\
\ \cdot & \ \cdot & \ \cdot & \\
r_{(K+Q)1} & r_{(K+Q)2} & \cdots & r_{(K+Q)(K+Q)}
\end{array}
$$

where $K = $ number of independent variables, and $Q = $ number of dependent variables.

STUD. RESID. The STUDENTIZED RESIDUAL is used to test if a single outlier is present:

$$\text{STUD. RESID.} = \frac{\text{residual}}{(\text{variance of residual})^{1/2}}$$

SUMS OF VARIABLES. The computer lists $\sum_{j=1}^{N} X_{ij}$ for each independent and dependent variable in the order of its appearance on data cards or as dictated by transformation instructions.

TAYLOR SERIES EXPANSIONS.

1. *Single independent variable.* Any function, $f(x)$, that possesses continuous derivatives $f^n(x)$ in the neighborhood of a can be expanded in the form:

$$f(x) = f(a) + (x - a)f^1(a) + \frac{(x - a)^2}{2!}f^2(a)$$

$$+ \cdots + \frac{(x - a)^{n-1}}{(n - 1)!}f^{n-1}(a) + R_n$$

where $R_n = (x - a)^n f^n(u)$, $\quad a < u < x$.

By analogy the following equation form can be viewed as the first three terms in a Taylor series expansion:

$$Y = b_0 + b_1(x - \bar{x}) + b_2(x - \bar{x})^2.$$

2. *Two independent variables.* Any function $f(x, z)$ that possesses continuous partial derivatives of all orders can be expanded about the point $(x = a, z = b)$ in the form:

$$f(x, z) = f(a, b) + \left(\frac{\partial f}{\partial x}\right)_{a,b} (x - a) + \left(\frac{\partial f}{\partial z}\right)_{a,b} (z - b)$$

$$+ \frac{1}{2!}\left[\left(\frac{\partial^2 f}{\partial x^2}\right)_{a,b} (x - a)^2 + 2\left(\frac{\partial^2 f}{\partial x \partial z}\right)_{a,b} (x - a)(z - b) + \left(\frac{\partial^2 f}{\partial z^2}\right)_{a,b} (z - b)^2\right]$$

$$+ \cdots.$$

The analogous second order fitting equation is:

$$Y = b_0 + b_1(x - \bar{x}) + b_2(z - \bar{z}) + b_3(x - \bar{x})^2 + b_4(x - \bar{x})(z - \bar{z})$$

$$+ b_5(z - \bar{z})^2.$$

Both the Taylor series and the analogous form may be extended to include any number of independent variables.

TOTAL SUM OF SQUARES.   The total sum of squares is a measure of the total variation in the dependent variable:

$$\text{Total sum of squares} = \sum_{j=1}^{N}(y_j - \bar{y}) .$$

T-VALUE.   The t-value is calculated as follows:

$$t_i = \frac{b_i}{s(b_i)} = \frac{\text{COEF. B(I)}}{\text{S. E. COEF.}} .$$

V(R)/RMS.   The ratio (variance of each residual)/(residual mean square) is dimensionless. The ratio (variance of the residual of *fitted Y*)/RMS is equal to $1 - v$. The ratio (variance of the residual of a *new observation*)/RMS is equal to $1 + v$.

V(Y)/RMS.   The ratio (variance of fitted Y)/(residual mean square) is dimensionless and is defined as v, the standardized distance from the centroid of data in factor space.

V(Y)/V(R).   This is the ratio of the variance of fitted Y to the variance of the residual. The ratio is equal to $v/(1 - v)$ for fitted observations and $v/(1 + v)$ for new observations.

WSSD.   The weighted squared standardized distance for point j from the centroid of all the observations:

$$\text{WSSD}_j = \sum_{i=1}^{K}\left[\frac{b_i(x_{ij} - \bar{x}_i)}{s_y}\right]^2$$

or between any two points j and j':

$$\text{WSSD}_{jj'} = \sum_{i=1}^{K}\left[\frac{b_i(x_{ij} - x_{ij'})}{s_y}\right]^2 .$$

The former is used to calculate the WSS DISTANCE values; the latter is used in the "near"-neighbor estimate of random error.

LINWOOD, A Computer Linear Least-Squares Curve-Fitting Program

Abstract

This computer program allows the user to transform data into an appropriate form, fits specified equations to the transformed data by linear least squares, and provides both statistics and plots to aid in evaluating the fit. A C_p-statistic search technique determines whether smaller sets of the variables will represent the data equally well.

The transformations which are available to the user include reciprocals, sums, differences, products, quotients, logarithms, and exponentials. Such transformations are used to convert the observed data to more convenient or more rational units, to add terms that are functions of the data variables, to stabilize variance, and to omit variables. The program allows the user to delete observations as well as to assign indicator variables to given observations.

In addition to the usual statistics, the program calculates the maximum and the minimum value of each variable, as well as its range, the relative influence of each variable, and the weighted squared standardized distance of each observation from the centroid of all observations. Near neighbors are used to estimate the standard deviation of the dependent variable. A table of component effects shows how each variable contributes to the fitted value of each observation. A table of functions related to the variance of the fitted value provides information on the influence of the location of each observation in x-space. Cross verification of coefficients can be made with a second sample of data.

Listings are made of the observed and fitted values of the dependent variable—both in the sequence in which observations were given to the computer, and in the order of the magnitude of the differences between the observed and fitted values. Plots are made to indicate (1) whether these differences are normally distributed and (2) how they are distributed over all the fitted values of the dependent variable. Plots of these differences, together with the component effects of each independent variable, can also be used (1) to choose the appropriate form of the equation, (2) to determine the distribution of the observations over the range of each independent variable, and (3) to ascertain the influence of each observation on each component of the equation.

Three versions of the program are available. One, A, will handle up to 105 variables before transformations, 80 after, and 1000 observations; the second, B, 65 variables before transformations, 40 after, and 1000 observations; the third, C, 35 variables before transformations, 10 after, and 200 observations. Multiple dependent variables are fitted one at a time, and multiple forms of specified linear equations can be fitted with one data loading.

Examples are given.

General Information

A. Program written and revised by: R. A. Brehmer, P. E. Piechocki, W. B. Traver, F. M. Jacobsen, F. M. Oliva, R. J. Toman, and F. S. Wood.

B. Instructions written and revised by: P. E. Piechocki, W. M Dudley, and F. S. Wood.

C. Form of equations:

$$Y = b_0 + b_1 x_1 + b_2 x_2 + \cdots + b_i x_i + \cdots + b_K x_K,$$

Computer Listing

where Y = fitted value of the dependent variable,   FITTED Y

$x_1, x_2, \ldots, x_i, \ldots, x_K$ = independent variables indexed by i from 1 to K,   IND. VAR. (I)

b_0 = estimated Y intercept when all $x_i = 0$,   COEF. B(I)

$b_1, b_2, \ldots, b_i, \ldots, b_K$ = estimated coefficients of the independent variables.   COEF. B(I)

Note: The dependent and independent variables may be transformations of the observed values, for example, $x_3 = (x_1)(x_2)$, $x_4 = 1/x_4'$, etc.

D. Restrictions of programs*

Number of variables	A	B	C
Maximum			
Before transformations	105	65	35
After transformations	80	40	10
Minimum			
Independent variable	1	1	1
Dependent variable	1	1	1
Number of observations			
Maximum	1000	1000	200
Minimum	One greater than the number of coefficients being estimated		

Control Cards and Data Entry

To submit a roblem, information is normally entered on two types of special forms trom which cards can be keypunched (see Figure 1 and Figure 2). The control card and data cards are always used; one or more transformation cards will be needed only if some variables of the entering data are to be transformed or omitted before calculations begin. Information cards may be used to display on the printout statements concerning the form of equation, the purpose of the run, or other pertinent information that will help identify the printout in the future. Variable name cards may be used to display the names of variables in English or alphanumeric code. Such displays help identify both the coefficients of the fitted equation and the variables in each of the candidate C_p-search equations. All possible combinations of the variables can be searched to ascertain whether a smaller set will fit the data equally well. If more than 18 variables are to be searched (262,144 equations), fractional factorial searches are required to find all candidate equations. In this case, the variables to be searched are read from factorial search cards.

To avoid obtaining garbled results, particularly on problems in which the transformation option is used, the user is advised to read all of this section before starting to fill out any forms. For routine use, the order of the cards and control card entries are summarized in the "User's Instructions" at the end of this manual.

A. Control card entry

One control card is necessary for each problem. This card contains information regarding the problem identification, the numbers of independent and

---

* These restrictions may be altered by changing the dimension statements of the computer program.

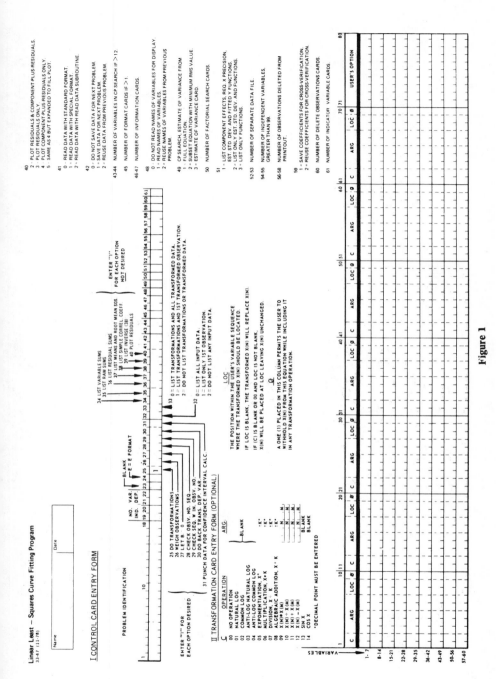

Figure 1

LINEAR LEAST - SQUARES CURVE FITTING PROGRAM

III STANDARD DATA-CARD ENTRY FORM

ENTER WEIGHTING FACTOR, IF ANY, AS LAST ENTRY
"END" CARD MUST BE LAST CARD.

Figure 2

358

dependent variables, and the operating and output options desired by the user. Column entries are as follows:

Cols. 1–18. Problem Identification. These columns may contain any grouping of letters, numbers, and blanks.

Cols. 19 and 20. Number of Independent Variables.

Cols. 21 and 22. Number of Dependent Variables.

Columns 19 and 20 contain either the number of independent variables that are read in as input or the number that will be correlated after transforming the input variables, whichever quantity is greater. Similarly, cols. 21 and 22 contain the larger number of dependent variables. For example, if the user wishes to read in five independent and four dependent variables and perform a transformation to increase the number of independent variables to seven, col. 20 must contain the number 7 and col. 22 the number 4. On the other hand, if seven independent variables and five dependent variables appear on the data cards and the user wishes to correlate only three independent and one dependent variable, while withholding four independent and four dependent variables, then the total number of each, seven and five, are the appropriate quantities to be entered for the number of independent and dependent variables. A subsequent adjustment for the number of variables to be used in correlation is performed within the program after the specified transformations and omissions are completed. Thus the program permits the user to try several different combinations of variables without repunching the data cards; only the control and transformation cards need to be changed in making successive runs.

Col. 23. Blank.

Col. 24. Method of Listing Residuals. When this column is blank, the residuals are printed with a digit field containing three places after the decimal. When the letter E is inserted, the residuals are printed in an E-fixed decimal-point format, for example, 4.5025E-03 is printed instead of 0.004.

Col. 25. Do transformations—Enter 1. When input variables are to be transformed, a one (1) must be placed in this column. This signals the program to read the appropriate number of transformation cards.

Col. 26. Weight Observations
1 When a user wishes to impute a greater degree of importance to certain observations as opposed to others, these observations may be "weighted". This column entry indicates that weighting is to be done. The actual weighting factor is entered on the data card as an entry following the last

dependent variable. When the weighting option is on, every observation must have a nonblank weighting factor, that is, a factor of 1 should be used on unweighted sets. *Note: The weighting factor is not considered a variable and must not be counted in the totals of the number of independent and dependent variables.*

2 When the weighting is by the reciprocal of the variance of the observation, a 2 is entered in this column and the variance of each observation is entered as the weighting factor.

Col. 27. Let $b_0 = 0$—Enter 1. In the fitted equation, $Y = b_0 + b_1x_1 + b_2x_2 + \cdots + b_Kx_K$, the b_0 term will be the calculated value of the intercept on the y-axis. If the user chooses to specify that b_0 be assumed equal to zero, he may indicate this by entering a 1 in col. 27. In most cases the value reported for b_0 under these circumstances will not be exactly zero, because of the nature of the calculations. When this option is used with this program, the means and the standard deviations of the independent and dependent variables require special interpretation, as $x = 0$, $y = 0$ is not included as an observed value but nonetheless $b_0 = 0$ is a restraint imposed upon the fit.

Col. 28. Check Observation Number Sequence—Enter 1. A check is made to ensure that the data card observation numbers are in ascending sequence as they are read into the program. Observations out of sequence are reported on the output listing. If one observation number out of sequence is found, the program bypasses all calculations on the problem but continues checking the observation numbers until either a total of five observations out of sequence have been found and reported, or all observations have been read in. It then proceeds to the next problem, if any.

Col. 29. Check Sequence Numbers within Observation Numbers—Enter 1. The check included here ensures that the sequencing of cards is consecutive and in ascending order, even when the number of variables included in one observation requires more than one data card per observation. The same conditions of operation apply as in checking sequencing of observation numbers (see col. 28 above).

On problems having a large number of observations, checking of observation numbers (and sequence numbers, if used) is desirable to prevent waste of machine time on data which are improperly set up. *Note pitfall:* when there is more than one data card per observation, each card in the sequence must carry the observation number as well as the sequence number. Otherwise, if checking of observation numbers is specified, a card with no observation number will be considered "out of order" and the whole problem will be rejected.

Col. 30. Do Back-Transformation of Dependent Variable—Enter 1. Dependent variables that have been transformed by a log or exponential operation may be back-transformed in the listing of the observed and fitted Y-values and their differences or residuals. Back-transformations are performed only on dependent variables and only for the log and exponential transformations. When back-transformations other than these are requested, the user is notified by a message that the back-transformation has not been effected and that the results listed are as the dependent variables were transformed.

When the fitted values have been back-transformed, the residual plots are not usable to determine the fit of the data—they too will have been transformed. Hence, the residual plots are automatically deleted.

Col. 31. Punch Data for Confidence Interval Calculations—Enter 1. Data required for confidence interval calculations (another computer program) is punched on cards when this option is chosen.

Col. 32. List Input Data.
0 (or blank) List all input data.
1 List only first observation.
2 Do not list any input data.

Col. 33. List Transformations and Transformed Data.
0 (or blank) List transformations and all transformed data.
1 List transformations and first transformed observation.
2 Do not list any transformations or transformed data.

Col. 34. Do not List the Sum of Each Variable—Enter 1.

Col. 35. Do Not List Raw Sums and Cross Products—Enter 1. When b_0 equals zero, the raw sums of squares and cross products of the variables are not listed.

Col. 36. Do Not List Residual Sums and Cross Products—Enter 1. When b_0 is a calculated value, the residual (deviations from the mean) sums of squares and cross products are not listed.

Col. 37. Do Not List the Mean and Root Mean Square of Each Variable—Enter 1.

Col. 38. Do Not List the Simple Correlation Coefficients—Enter 1.

Col. 39. Do Not List the Inverse of the Simple Correlation Coefficients—Enter 1.

Col. 40. 1 Do Not Print Plots of Residuals. The cumulative distribution plot of the residuals and the plot of the residuals versus their corresponding fitted Y-values are not printed. 2 Plot a residuals and b component and component-plus-residuals vs. each independent variable. 3 Plot a only. 4 Plot b only. 5 Plot b only but expand increments to fill each plot.

Col. 41. Method of Reading Data.

0 (or blank) Read data with standard format—cards keypunched from the standard data entry form.

1 Read data with special format. The format card containing this information immediately follows the control card. If more than 72 columns are needed, up to 8 additional format cards can be used. The total number of format cards, if more than 1, is entered in col. 45. The standard format, written in FORTRAN, is (A6, I4,I2, 10F6.3). This means there are 6 columns of alphanumeric information for identification, 4 columns of integers for observation numbers, 2 columns of integers for sequence numbers, and 10 fields of 6 columns for fixed-point numbers—each with 3 digits after the decimal point unless the decimal point is punched in some other position. The fields for identification, observation numbers, and sequence numbers are required even though they are not used. If each observation had data on 2 cards with the last variable in floating-point notation, the format card might read (A6, I2, I1, 3F4.1/12x, E 12.5). The / mark indicates that another card should be read and 12 columns omitted before reading the fourth variable. The use of such a format requires that two END cards follow the data cards (unless data are read from a separate file; see cols. 52–53).

2 Read data, using a read data subroutine. (See printout of the read data subroutine, Figure 3, for available common dimensions and requirements for returning data to main program.)

Col. 42. Save or Reuse Data.
0 (or blank) Do not save data for next problem.
1 Save data for next problem.
2 Reuse data from previous problem.

Cols. 43 and 44. Number of Variables in C_p Search if Greater than 12.

Col. 45. Number of Format Cards If Greater than 1 (8 Max.).

Cols. 46 and 47. Number of Information Cards. If desired, the user can specify the number of 72-column information cards that are to be read for

COMPUTER LINEAR LEAST-SQUARES CURVE-FITTING PROGRAM

LINEAR READ DATA SUBROUTINE

```
      SUBROUTINE REDATA
C
C ---     SUBROUTINE TO READ DATA. CAN BE REVISED TC SUIT SPECIAL NEECS.
C
C ---     CCNTRCL CARD SHOULD INDICATE THE NUMBER CF INDEPENDENT
C      VARIABLES BEING RETURNED IN ARRAY LABELLED DATA(1C6).
C      IDENTIFICATION CAN EE RETURNED USING A6 FORMAT IN IDENT.
C ---     NUMBER OF OBSERVATICNS (NOOBSV) ARE COUNTED IN CF2.
C      THE OBSERVATION NUMBER CAN EE RETURNED USING I4 FORMAT IN IOBSV.
C      THE SEQUENCE NUMBER CAN BE RETURNED USING I2 FORMAT IN ISEQ IF
C      DESIRED.
C ---     DATA CAN EE READ FROM NFILE SPECIFIED IN COLUMNS 52-53 OF
C --- CONTROL CARD.
C ---     ARRAYS THAT CAN BE USED IN REDATA TC SAVE CCMPUTER CCRE SPACE
C      ARE CCM4, LSORT AND IDZB. E.G.
C      EQUIVALENCE (COM4,NEWARY)  WHERE NEWARY IS USED IN REDATA.
C ---     KSL2 CAN BE USED AS A SWITCH.  KSL_2 IS SET = 1 IN MAIN AFTER
C      CONTROL CARD IS READ.  IF KSL2 IS SET = 2 IN REDATA AFTER READING
C      FIRST OBSERVATION. SUBSEQUENT OBSERVATICNS CAN EE READ IN
C      DIFFERENT MANNER USING KSL2 AS A SWITCH.
C
C DP  IMPLICIT REAL*8(A-H,O-Z)
C IBM DOUBLE PRECISION IDENT, IEND
C IBM DOUBLE PRECISION PR1, PR2, PR3
C IBM DOUBLE PRECISION BI, EQU
      DIMENSICN AVATR( 65), CATA( 66), FMT( 66), ICMIT( 65), ITLOC( 65)
      CCMMCN /AAAAAA/ CCM1(1000)
      CCMMCN /EEEEEE/ CCM2(186)
      COMMON /CCCCCC/ CCM3(240)
C DP  COMMON /CCCCCC/ CCM4(3500)
C SP  COMMON /CCCCCC/ CCM4(4796)
      COMMON /EEEEEE/ LSORT(1000)
      CCMMCN /FFFFFF/ VARTR( 65), ITRFM( 65)
      COMMON /HHHHHH/ BETA(41), C(40,40), Z(41,41), NNNN, MXSIZE
      CCMMON /JJJJJJ/ ICZE(1000)
      CCMMON /KKKKKK/ IG(40), PR1, PR2, PR3, K, KFM, NODEP, NOIND,
     1                NCOBSV, ICURY
C
      COMMON /MMMMMM/ IDENT, IEND, BZRO, IOBSV, ISEQ, JOBSV, KSL2,
     1                NCERR, NVP1, NOEQ
      CCMMON /CCCCCC/ KNCDEP, KNCIND, KNOVAR, KNVP1, L, NOVAR
      COMMON /GGGGGG/ KTIN, KTOU, KTPCH, KKK, KONE, KTWO, MNO, NFILE
      CCMMCN /FFFFFF/KTEIN1, KTBIN2, KTBIN3, KTBIN4, KTBINS
C ---     EQUIVALENCE STATEMENTS IN CF 4 ARE THE SAME AS IN CF 1
      EQUIVALENCE (COM1(1),AVATR(1)), (COM1(10C),ITLOC(1)),
     1            (CCM1(321),ICMIT(1)),
      EQUIVALENCE (COM2(1),DATA(1))
      EQUIVALENCE (COM3(1),FMT(1))
C
      READ(NFILE,FMT,END=99) IDENT,ICBSV,ISEQ,(DATA(JZ), JZ=1,40)
      IF(IDENT - IEND ) 4,99,4
    4 IA = IA
      RETURN
   99 IDENT = IEND
      RETURN
      END
```

Figure 3

363

display on the computer printout. This information can include the form of equation used, a statement of the purpose of the run, or other pertinent information that will help identify the printout in the future. The information cards follow the transformation cards.

Col. 48. Names of Variables.

0 (or blank) Do not read names of variables for display on printout.

1 Read names of variables from cards. The names or abbreviations of the names of the variables, independent and dependent, are keypunched in the first six positions of fields in the same locations as the variables in the transformation cards. The variables listed and the order in which they are placed depend on the transformations. If no transformations are made, all variable names are displayed. The variable name cards follow the information cards. English or alphanumeric code names aid in identifying both the coefficients of the fitted equation and the variables in each of the candidate C_p-search equations.

2 Reuse names of variables from previous problem. Again, the names displayed depend on the transformations.

Col. 49. Search for Candidate Equations via C_p. Use estimate of variance (RMS) from:

1 Full equation.

2 Subset equation with minimum RMS value.

3 Estimate of variance card with an F16.8 format.

Unless a factorial search is requested in col. 50, a directed search will be made for the combination of variables whose C_p values are less than those of the full equation. If greater than 12, the limit to the number of variables to be searched after a basic set has been chosen, is placed in cols. 43–44. If more variables remain, a factorial search will be required to look for all possible candidate equations.

Col. 50. Number of C_p Fractional Factorial Search Cards Read for Each Dependent Variable.* For a factorial search, the identification numbers of the variables to be searched are listed numerically in ascending order on a factorial search card with a 12(2X, 12) format. All variables not listed are included in the basic equation. The first search card follows the variable name cards. Additional factorial search cards may be placed after the END card(s) of the data set. Each dependent variable requires its own factorial search card(s). An entry in col. 49 is *not* needed to request a factorial search.

---

*After transformations.

Col. 51. Calculate the Component Effect of Each Variable on Each Observation, Calculate the Required Precision of the Independent Variables, Estimate the Standard Deviation of y from Near Neighbors, and List Functions Related to the Variance of Each Fitted Y.

The component effect of each variable on each observation is listed in units of Y. The variables are ordered by their relative influences in the full equation. The observations in turn are ordered by the magnitude of the effect of the most influential variable.

The required precision of each independent variable is back calculated from the perturbation or rounding index suggested by Beaton et al. Knowing the required x_i precision allows the user to determine for the particular fitted equation if the data have been recorded (1) to a sufficient number of places or (2) to more places than the source warrants (resulting in a questionable fitted equation).

The program determines the weighted squared standardized distances between pairs of observations with nearly adjacent fitted Y values. They are sorted on this distance. Starting from the smallest, the absolute differences between the residuals of each pair of near-neighbors are summed, averaged, and multiplied by 0.886 times the square root of the ratio of the total number of observations to the residual degrees of freedom of the fitted equation. These values provide an estimate of the standard deviation of y.

The functions related to the variance of fitted Y include for each observation the studentized residual, the variance of fitted Y, the variance of the residual, and the ratio of these variances.

1 All.

2 Estimate standard deviation and function table only.

3 List function table only.

Cols. 52 and 53. Number of File if Data are Read from a Separate File. End card(s) are not needed if only one set of data is on that file.

Cols. 54 and 55. Number of Independent Variables Greater than the 99 allowed in Cols. 19 and 20.

Cols. 56–58. Number of Central Observations and Residuals Not to be Printed with E Format in Col. 24. This is desirable when there are many observations or when results are printed on a time-share terminal.

Col. 59. Cross Verification of Model and Fitted Coefficients with a Second Sample of Data.

1 Write coefficients on fourth file in one problem using the first set of observations. (Transformations are not saved.)

2 Read coefficients saved on fourth file in next problem using a second set of observations. This provides a relatively rigorous test of the data, the model, and the fitted coefficients.

Col. 60. Number of Delete-Observations Cards. Enter observation number of data to be deleted in the order read by computer in columns 1–4, 11–14, ..., 61–64 right justified. The delete-observations cards follow the format cards.

Col. 61. Number of Indicator-Variable-Observations Cards. Enter observation number of data in order read by the computer and the assigned indicator variable location in columns 1–4 and 7–9, 11–14 and 17–19, ..., 61–64 and 67–69 right justified. For each observation, the program places a 1 in the assigned (blank or 0) variable location before transformations. These cards follow the delete-observations cards.

B. Standard data card entry

Without a special format card, up to ten variables can be entered on one data card. Therefore, the number of cards read per observation must equal the sum of the number of independent and dependent variables, either before or after transformations (whichever is larger), divided by 10; for example, before transformations 4 ind. + 2 dep. = 6, after transformations 7 ind. + 4 dep. = 11, 11/10 = 1.1 or 2 cards per observation are required.

The data entry card allows six card columns for each variable. If the decimal point is not entered, it is understood to lie between the third and fourth card columns of each variable field. If more than three digits are required before or after the decimal point, the point must be entered in one of the six spaces. Negative numbers are designated by a minus sign preceding the number as in standard algebraic notation. Plus signs need not be entered. Blanks are interpreted as zeros, whether or not the position is significant.

Normally the data card entry fields are interpreted by the program as numbers of the following magnitude:

-50.423	is entered as	$-$	5	0	4	2	3
121.242	is entered as	1	2	1	2	4	2
50.0	is entered as		5				
-0.4	is entered as			$-$	4		
-0.1424	is entered as	$-$	.	1	4	2	4
5000.	is entered as	5	0	0	0	.	
-121.242	is entered as	$-$	1	2	1	.	2

Thus, scaling is accomplished automatically when the decimal point is entered.

Column entries of a data card are as follows:

Columns	Entry
1–6	Card or observation identification
7–10	Observation number
11–12	Sequence number of card within observation
13–18, etc.	Data
73–80	Not used by program. May be used for identification of data sets.

The identification contained in cols. 1–6 may be any combination of letters, numbers, and blanks.

The observation number entered in cols. 7–10 and the sequence number in cols. 11 and 12 *must* be numeric. If no transformations are to be used, independent variables must precede the dependent variables as data card entries. If transformations are being done, there are different restrictions (see below).

Weighting factor, if any, must be the last data entry of an observation sequence.

As an example, suppose that data on four independent and two dependent variables are to be read in, and no transformations or omissions are desired. The independent variables must occupy the fields headed x_1, x_2, x_3, and x_4 on the data card entry sheet; the order of the four variables relative to each other is immaterial. The dependent variables must occupy the x_5 and x_6 fields; x_5 will be identified on the printout as "Dependent Variable 1," and x_6 as "Dependent Variable 2." If the weighting option is used, a weighting factor must appear on *every* data set, in the field to the right of the last dependent variable. In the above example, weighting factors, if used, would appear in field 7. They are not included in the variable count.

The number of observations is limited to 1000 (Programs A and B) or 200 (Program C). The last card of every problem must be a card with END punched in cols. 1–3 unless data are read from a separate file (see control cards, cols. 52 and 53).

C. Problems using transformation option

Transformation and data cards

In trying to discover what relationships exist in a body of data, or how to simplify relationships which are already apparent, it is frequently desirable to make simple transformations of the original variables. New forms such as sums, differences, products, ratios, logarithms, or exponentials of variables with other variables or constants can be generated by the program, thus saving time and avoiding arithmetical mistakes.

Whether the program regards a certain field of the data cards as an independent or a dependent variable can be controlled by suitably arranged control and transformation cards.

When the mass of data is large, the data need be punched into a set of data cards only once. In running a series of problems using the same data cards (but different control and transformation cards), the individual variables may be worked over and combined or omitted in any desired pattern. *All must use the same format.*

Each transformation card can carry instructions for transformations on as many as seven variables (see transformation card keypunch sheet, shown in Figure 1). The directions for one transformation occupy a field of ten digits. Thus the seven transformations occupy cols. 1–70 of a transformation card; cols. 71–80 are available for sorting or identification.

Each of the operation fields applies to a specific variable whose data entries are in corresponding locations on the data cards. Thus the first transformation card applies to variables 1–7; the second, to 8–14; etc.

Each ten-digit transformation field is divided into four subfields:

Columns of Transformation Field	Column Heading	Subfield Description
1–2	C	Transformation code
3–7	ARG.	Transformation argument
8–9	LOC.	Where to put results
10	ϕ	"Omit" instruction

For a printout of transformed data without a fit of the data, 1 is placed in col. 25, 1 in col. 27 ($b_0 = 0$), and 0 in col. 33, and a blank field is indicated for the independent variable.

The transformations included are as follows:

Code	Transformation
0, blank	None
1	Natural log
2	Common log
3	Antilog of natural log
4	Antilog of common log
5	Exponentiation by indicated entry
6	Multiplication by indicated entry
7	Division by indicated entry
8	Algebraic addition by indicated entry

9	Replacement of variable by product of variables M and N
10	Replacement of variable by ratio of variables M and N
11	Replacement of variable by sum of variables M and N
12	Replacement of variable by difference of variables M and N
13	Replacement of variable by sine of variable, in radians
14	Replacement of variable by cosine of variable, in radians
15	Indicated entry divided by variable
16	Natural log of natural log of variable
17	Common log of common log of variable
18	Common log (indicated entry added to variable)
19	Antilog of antilog of natural log
20	Antilog of antilog of common log

The program is not capable of rounding either input data or transformed data.

The transformation code digits are entered into cols. 1 and 2 of the transformation field, right-justified. Zeros may be omitted.

The transformation argument field occupying cols. 3–7 of the transformation field may contain three different kinds of entries: blanks, constants, or variable subscripts.

Transformation codes 1–4 and 13 and 14 indicate log, antilog, or trigonometric transformations on the appropriate variable. These codes require only the variable itself as an argument and thus require no entry in the argument subfield. For example, to replace x_1 by $\log_e x_1$:

$$x_1$$

C	ARG					LOC	ϕ
	1						

Transformation codes 5–8 indicate operations performed on the appropriate variables by a constant entered in the ARG subfield. Entries in the ARG subfield must be entered right-justified. Decimal values must include a decimal point. Some examples of transformation codes using constants as arguments are as follows:

(1) Replace x_1 by $x_1{}^2$. (*Note:* If the 2 had been placed in the 5th position a decimal would not have been required, as the format of the field is F5.0.)

$$x_1$$

C	ARG				LOC	ϕ
5		2	.			

(2) Replace x_3 by $x_3/1.5$.

x_3

C		ARG				LOC	ϕ
	7			1	.	5		

(3) Replace x_8 by $x_8 + (-5.2)$.

Second card:

x_8

C		ARG				LOC	ϕ
	8		−	5	.	2		

In any of the preceding cases, the transformed result can be moved to another position in the series of variables by writing the location, right justified, under LOC.

Transformation codes 9–12 indicate an operation performed using two variables to develop a transformed variable. The values inserted in the argument subfields are the subscripts of the variables to be used in the indicated operation; the value inserted in LOC tells where the result should be placed in the user's series of variables.

If one of the two variables entering the transformation is to be omitted from a particular equation, the transformation may be written in that field of the transformation card entry, with a 1 in the ϕ column. As an alternative, the transformation may be written elsewhere; the omission can be accomplished by a no-operation code in the desired field, with a 1 in the ϕ column. An omission code cannot be made to refer to any variable other than the one in whose field it appears. The omissions are not made until after all transformations have been completed; for example, an omit code in field 2 does not prevent the use of x_2 in a transformation appearing in its own or any other field.

Some examples of transformation codes using variable subscripts as arguments are as follows:

(1) Form product x_1 times x_2 and place result at x_4.

x_1

C		ARG				LOC	ϕ
	9			1		2	4	

(2) Form product x_3 times x_5 and place result at x_5; omit x_3 from the equation. (Note that space is always available for two-digit subscripts. Single-digit subscripts should be right-justified as shown.)

x_3

C	ARG	LOC	ϕ
9	3	5	5 1

(3) Form x_4/x_2 and place at x_4. Include x_2 in the equation. (Note that numerator is left entry; divisor is right entry.)

x_2

C	ARG	LOC	ϕ
1 0	4 2	4	

(4) Form sum of x_{13} and x_{14} and place at x_{15}.

C	ARG	LOC	ϕ
1 1	1 3 1 4	1 5	

(5) Form x_3-x_9 and place at x_{10}. (Note that minuend is left entry, subtrahend right entry.)

x_{10}

C	ARG	LOC	ϕ
1 2	3 9		

LOC may be left blank if the transformation is written in field 10, which is the third field of the second transformation card.

When transformations are to be done in a problem, a 1 must appear in col. 25 of the control card. This tells the program to look for transformation cards following the control card, and to figure out how many such cards to expect. It does this by adding up the total number of independent and dependent variables listed in the control card, dividing by seven operations per card, and taking the next larger whole number. For example, if there are four independent and two dependent variables, there will be one transformation card; but if there are six independent and two dependent variables, there will be two transformation cards. If the number of transformation cards actually placed in the deck is either greater or less than the expected number, the calculated results will be garbled, because the program will be trying to read transformation cards as data cards, or vice versa. Thus, in the second example above, even if only the first three independent variables are to be transformed, two transformation cards must be put in; the first one carries the three transformations, and the second one is blank. When a certain field on a transformation card is blank, the program assumes that no transformation of the corresponding variable is desired.

In problems large enough to approach the capacity of the program (105 input variables, 80 to be used in the equation), the above arrangement limits the number of transformations to an average of about one per variable. In smaller problems, transformations requiring several steps can be done by having the results of intermediate steps listed as "variables" and later omitting them from the equation.

When the transformation option is used, none of the input tells the program directly how many independent or dependent variables are to be included in the calculations. The program develops this information by starting with the number of independent or dependent variables listed in the control card, and subtracting the specified omissions. The various types of situation that may occur are best understood by means of examples.

1. The user has placed data for six variables in fields 1–6 of a set of data cards. He wants to use 1, 2, and 4 as independent variables, and 5 as a single dependent variable, all without any transformations. In the control card he lists 4 independent variables and 1 dependent variable, and enters a 1 in col. 25, calling for transformations. In the transformation card entry, he orders omission of x_3. In the printed output the program reports under "Data Transformations" that "the equation has 3 independent variables and 1 dependent variable"; the transformation instructions will be listed as written, followed by a repetition of the data for x_1, x_2, x_4, and x_5 in the first four columns, x_3 and x_6 having been omitted.

2. With the same raw data as before, the user now wishes to consider x_2 and x_5 as independent variables, and x_3 as the dependent variable. He can use the old data cards by listing "6 independent, 1 dependent" on the control card, and entering the following transformations:

> x_1 Mark for "omit"
> x_2 No operation (leave blank)
> x_3 Move to location 7 and "omit" in 3
> x_4 Mark for "omit"
> x_5 Leave blank
> x_6 Mark for "omit"
> x_7 Leave blank

The program, having been told that data fields 1–6 are now reserved for six independent variables, and field 7 for a dependent variable, and that x_1, x_3, x_4, and x_6 are to be omitted, will report the transformed equation as "2 independent and 1 dependent" and proceed, using field 7 (which now contains the data from field 3) as the dependent variable, and cols. 2 and 5 as the independents. The user should note that, when a transformation in field 3 calls for "move to field 7 and omit," it is x_3 that will be omitted, not x_7, which can be omitted only by an omit instruction in field 7.

3. The user has data for five independent variables $x_1 \cdots x_5$ and a dependent variable x_6. He wants to perform a fit of x_6 against $x_1, x_2, x_3 - x_4$, and $x_5 - 100$. On the data sheet, he enters $x_1 \cdots x_6$ in the first six fields. The control card will list "5 independents, 1 dependent." Transformations will be as follows:

x_1 No operation (leave blank)

x_2 No operation

x_3 Take $x_3 - x_4$

x_4 Mark for "omit"

x_5 Take $x_5 - 100$, leave "location" blank

x_6 No operation

The program reports the transformed equation as "4 independent, 1 dependent." The transformed data $x_1, x_2, x_3 - x_4, x_5 - 100, x_6$ will appear in the first five columns.

4. The user has data for $x_1 \cdots x_5$. He wants to perform a fit of x_4/x_5 as dependent variable against $x_3(x_1 - x_2)$ as independent variable. On the data sheet, he enters $x_1 \cdots x_5$ in the first five fields. The control card will list "6 independent, 1 dependent." Transformations will be as follows:

x_1 Take $x_1 - x_2$, place in LOC 6, omit

x_2 Mark for "omit"

x_3 Take x_3 times x_6, place in LOC 6, omit

x_4 Take x_4/x_5, LOC 7, omit

x_5 Mark for "omit"

x_6 No operation

x_7 No operation

The program reports that the fitted equation has one independent and one dependent variable.

The examples illustrate that, when a control card lists seven independent and three dependent variables, the program assumes that:

1. The combination of raw data and transformations is such that, after the transformations are completed, there will be data in the first ten fields, with independent variables in fields 1–7, and dependents in 8–10.

2. The number of independent variables to be used is 7 minus the total omissions listed for fields 1–7.

3. The number of dependent variables is 3 minus the omissions listed for fields 8–10.

Most problems can be handled in more than one way, since a transformed variable can either replace one of the variables from which it came or, if space is available, go to a new field. When a variable is to be omitted, its position in the series identifies it as either independent or dependent; the program adjusts the makeup of the equation accordingly.

Order of cards

First Problem

1.	Control card	(0 or blank in col. 42)
2.	Format card(s)	(If special format is used, 1 col. 41)
3.	Delete observations card(s)	(If desired, number in col. 60)
4.	Indicator variable card(s)	(If desired, number in col. 61)
5.	Transformation card(s)	(If transformations are used, 1 col. 25)
6.	Information card(s)	(If desired, number in cols. 46 and 47)
7.	Variable name card(s)	(If desired, 1 col. 48)
8.	Factorial search card	(If desired, number in col. 50)
9.	Data cards	
10.	END card(s)	(The word END in cols. 1–3 of data card identification. The number of END cards *must* equal the number of data cards per observation.)
11.	Estimate of variance card	(If desired, 3 in col. 49)
12.	Additional factorial search card(s)	(If desired, number in col. 50)

Second Problem (Different Data)

1.	Control card	(1 in col. 42)
2.	Format card(s)	(If special format is used, 1 col. 41)
3.	Delete observations card(s)	(If desired, number in col. 60)
4.	Indicator variable card(s)	(If desired, number in col. 61)
5.	Transformation card(s)	(If transformations are used, 1 col. 25)
6.	Information card(s)	(If desired, number in cols. 46 and 47)

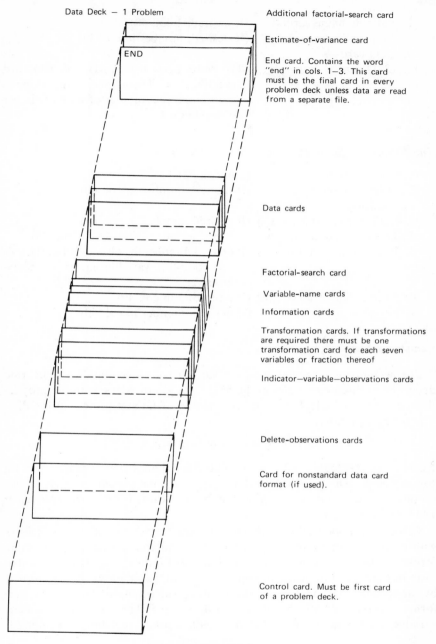

Data Deck — 1 Problem

Additional factorial-search card

Estimate-of-variance card

END

End card. Contains the word "end" in cols. 1—3. This card must be the final card in every problem deck unless data are read from a separate file.

Data cards

Factorial-search card

Variable-name cards

Information cards

Transformation cards. If transformations are required there must be one transformation card for each seven variables or fraction thereof

Indicator—variable—observations cards

Delete-observations cards

Card for nonstandard data card format (if used).

Control card. Must be first card of a problem deck.

Figure 4 Order of cards

375

7. Variable name card(s)   (If desired, 1 col. 48)
8. Factorial search card   (If desired, number in col. 50)
9. Data cards
10. END card(s)   (The word END in cols. 1–3 of data card
                              identification. The number of END cards
                              must equal the number of data cards per
                              observation.)

Third Problem (Same Data as in Second Problem)

1. Control card   (2 in col. 42)
2. Delete observation   (If desired, number in col. 60)
   card(s)
3. Indicator variable card(s) (If desired, number in col. 61)
4. Transformation card(s)   (If transformations are used, 1 col. 25)
5. Information card(s)   (If desired, number in cols. 46 and 47) (Enter
                              2 in col. 48 if variable names are to be
                              reused.)
6. Estimate of variance card (If desired, 3 in col. 49)
7. Factorial search card   (If desired, number in col. 50)

Sample problem*

The accompanying table gives, for various countries, data on death rate
due to heart disease in males in the 55–59 age group, along with the propor-
tionate number of telephones and the calories of fat and protein in the diet.
The fitting equation is

$$Y = b_0 + b_1 x_1 + b_2 x_2 + b_3 x_3,$$

where $y = 100(\log y' - 2)$, $x_1 = 1000$ (telephones per head), $x_2 = $ fat
calories as percentage of total calories, $x_3 = $ animal protein calories as percentage of total calories, and $y' = $ number of deaths from heart disease per
100,000 for males in 55–59 age group.

This example shows the mechanics of cross verification of the coefficients
with a second sample of data. The control and transformation card entry
forms for the *first* and *next* pass are shown in Figure 5; the data card entry
form in Figure 6. Following Brownlee, the death rate data were transformed
by the formula $100(\log y' - 2)$, before fitting the equation. The three transformations required are shown in fields 4, 5, and 6. The resulting dependent
variable is left in field 6, while the intermediate results in 4 and 5 are omitted.

* From K. A. Brownlee (pages 462–464).

The independent variables x_1, x_2, and x_3 did not require any transformations or omissions.

	x_1	x_2	x_3	y'
Australia	124	33	8	646
Austria	49	31	6	355
Canada	181	38	8	631
Ceylon	4	17	2	174
Chile	22	20	4	603
Denmark	152	39	6	331
Finland	75	30	7	759
France	54	29	7	282
Germany	43	35	6	316
Ireland	41	31	5	490
Israel	17	23	4	457
Italy	22	21	3	282
Japan	16	8	3	174
Mexico	10	23	3	269
Netherlands	63	37	6	240
New Zealand	170	40	8	525
Norway	125	38	6	257
Portugal	15	25	4	240
Sweden	221	39	7	331
Switzerland	171	33	7	331
United Kingdom	97	38	6	457
United States	254	39	8	776

To show how the indicator-variable-observations card is used, a fictitious observation, entitled the U.N., has been added to the data set. It contains the average values for x_1, x_2, and x_3 and, since members of the U.N. are living in the United States, it was erroneously given the same value for y as the United States. To see if the response of this observation differs significantly from the remainder of the data (is an outlier), an indicator variable was assigned to the observation. To allow room for this new variable, an extra (blank) field was read by the format and the name UN IV given to the fourth variable. A 1 is placed in col. 61 of the control card (the number of indicator-variable-observations cards); a 23 (the observation number of the UN data) is placed in cols. 3 and 4, and a 4 (the variable number) in col. 9 of the indicator-variable-observations card. The numbers of independent and dependent variables on the control card were listed as 4 and 3.

To show the procedure of cross-verification, all of the data except the observation from the United States are fitted in the *First Pass*. This observation is deleted by a 1 in col. 60 (the number of delete-observations cards) and

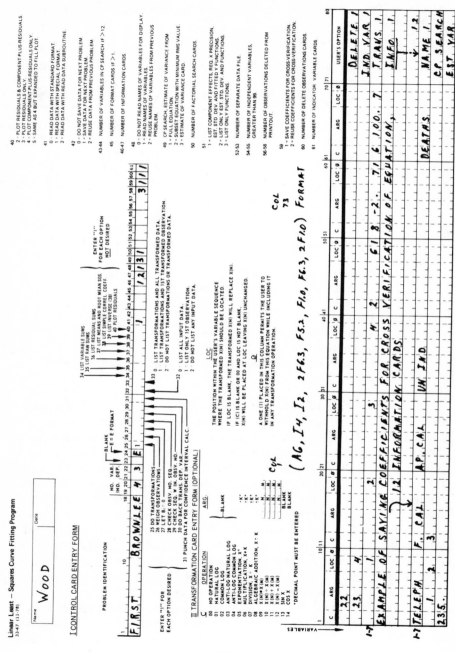

Linear Least — Squares Curve Fitting Program
33-47 (11-78)

378

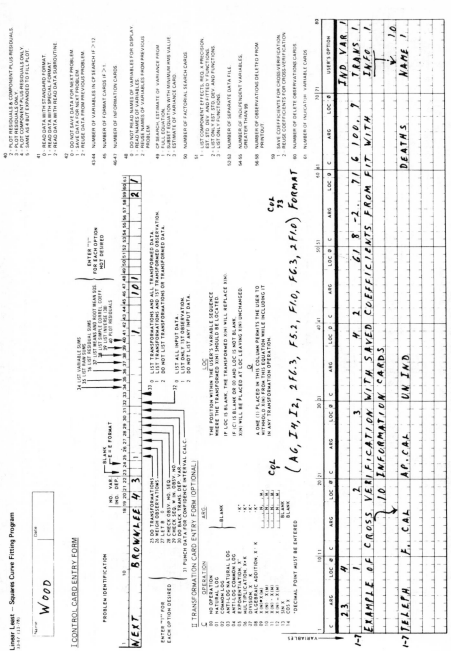

Figure 5

379

LINEAR LEAST – SQUARES CURVE-FITTING PROGRAM

III STANDARD DATA-CARD ENTRY FORM

ENTER WEIGHTING FACTOR, IF ANY, AS LAST ENTRY.
"END" CARD MUST BE LAST CARD.

IDENT.	OBSV. NO.	SEQ NO.	x1–11–21	x2–12–22	x3–13–23	x4–14–24	x5–15–25	x6–16–26	x7–17–27	x8–18–28	x9–19–29	x10–20–30	USER'S OPTION
1 2 3 4 5 6	7 8 9 0	1 2	3 4 5 6 7 8	9 0 1 2 3 4	5 6 7 8 9 0	1 2 3 4 5 6	7 8 9 0 1 2	3 4 5 6 7 8	9 0 1 2 3 4	5 6 7 8 9 0	1 2 3 4 5 6	7 8 9 0 1 2	3 4 5 6 7 8 9 0
AUSTRL	1		124	33	8	646							
ASTRIA	2		49	31	6	355							
CANADA	3		181	38	8	631							
CEYLON	4		4	17	2	174							
CHILE	5		22	20	4	603							
DENMK	6		152	39	6	331							
FINLND	7		75	30	7	759							
FRANCE	8		54	29	7	282							
GMANY	9		43	35	6	316							
IRELND	10		41	31	5	490							
ISRAEL	11		17	23	4	457							
ITALY	12		22	21	3	282							
JAPAN	13		16	8	3	174							
MEXICO	14		10	23	3	269							
NTHLDS	15		63	37	6	240							
NWZEAL	16		170	40	8	525							
NORWAY	17		125	38	6	257							
PORTGL	18		15	25	4	240							
SWEDEN	19		221	39	7	331							
SWITZD	20		171	33	7	331							
UNKING	21		97	38	6	457							
U.S.A.	22		254	39	8	776							
U.N.	23		87	30	5	776							
END													
1 2 3 4 5 6	7 8 9 0	0 1	2 3 4 5 6 7 8 9	0 1 2 3 4	5 6 7 8 9	0 1 2 3 4 5 6 7 8 9	0 1 2 3 4 5 6 7 8 9		0 1 2 3 4 5 6 7 8 9		0 1 2 3 4 5 6 7 8 9	0 1 2 3 4 5 6 7 8 9	0
		1		2		3	4		5		6	7	8

FORM B-56 REV. 3-69

Figure 6

a 22 (the observation number of U.S. data) in cols. 3 and 4 on the delete-observations card. A 1 in col. 59 of the control card indicates that the coefficients are to be saved. In the *Next Pass* all the data are used and new coefficients calculated. With a 2 in col. 59, the coefficients from the first pass are read and displayed. The residuals are calculated using the fitted values from the retained coefficients. Component and component-plus-residual plots (not shown) are often useful to see the relationship between the residuals from the second set (or sample) of data and the components from the first set of data. If outliers occur, finding their cause may lead to recognizing new variables not observed in the first set of data.

Computer printouts from this problem are shown in Figures 7–12.

Example of weighted observations*

In this relatively simple problem, four samples have been titrated. Single determinations have been obtained on the first, third, and fourth sample; four determinations have been averaged to give a value for the second sample. The measurements are as follows:

Sample and Observation Number	x	y
1	10	33.8
2	20	62.2
3	30	92.0
4	40	122.4

The fitted equation is $Y = b_0 + bx$. We wish to determine the values of the coefficients of the equation, as well as a measure of the variance of titration error.

The control card entry form, with a 1 in col. 26 for weighted observations, is shown in Figure 13. In Figure 14 the samples have been identified in cols. 1–6 of each data card, the observation numbers in cols. 7–10, the x-values entered in cols. 13–18, and the y-values in cols. 19–24 so that a standard format could be used (this is convenient but not necessary). The weight to be given to each y-observation is entered in the next field, cols. 25–30. In this case, observations 1, 3, and 4 are given a weight of 1.0 to represent single observations. Observation 2 is given a weight of 4.0 for the four determinations that were averaged. The weighting factor switch on the control card

* From John Mandel, Section 7.6.   (*continued on page 396*)

LINEAR LEAST-SQUARES CURVE-FITTING PROGRAM

1979 VERSION OF THE LINWOOD 40 VARIABLE, 1000 OBSERVATION PROGRAM

REFER TO FITTING EQUATIONS TO DATA BY DANIEL AND WOOD, SECOND EDITION, WILEY PUBLISHER, FOR GLOSSARY OF TERMS, USER'S MANUAL, DETAILS OF CALCULATIONS AND INTERPRETATION OF RESULTS.

FIRST   BROWNLEE

CONTROL CARD INFORMATION

COL.	INPUT	MAX.	ITEM
1-18			(NOTE: BLANK ON CARD = 0)
			PROBLEM IDENTIFICATION.
19-20	4		NUMBER OF INDEPENDENT VARIABLES READ IN
			(ALLOW SPACE FOR TRANSFORMATIONS).
21-22	3	65	NUMBER OF DEPENDENT VARIABLES READ IN.
24	E		TOTAL NUMBER OF VARIABLES (BEFORE TRANSFORMATIONS, IF ANY).
			E CAUSES Y, FITTED Y AND RESIDUALS TO BE
			LISTED WITH AN E RATHER THAN AN F FORMAT.
25	1		DO TRANSFORMATIONS.
26	0		WEIGHT OBSERVATIONS. WEIGHT ENTERED IN LAST POSITION.
			2 WEIGHT BY 1/VAR. OF OBSERVATION. VARIANCE ENTERED.
27	0		LET B(0) = ZERO. OTHERWISE B(0) IS CALCULATED VALUE.
28	0		CHECK OBSERVATION NUMBER SEQUENCE.
29	0		CHECK SEQUENCE WITHIN OBSERVATION NUMBER.
30	0		DO BACK TRANSFORMATION OF DEPENDENT VARIABLE.
31	0		PUNCH DATA FOR CONFIDENCE INTERVAL PROGRAM.
32	0		LIST ALL INPUT DATA.
			0 LIST ONLY 1ST OBSERVATION.
			2 DO NOT LIST ANY INPUT DATA.
33	0		0 LIST TRANSFORMATIONS AND ALL TRANSFORMED DATA.
			1 LIST TRANSFORMATIONS AND 1ST TRANSFORMED OBSERVATION.
			2 DO NOT LIST TRANSFORMATIONS OR TRANSFORMED DATA.
34	0		1 DO NOT LIST SUMS OF VARIABLES.
35	0		1 DO NOT LIST RAW SUMS AND CROSS PRODUCTS WHEN B(0)=0.
36	0		1 DO NOT LIST RESIDUAL SUMS AND CROSS PRODUCTS.
37	0		1 DO NOT LIST MEANS AND ROOT MEAN SQUARES OF VARIABLES.
38	0		1 DO NOT LIST SIMPLE CORRELATION COEFFICIENTS.
39	0		1 DO NOT LIST INVERSE MATRIX.
40	0		1 DO NOT PRINT PLOTS OF RESIDUALS VS. FITTED Y.
			2 PLOT (A) RESIDUALS AND (B) COMPONENT AND COMPONENT-
			PLUS-RESIDUALS VS. EACH INDEPENDENT VARIABLE.
			3 PLOT (A) RESIDUALS ONLY.
			4 PLOT (B) COMPONENT AND COMPONENT-PLUS-RESIDUALS ONLY.
			5 PLOT (B) BUT EXPAND SCALE TO FILL EACH PLOT.
41	1		0 READ DATA WITH STANDARD FORMAT (A6, I4, I2, 10F6.3).
			1 READ DATA WITH FORMAT TO BE READ. 1 CARD (72COL) ASSUMED.
			2 READ DATA WITH READ DATA (REDATA) SUBROUTINE.
42	0		0 DO NOT SAVE DATA FOR NEXT PROBLEM.
			1 SAVE DATA FOR NEXT PROBLEM.
			2 REUSE DATA FROM PREVIOUS PROBLEM.
43-44	0	18	NUMBER OF VARIABLES IN CP SEARCH IF GREATER THAN 12.
45	0	8	NUMBER OF FORMAT CARDS IF GREATER THAN 1.
46-47	12	12	NUMBER OF INFORMATION CARDS TO BE READ FOR DISPLAY ON
			PRINTOUT IF DESIRED, 72 CPL EACH.
48	1		1 READ NAMES OF VARIABLES FOR DISPLAY ON PRINTOUT.
			NAMES IN SAME POSITIONS AS VARIABLES ON TRANSFORMATION
			CARDS - CCLS. 1-6, 11-16, - - - , 61-66.
			NAMES ARE NOT MOVED. ONLY LISTED CR DELETED.
			2 REUSE NAMES OF VARIABLES FROM PREVIOUS PROBLEM.
49	3		SEARCH FOR CANDIDATE EQUATIONS VIA CP. USE ESTIMATE
			OF VARIANCE (RMS) FROM:
			1 FULL EQUATION.
			2 SUBSET EQUATION WITH MINIMUM RMS VALUE. CR
			3 ESTIMATE OF VARIANCE CARD. F16.8 FORMAT.

ORDER OF CARDS

FIRST PROBLEM
1 CONTROL CARD.
2 FORMAT CARD(S), IF ANY.
3 DELETE OBSERVATIONS CARD(S), IF ANY.
4 INDICATOR-VARIABLE-OBSERVATIONS CARD(S), IF ANY.
5 TRANSFORMATION CARD(S), IF ANY.
6 INFORMATION CARD(S) FOR PRINTOUT, IF ANY.
7 NAMES OF VARIABLES CARD(S) FOR PRINTOUT, IF ANY.
8 CP FACTORIAL SEARCH CARD, IF ANY.
9 DATA CARDS (IF NOT READ FROM A FILE).
10 END CARD (END IN COLUMNS 1 - 3 OF IDENTIFICATION
   FIELD). THE NUMBER OF END CARDS MUST EQUAL THE
   NUMBER OF CARDS PER OBSERVATION. (END CARDS
   ARE NOT NEEDED IF DATA ARE READ FROM A FILE).
11 ESTIMATE OF VARIANCE CARD, IF ANY.
12 ADDITIONAL CP FACTORIAL SEARCH AND ESTIMATE OF
   VARIANCE CARDS, IF ANY.

SECOND PROBLEM
IF DATA ARE REUSED FROM FIRST PROBLEM,
   DELETE THE FORMAT, DATA AND END CARDS.
IF NAMES ARE REUSED,
   DELETE THE NAME CARDS.
IF DIFFERENT INPUT DATA AND NAMES,
   REPEAT 1 - 12 ABOVE.

50   1   NUMBER OF CP FACTORIAL SEARCH CARDS TO BE READ PER
              DEPENDENT VARIABLE. KEYPUNCH IN ASCENDING ORDER THE
              IDENTIFING NUMBER OF THE VARIABLES (AFTER TRANSFOR-
              MATIONS) TO BE USED IN THAT SEARCH. 18(2X.I2) FORMAT.
              ALL OTHER VARIABLES WILL BE PLACED IN BASIC EQUATION.

51   0   1 LIST COMPONENT-EFFECT TABLE, REQUIRED X(I) PRECISION.
              ESTIMATE OF STANDARD DEVIATION FROM NEAR NEIGHBORS,
              AND FUNCTIONS OF FITTED Y.
           2 LIST ONLY ESTIMATE OF STD. DEVIATION AND FUNCTIONS.
           3 LIST ONLY FUNCTIONS OF FITTED Y.

52-53 0   NUMBER OF FILE IF DATA ARE READ FROM SEPARATE FILE.

54-55 0   NO END CARD IF ONLY ONE SET OF DATA IS ON EACH FILE.
           NUMBER OF INDEPENDENT VARIABLES GREATER THAN THE 99
           ALLOWED IN COLS 19-20.

56-58 3   TO REDUCE PRINTOUT WITH MANY OBSERVATIONS,
           NUMBER OF CENTRAL OBSERVATIONS AND RESIDUALS NOT
           TO BE PRINTED. THIS REQUIRES AN E IN COL 24.

59   1   CROSS VERIFICATION OF MODEL AND COEFFICIENTS WITH A
           SECOND SAMPLE OF DATA.
           1 WRITE B(0) AND B(I)S ON FILE IN ONE PROBLEM
             (TRANSFORMATIONS ARE NOT SAVED).
           2 READ B(0) AND B(I)S FROM FILE IN NEXT PROBLEM
             WITH SECOND SET OF OBSERVATIONS.

60   1   5 NUMBER OF DELETE-OBSERVATIONS CARDS. KEYPUNCH
           OBSERVATION NUMBER OF DATA TO BE DELETED IN ORDER READ
           BY COMPUTER USING 7(I4.6X) FORMAT.

61   1   5 NUMBER OF INDICATOR-VARIABLE-OBSERVATIONS CARDS.
           USING A 7(I4.2X.I3.1X) FORMAT.KEYPUNCH OBSERVATION
           NUMBER (IN ORDER READ) AND VARIABLE NUMBER (BEFORE
           TRANSFORMATIONS) INTO WHICH A 1 IS TO BE INSERTED.

PROBLEM HAS ONE EQUATION

DATA READ WITH SPECIAL FORMAT

FORMAT CARD 1 (A6. I4. I2. 2F6.3. F5.2. F1.0. F6.3. 2F1.0)

BZERO = CALCULATED VALUE

OBSERVATIONS TO BE DELETED
    22

OBSERVATIONS AND INDICATOR VARIABLES INTO WHICH A 1 IS TO BE INSERTED
    23   4

INFORMATION CARDS
    EXAMPLE OF SAVING COEFFICIENTS FOR CROSS VERIFICATION OF EQUATION.
    REDUCING PRINTOUT (CENTRAL RESIDUALS AND OBSERVATIONS 10-12), E FORMAT.
    DELETED OBSERVATION. INDICATOR-VARIABLE-OBSERVATION. AND CP FACTORIAL
    SEARCH WITH GIVEN ESTIMATE OF VARIANCE.
        OBSERVATION 22 DELETED FROM FIRST SET OF DATA. INCLUDED IN NEXT SET.
    $Y = B(0) + B(1)X1 + B(2)X2 + B(3)X3 + B(4)X4$
    $Y = 100($ LOG(NUMBER OF DEATHS FROM HEART DISEASE PER 100,000 FOR MALES
            IN THE 55 TO 59 AGE GROUP) $- 2)$
    $X1 = 1000($TELEPHONES PER PERSON$)$
    $X2 =$ FAT CALORIES AS PER CENT OF TOTAL CALORIES
    $X3 =$ ANIMAL PROTEIN CALORIES AS PER CENT TOTAL CALORIES
    $X4 =$ INDICATOR VARIABLE, 1 FOR OBSERVATION 23 (U.S.A.), 0 OTHERWISE.

DATA TRANSFORMATIONS

POSITION	CODE	OPERATION	CONSTANT	LOCATION	OMIT	VARIABLE	NAME
1		NONE		1	0	1	TELEPH
2		NONE		2	0	2	F. CAL
3		NONE		3	0	3	AP.CAL
4		NONE		4	0	4	UN IND
5	2	COMMON LOG		6	1		
6	8	ADD CONSTANT	-2.	7	1		
7	6	MULTIPLY BY CONSTANT	100.	7	0	5	DEATHS

VARIABLES ON FIRST CP FACTORIAL SEARCH CARD
    1   2   3

Figure 7

LINEAR LEAST-SQUARES CURVE-FITTING PROGRAM

FIRST   BROWNLEE

DATA INPUT   4 INDEPENDENT VARIABLES   3 DEPENDENT VARIABLE(S)

OBSV.	SEQ.	1-11-21	2-12-22	3-13-23	4-14-24	5-15-25	6-16-26	7-17-27	8-18-28	9-19-29	10-20-30
1	0	124.000	33.000	8.000	0.0	646.000	0.0	0.0			
2	0	49.000	31.000	6.000	0.0	355.000	0.0	0.0			
3	0	181.000	38.000	8.000	0.0	631.000	0.0	0.0			
4	0	4.000	17.000	2.000	0.0	174.000	0.0	0.0			
5	0	22.000	20.000	4.000	0.0	603.000	0.0	0.0			
6	0	152.000	39.000	6.000	0.0	331.000	0.0	0.0			
7	0	75.000	30.000	7.000	0.0	759.000	0.0	0.0			
8	0	54.000	29.000	7.000	0.0	282.000	0.0	0.0			
9	0	43.000	35.000	6.000	0.0	316.000	0.0	0.0			
10	0	41.000	31.000	5.000	0.0	490.000	0.0	0.0			
11	0	17.000	23.000	4.000	0.0	457.000	0.0	0.0			
12	0	22.000	21.000	3.000	0.0	282.000	0.0	0.0			
13	0	16.000	8.000	3.000	0.0	174.000	0.0	0.0			
14	0	10.000	23.000	3.000	0.0	269.000	0.0	0.0			
15	0	63.000	37.000	6.000	0.0	240.000	0.0	0.0			
16	0	170.000	40.000	8.000	0.0	525.000	0.0	0.0			
17	0	125.000	38.000	6.000	0.0	257.000	0.0	0.0			
18	0	15.000	25.000	4.000	0.0	240.000	0.0	0.0			
19	0	221.000	39.000	7.000	0.0	331.000	0.0	0.0			
20	0	171.000	33.000	7.000	0.0	331.000	0.0	0.0			
21	0	97.000	38.000	6.000	0.0	457.000	0.0	0.0			
23	0	87.000	30.000	5.000	1.000	776.000	0.0	0.0			

A VALUE OF 1.0 HAS BEEN PLACED IN VARIABLE 4 OF OBSERVATION 23

DATA TRANSFORMATIONS

POSITION	CODE	OPERATION	CONSTANT	LOCATION	OMIT	VARIABLE	NAME
1		NONE		1	0	1	TELEPH
2		NONE		2	0	2	F. CAL
3		NONE		3	0	3	AP.CAL
4		NONE		4	0	4	UN IND
5	2	COMMON LOG		6	1		
6	8	ADD CONSTANT	-2.	7	1		
7	6	MULTIPLY BY CONSTANT	130.	5		5	DEATHS

DATA AFTER TRANSFORMATIONS   THE FITTED EQUATION HAS 4 INDEPENDENT VARIABLES, 1 DEPENDENT VARIABLE(S)

```
OBSV.      TELEPH       F. CAL       AP.CAL       UN IND       DEATHS
           1-11-21      2-12-22      3-13-23      4-14-24      5-15-25      6-16-26   7-17-27   8-18-28   9-19-29   10-20-30
  1    1.24000D+02  3.30000D+01  8.00000D+00  0.0          8.10233D+01
  2    4.90000D+01  3.10000D+01  6.00000D+00  0.0          5.50280D+01
  3    1.81000D+02  3.80000D+01  8.00000D+00  0.0          8.00029D+01
  4    4.00000D+00  1.70000D+01  2.00000D+00  0.0          2.40549D+01
  5    2.20000D+01  2.00000D+01  4.00000D+00  0.0          7.80317D+01
  6    1.52000D+02  3.90000D+01  6.00000D+00  0.0          5.19828D+01
  7    7.50000D+01  3.00000D+01  7.00000D+00  0.0          8.80420D+01
  8    5.40000D+01  2.90000D+01  7.00000D+00  0.0          4.50249D+01
  9    4.30000D+01  3.50000D+01  6.00000D+00  0.0          4.99687D+01
 10    4.10030D+01  3.10000D+01  5.00000D+00  0.0          6.90196D+01
 11    1.70000D+01  2.30000D+01  4.00000D+00  0.0          6.59916D+01
 12    2.20000D+01  2.10000D+01  3.00000D+00  0.0          4.50249D+01
 13    1.60000D+01  8.00000D+00  3.00000D+00  0.0          2.40549D+01
 14    1.00030D+01  2.30000D+01  3.00000D+00  0.0          4.29752D+01
 15    6.30000D+01  3.70000D+01  6.00000D+00  0.0          3.80211D+01
 16    1.70000D+02  4.00000D+01  8.00000D+00  0.0          7.20159D+01
 17    1.25000D+02  3.80000D+01  6.00000D+00  0.0          4.09933D+01
 18    1.50000D+01  2.50000D+01  4.00000D+00  0.0          3.80211D+01
 19    2.21000D+02  3.90000D+01  7.00000D+00  0.0          5.19828D+01
 20    1.71000D+02  3.30000D+01  7.00000D+00  0.0          5.19828D+01
 21    9.70000D+01  3.80000D+01  6.00000D+00  0.0          6.59916D+01
 23    8.70000D+01  3.00000D+01  5.00000D+00  1.00000D+00  6.89662D+01


SUMS OF VARIABLES
   1.75900D+03  6.58000D+02  1.21000D+02  1.00000D+00  1.24820D+03

RESIDUAL SUMS OF SQUARES + CROSS PRODUCTS
   1   9.04810D+04
   2   8.93191D+03  1.50582D+03
   3   1.95450D+03  2.61000D+02  6.75000D+01
   4   7.04545D+00  9.09091D-02 -5.00000D-01  9.54545D-01
   5   8.93015D+03  1.27891D+03  3.72718D+02  3.22499D+01  7.82405D+03

OBSERVATIONS DELETED
        22

MEANS OF VARIABLES
   7.99545D+01  2.99091D+01  5.50000D+00  4.54545D-02  5.67362D+01

ROOT MEAN SQUARES OF VARIABLES
   6.56401D+01  8.46792D+00  1.79284D+00  2.13201D-01  1.93022D+01

SIMPLE CORRELATION COEFFICIENTS, R(I,I PRIME)
   1   1.000
   2   0.765   1.000
   3   0.791   0.819   1.000
   4   0.024   0.202  -0.062   1.000
   5   0.336   0.373   0.513   0.373   1.000

INVERSE, C(I,I PRIME)
   1   3.035
   2  -1.057   3.427
   3  -1.546  -1.976   3.858
   4  -0.166  -0.106   0.282   1.022
```

Figure 8

385

LINEAR LEAST-SQUARES CURVE-FITTING PROGRAM

FIRST   BROWNLEE   DEP VAR 1: DEATHS   MIN Y = 2.405D+01   MAX Y = 8.899D+01   RANGE Y = 6.493D+01

EXAMPLE OF SAVING COEFFICIENTS FOR CROSS VERIFICATION OF EQUATION.
REDUCING PRINTOUT (CENTRAL RESIDUALS AND OBSERVATIONS 10-12), E FORMAT.
DELETED OBSERVATION, INDICATOR-VARIABLE-OBSERVATION, AND CP FACTORIAL
SEARCH WITH GIVEN ESTIMATE OF VARIANCE.
OBSERVATION 22 DELETED FROM FIRST SET OF DATA, INCLUDED IN NEXT SET.
Y = B(0) + B(1)X1 + E(2)X2 + B(3)X3 + B(4)X4
Y = 100(LOG(NUMBER OF DEATHS FROM HEART DISEASE PER 100,000 FOR MALES
      IN THE 55 TO 59 AGE GROUP) - 2)

X1 = 100(TELEPHONES PER PERSON)
X2 = FAT CALORIES AS PER CENT OF TOTAL CALORIES
X3 = ANIMAL PROTEIN CALORIES AS PER CENT TOTAL CALORIES
X4 = INDICATOR VARIABLE, 1 FOR OBSERVATION 23 (U.N.), 0 OTHERWISE.
OBSERVATIONS DELETED: 22

IND.VAR(I)	NAME	COEF.B(I)	S.E. COFF.	T-VALUE	R(I)SQRD	MAX X(I)	MIN X(I)	RANGE X(I)	REL.INF.X(I)
0		2.0254D+01							
1	TELEPH	-6.7601D-02	9.13D-02	0.7	0.6705	2.210D+02	4.000D+00	2.170D+02	0.23
2	F.CAL	-2.9854D-01	7.52D-01	0.4	0.7082	4.030D+01	8.000D+00	3.230D+01	0.15
3	AP.CAL	8.9223D+00	3.77D+00	2.4	0.7408	8.030D+00	2.000D+00	6.030D+00	0.82
4	UN IND	3.8986D+01	1.63D+01	2.4	0.0214	1.000D+00	0.0	1.000D+00	0.60

NO. OF OBSERVATIONS   22
NO. OF IND. VARIABLES   4
RESIDUAL DEGREES OF FREEDOM   17
F-VALUE   3.6
RESIDUAL ROOT MEAN SQUARE   15.76802212
RESIDUAL MEAN SQUARE   248.63052149
RESIDUAL SUM OF SQUARES   4226.7188529
TOTAL SUM OF SQUARES   7824.04773566
MULT. CORREL. COEF. SQUARED   .4598

*** COEFFICIENTS SAVED AS REQUESTED ***

------ORDERED BY COMPUTER INPUT------

IDENT.	OBS. NO.	OBS. Y	FITTED Y	RESIDUAL
AUSTRL	1	8.1023D+01	7.3370D+01	7.6536D+00
ASTRIA	2	5.5023D+01	6.1192D+01	-6.1694D+00
CANADA	3	8.0003D+01	6.8024D+01	1.1979D+01
CEYLON	4	2.4055D+01	3.2724D+01	-8.6696D+00
CHILE	5	7.8332D+01	4.8457D+01	2.9575D+01
DENMK	6	5.1983D+01	5.1841D+01	1.4190D-01
FINLND	7	8.8024D+01	6.8655D+01	1.9369D+01
FRANCE	8	4.5025D+01	7.0374D+01	-2.5349D+01
GMANY	9	4.9969D+01	6.0404D+01	-1.0435D+01
JAPAN	13	2.4055D+01	4.3522D+01	-1.9460D+01
MEXICO	14	4.2975D+01	3.9450D+01	3.5253D+00
NTHLDS	15	3.8021D+01	5.8455D+01	-2.0433D+01
NWZEAL	16	7.2016D+01	6.8170D+01	3.8457D+00
NORWAY	17	4.0993D+01	5.3965D+01	-1.2971D+01
PORTGL	18	3.8021D+01	4.7437D+01	-9.4161D+00
SWEDEN	19	5.1983D+01	5.6099D+01	-4.1159D+00
SWITZD	20	5.1983D+01	6.1270D+01	-9.2873D+00
UNKING	21	6.5920D+01	5.8580D+01	1.0134D+01
U.N.	23	8.8986D+01	8.8986D+01	0.0

------ORDERED BY RESIDUALS------

OBS. NO.	OBS. Y	FITTED Y	ORDERED RESID.	STUD. RESID.	SEQ
5	7.8032D+01	4.8457D+01	2.9575D+01	2.0	1
11	8.8024D+01	6.8655D+01	1.9369D+01	1.4	2
10	6.5992D+01	4.7499D+01	1.8093D+01	1.2	3
3	6.9020D+01	5.2811D+01	1.6200D+01	1.1	4
21	8.0003D+01	6.8024D+01	1.1979D+01	0.8	5
12	6.5992D+01	5.5858D+01	1.0134D+01	0.7	6
16	8.1023D+01	7.3370D+01	7.6536D+00	0.7	7
19	7.2016D+01	6.8170D+01	5.7391D+00	0.4	8
2	5.1983D+01	3.9236D+01	3.8457D+00	0.3	9
4	2.4055D+01	3.2724D+01	-4.1159D+00	-0.3	10
20	2.4055D+01	3.2724D+01	-6.1694D+00	-0.4	11
18	5.0230D+01	5.8455D+01	-8.6696D+00	-0.6	12
9	5.1983D+01	6.1270D+01	-9.2873D+00	-0.7	13
17	3.8021D+01	4.7437D+01	-9.4161D+00	-0.7	14
13	4.9969D+01	6.0404D+01	-1.0435D+00	-0.7	15
8	4.0993D+01	5.3965D+01	-1.2971D+01	-0.9	16
	2.4055D+01	4.3522D+01	-1.2971D+01	-0.7	17
	3.8021D+01	4.3522D+01	-1.9468D+01	-1.7	18
	5.1983D+01	3.5965D+01	-1.2971D+01	-0.9	19
	2.4055D+01	4.3522D+01	-1.9468D+01	-1.7	20
	3.8021D+01	5.8455D+01	-2.0433D+01	-1.4	21
	4.5025D+01	7.0374D+01	-2.5349D+01	-1.9	22

Figure 9

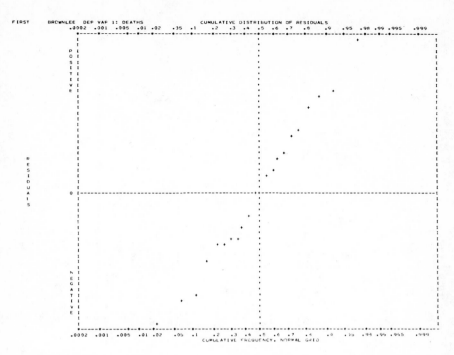

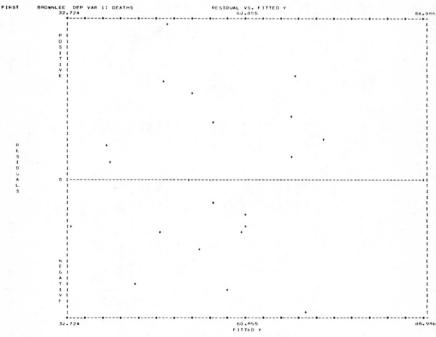

Figure 10

387

LINEAR LEAST-SQUARES CURVE-FITTING PROGRAM

NEXT   BROWNLEE

DATA INPUT   4 INDEPENDENT VARIABLES   3 DEPENDENT VARIABLE(S)

OBSV.	SEQ.	1-11-21	2-12-22	3-13-23	4-14-24	5-15-25	6-16-26	7-17-27	8-18-28	9-19-29	10-20-30
1	0	124.000	33.000	8.000	0.0	646.000	0.0	0.0			
2	0	49.000	31.000	6.000	0.0	355.000	0.0	0.0			
3	0	181.000	38.000	8.000	0.0	631.000	0.0	0.0			
4	0	4.000	17.000	2.000	0.3	174.000	0.0	0.0			
5	0	22.000	20.000	4.000	0.0	603.000	0.0	0.0			
6	0	152.000	39.000	6.000	0.0	331.000	0.0	0.0			
7	0	75.000	30.000	7.000	0.0	759.000	0.0	0.0			
8	0	54.000	29.000	7.000	0.0	282.000	0.0	0.0			
9	0	43.000	35.000	6.000	0.0	316.000	0.0	0.0			
10	0	41.000	31.000	5.000	0.0	450.000	0.0	0.0			
11	0	17.000	23.000	4.000	0.0	457.000	0.0	0.0			
12	0	22.000	21.000	3.000	0.0	282.000	0.0	0.0			
13	0	16.000	8.000	3.000	0.0	174.000	0.0	0.0			
14	0	10.000	23.000	3.000	0.0	269.000	0.0	0.0			
15	0	63.000	37.000	6.000	0.0	240.000	0.0	0.0			
16	0	170.000	40.000	8.000	0.0	525.000	0.0	0.0			
17	0	125.000	38.000	6.000	0.0	257.000	0.0	0.0			
18	0	15.000	25.000	4.000	0.0	240.000	0.0	0.0			
19	0	221.000	39.000	7.000	0.0	331.000	0.0	0.0			
20	0	171.000	33.000	7.000	0.0	331.000	0.0	0.0			
21	0	97.000	38.000	6.000	0.0	457.000	0.0	0.0			
22	0	254.000	39.000	8.000	0.0	776.000	0.0	0.0			
23	0	87.000	30.000	5.000	1.000	776.000	0.0	0.0			

A VALUE OF 1.0 HAS BEEN PLACED IN VARIABLE 4 OF OBSERVATION 23

DATA TRANSFORMATIONS

POSITION	CODE	OPERATION	CONSTANT	LOCATION	OMIT	VARIABLE	NAME
1		NONE		1	1	1	TFLEPH
2		NONE		2	0	2	F. CAL
3		NONE		3	0	3	AP.CAL
4		NONE		4	1	4	UN IND
5	2	COMMON LOG		6	1		
6	8	ADD CONSTANT	-2.	7	0	5	DEATHS
7	6	MULTIPLY BY CONSTANT	100.				

DATA AFTER TRANSFORMATIONS   THE FITTED EQUATION HAS 4 INDEPENDENT VARIABLES, 1 DEPENDENT VARIABLE(S)

OBSV.	TELEPH 1-11-21	F.CAL 2-12-22	AP.CAL 3-13-23	UN IND 4-14-24	DEATHS 5-15-25	6-16-26	7-17-27	8-18-28	9-19-29	10-20-30
1	1.24000D+02	3.30000D+01	8.00000D+00	0.0	8.10233D+01					
2	4.90000D+01	3.10000D+01	6.00000D+00	0.0	5.50228D+01					
3	1.81000D+02	3.80000D+01	2.00000D+00	0.0	8.00029D+01					
4	4.00000D+00	1.70000D+01	0.0	0.0	2.40549D+01					
5	2.20000D+01	2.00000D+01	4.00000D+00	0.0	7.80317D+01					
6	1.52000D+02	3.90000D+01	6.00000D+00	0.0	5.19828D+01					
7	7.50000D+01	3.00000D+01	7.00000D+00	0.3	8.80242D+01					
8	5.40000D+01	2.90000D+01	7.00000D+00	0.0	4.50249D+01					
9	4.30000D+01	3.50000D+01	0.0	0.0	4.96687D+01					
10	4.10000D+01	3.10000D+01	5.00000D+00	0.0	6.90196D+01					
11	1.70000D+01	2.30000D+01	4.00000D+00	0.0	6.59516D+01					
12	2.20000D+01	2.10000D+01	3.00000D+00	0.0	4.50249D+01					
13	1.60000D+01	8.00000D+00	3.00000D+00	0.0	2.40549D+01					
14	1.00000D+01	2.30000D+01	3.00000D+00	0.0	4.29752D+01					
15	6.30000D+01	3.70000D+01	0.0	0.0	3.80211D+01					
16	1.70000D+02	4.00000D+01	8.00000D+00	0.0	7.20159D+01					
17	1.25000D+02	3.80000D+01	6.00000D+00	0.0	4.09933D+01					
18	1.50000D+01	2.50000D+01	4.00000D+00	0.0	3.80211D+01					
19	2.21000D+02	3.90000D+01	7.00000D+00	0.0	5.19828D+01					
20	1.71000D+02	3.30000D+01	7.00000D+00	0.0	5.19828D+01					
21	9.70000D+01	3.80000D+01	6.00000D+00	0.0	6.59916D+01					
22	2.54000D+02	3.90000D+01	8.00000D+00	0.0	8.89620D+01					
23	8.70000D+01	3.00000D+01	5.00000D+00	1.00000D+00	8.89620D+01					

SUMS OF VARIABLES

2.01300D+03 6.97000D+02 1.29000D+02 1.00000D+00 1.33718D+03

RESIDUAL SUMS OF SQUARES + CROSS PRODUCTS

1 1.19456D+05
2 1.04453D+04 1.58487D+03
3 2.37070D+03 2.82739D+02 7.34783D+01
4 -5.21739D-01 -3.04348D-01 -6.08696D-01 9.56522D-01
5 1.42991D+04 1.55935D+03 4.49837D+02 3.08478D+01 8.81889D+03

MEANS OF VARIABLES

8.75217D+01 3.03043D+01 5.60870D+00 4.34783D-02 5.81384D+01

ROOT MEAN SQUARES OF VARIABLES

7.36872D+01 8.48761D+00 1.82755D+00 2.08514D-01 2.00214D+01

Figure 11

LINEAR LEAST-SQUARES CURVE-FITTING PROGRAM

NEXT   BROWNLEE   DEP VAR 1: DEATHS   MIN Y = 2.405D+01   MAX Y = 8.899D+01   RANGE Y = 6.493D+01

EXAMPLE OF CROSS VERIFICATION WITH SAVED COEFFICIENTS FROM FIT WITH
FIRST SET OF DATA. THIS SECOND SET OF DATA CONTAINS USA. OBSERVATION
22. DELETED FROM FIRST SET OF DATA.
Y = B(0) + B(1)X1 + B(2)X2 + B(3)X3 + B(4)X4
Y = 100(LOG(NUMBER OF DEATHS FROM HEART DISEASE PER 100.000 FOR MALES
        IN THE 55 TO 59 AGE GROUP) - 2)
X1 = 1000(TELEPHONES PER PERSON)
X2 = FAT CALORIES AS PER CENT OF TOTAL CALORIES
X3 = ANIMAL PROTEIN CALORIES AS PER CENT TOTAL CALORIES
X4 = INDICATOR VARIABLE, 1 FOR OBSERVATION 23 (U.N.), 0 OTHERWISE.

IND.VAR(I)	NAME	COEF.B(I)	S.E. COEF.	T-VALUE	R(I)SQRD	MIN X(I)	MAX X(I)	RANGE X(I)	REL.INF.X(I)
0		2.39806D+01							
1	TELEPH	-6.87054D-03	8.14D-02	0.1	0.6715	4.000D+00	2.540D+02	2.500D+02	0.03
2	F. CAL	-4.80840D-01	7.57D-01	0.6	0.7136	8.000D+00	4.000D+01	3.200D+01	0.24
3	AP.CAL	8.50465D+00	3.84D+00	2.2	0.7603	2.000D+00	8.000D+00	6.000D+00	0.79
4	UN IND	3.75052D+01	1.67D+01	2.3	0.0190	0.0	1.000D+00	1.000D+00	0.58

NO. OF OBSERVATIONS   23
NO. OF IND. VARIABLES   4
RESIDUAL DEGREES OF FREEDOM   18
F-VALUE   4.0
RESIDUAL ROOT MEAN SQUARE   16.13185806
RESIDUAL MEAN SQUARE   260.23684450
RESIDUAL SUM OF SQUARES   4684.26320100
TOTAL SUM OF SQUARES   8818.88546632
MULT. CORREL. COEF. SQUARED   .4688

IND.VAR(I)	NAME	COEF.B(I)				MIN X(I)	MAX X(I)	RANGE X(I)	REL.INF.X(I)
*** COEFFICIENTS FROM PREVIOUS FIT ***

0		2.02254D+01							
1	TELEPH	-6.76016D-02				4.000D+00	2.540D+02	2.500D+02	0.26
2	F. CAL	-2.98541D-01				8.000D+00	4.000D+01	3.200D+01	0.15
3	AP.CAL	8.92234D+00				2.000D+00	8.000D+00	6.000D+00	0.82
4	UN IND	3.89866D+01				0.0	1.000D+00	1.000D+00	0.60

390

LINEAR LEAST-SQUARES CURVE FITTING PROGRAM

NEXT   BROWNLEE DEP VAR I: DEATHS

FIT OF SECOND SAMPLE USING CCEFFICIENTS FROM FIRST SAMPLE OF DATA

NO. CF CBSERVATICNS	23
NO. OF PARAMETERS	5
RESIDUAL DEGREES OF FREEDOM	18
RESIDUAL ROCT MEAN SQUARE	16.52094110
RESIDUAL MEAN SQUARE	272.94149490
RESIDUAL SUM OF SQUARES	4912.94650819
TOTAL SUM OF SQUARES	8818.88546632
MULT. CCRREL. COEF. SQUARED	.4429

IDENT.	OBSV.	WSS DISTANCE	OBS. Y	FITTED Y	RESIDUAL
AUSTRL	1	2.	81.023	73.370	7.654
ASTRIA	2	0.	55.023	61.192	-6.169
CANADA	3	2.	80.303	68.024	11.979
CEYLON	4	4.	24.055	32.724	-8.670
CHILE	5	1.	78.032	48.457	29.575
DENMK	6	0.	51.983	51.841	0.142
FINLND	7	1.	88.024	68.655	19.369
FRANCE	8	1.	45.025	70.374	-25.349
GMANY	9	0.	49.969	60.404	-10.435
IRELND	10	0.	69.020	52.811	16.209
ISRAEL	11	1.	65.992	47.899	18.093
ITALY	12	2.	45.025	39.236	5.789
JAPAN	13	2.	24.055	43.522	-19.468
MEXICO	14	2.	42.975	39.450	3.525
NTHLDS	15	0.	38.021	58.455	-20.433
NWZEAL	16	2.	72.016	68.170	3.846
NORWAY	17	0.	40.993	53.965	-12.971
PORTGL	18	1.	38.021	47.437	-9.416
SWEDEN	19	1.	49.969	56.099	-4.116
SWITZD	20	1.	40.993	61.270	-9.287
UNKING	21	0.	65.992	55.858	10.134
U.S.A.	22	2.	88.986	62.790	26.196
U.N.	23	5.	88.986	88.986	0.0

ORDERED BY COMPUTER INPUT

---ORDERED BY RESIDUALS---

OBSV.	OBS. Y	FITTED Y	ORDERED RESID.	STUD. RESID.	SEQ
5	78.032	48.457	29.575	1.8	1
22	88.986	62.790	26.196	1.4	2
7	88.024	68.655	19.369	1.1	3
11	65.992	47.899	18.093	1.1	4
10	69.020	52.811	16.209	1.0	5
3	80.003	68.024	11.979	0.7	6
21	65.992	55.858	10.134	0.6	7
1	81.023	73.370	7.654	0.4	8
12	45.025	39.236	5.789	0.3	9
16	72.016	68.170	3.846	0.2	10
14	42.975	39.450	3.525	0.2	11
6	51.983	51.841	0.142	0.0	12
23	88.986	88.986	0.0	0.0	13
19	51.983	56.099	-4.116	-0.2	14
2	55.023	61.192	-6.169	-0.4	15
4	24.055	32.724	-8.670	-0.5	16
20	51.983	61.270	-9.287	-0.5	17
18	38.021	47.437	-9.416	-0.6	18
9	49.969	60.404	-10.435	-0.6	19
17	40.993	53.965	-12.971	-0.8	20
13	24.055	43.522	-19.468	-1.0	21
15	38.021	58.455	-20.433	-1.2	22
8	45.025	70.374	-25.349	-1.4	23

Figure 12

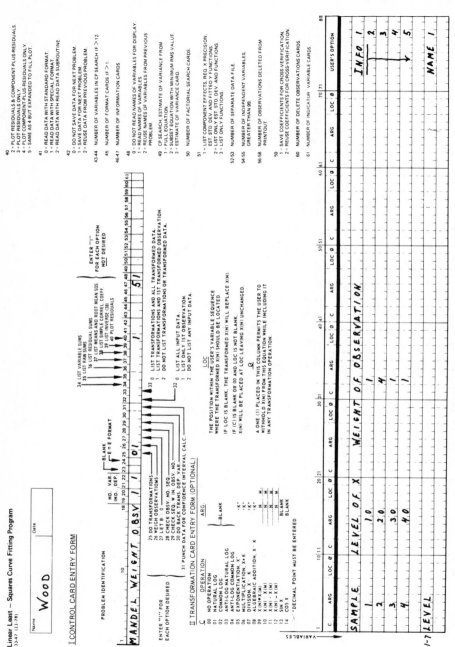

Figure 13

392

III STANDARD DATA-CARD ENTRY FORM

ENTER WEIGHTING FACTOR, IF ANY, AS LAST ENTRY.
"END" CARD MUST BE LAST CARD.

IDENT.	OBSV. NO.	SEQ. NO.	x1-11-21	x2-12-22	x3-13-23	x4-14-24	x5-15-25	x6-16-26	x7-17-27	x8-18-28	x9-19-29	x10-20-30	USER'S OPTION
MANDEL DATA, IF WEIGHT OF OBSERVATION IS USED, COL 26 = 1													
	1		10.,	33.8,	1.,								
	2		20.,	62.2,	4.,								
	3		30.,	92.0,	1.,								
	4		40.,	122.4,	1.,								
END.													
MANDEL DATA, IF VARIANCE OF OBSERVATION IS USED, COL 26 = 2													
	1		10.,	33.8,	1.,								
	2		20.,	62.2,	0.,25								
	3		30.,	92.0,	1.,								
	4		40.,	122.4,	1.,								
END.													

Figure 14

FORM B-56 REV. 3-69

LINEAR LEAST-SQUARES CURVE-FITTING PROGRAM

MANDEL WEIGHT OBSV

DATA INPUT   1 INDEPENDENT VARIABLES   1 DEPENDENT VARIABLE(S)   WEIGHT OF OBSERVATION

OBSV.	SEQ.	1-11-21	2-12-22	3-13-23	4-14-24	5-15-25	6-16-26	7-17-27	8-18-28	9-19-29	10-20-30
1	0	10.000	33.800	1.000							
2	0	20.000	62.200	4.000							
3	0	30.000	92.000	1.000							
4	0	40.000	122.400	1.000							

SUMS OF VARIABLES
1.60000D+02 4.97000D+02

RESIDUAL SUMS OF SQUARES + CROSS PRODUCTS
1   5.42857D+02
2   1.61000D+03 4.77656D+03

MEANS OF VARIABLES
2.28571D+01 7.10000D+01

ROOT MEAN SQUARES OF VARIABLES
9.51190D+00 2.82151D+01

SIMPLE CORRELATION COEFFICIENTS, R(I,I PRIME)
1   1.000
2   1.000   1.000

INVERSE, C(I,I PRIME)
1   1.000

394

Figure 15

LINEAR LEAST-SQUARES CURVE FITTING PROGRAM

MANDEL WEIGHT OBSV DEP VAR 1:

MIN Y = 3.380D+01   MAX Y = 1.224D+02   RANGE Y = 8.860D+01

SAMPLE	LEVEL OF X	WEIGHT OF OBSERVATION
1	10	1
2	20	4
3	30	1
4	40	1

IND.VAR(I)	NAME	COEF.B(I)	S.E. COEF.	T-VALUE	R(I)SQRD	MIN X(I)	MAX X(I)	RANGE X(I)	REL.INF.X(I)
0		3.21053D+00							
1	LEVEL	2.96579D+00	3.89D-02	76.3	0.0	1.000D+01	4.000D+01	3.000D+01	1.00

NO. OF OBSERVATIONS   4
NO. OF IND. VARIABLES   1
RESIDUAL DEGREES OF FREEDOM   2
F-VALUE   5826.8
RESIDUAL ROOT MEAN SQUARE   0.90524786
RESIDUAL MEAN SQUARE   0.81947368
RESIDUAL SUM OF SQUARES   1.63894737
TOTAL SUM OF SQUARES   4776.56000000
MULT. CORREL. COEF. SQUARED   .9997

IDENT.	OBSV.	WSS DISTANCE	OBS. Y	FITTED Y	RESIDUAL	OBSV.	OBS. Y	FITTED Y	ORDERED RESID.	STUD.RESID.	SEQ
1	1774.		33.800	32.868	0.932	1	33.800	32.868	0.932	1.5	1
2	88.		62.200	62.526	-0.326	4	122.400	121.842	0.558	1.3	2
3	548.		92.000	92.184	-0.184	3	92.000	92.184	-0.184	-0.3	3
4	3154.		122.400	121.842	0.558	2	62.200	62.526	-0.326	-2.9	4

----------ORDERED BY COMPUTER INPUT----------   ----------ORDERED BY RESIDUALS----------

Figure 16

395

identifies the values in the field following the last dependent variable as the weighting factor. These weights are then applied to all of the dependent variables that are present.

The printout is shown in Figures 15 and 16.

The residual mean square is a measure of fit, but because of the weighting factor sample 2 contributed only $\frac{1}{4}$ to the variance of titration error. Whenever possible, more is learned by fitting all of the values at once (include all determinations of sample 2 rather than its average). The plot of the residuals will show the distribution of replicates about the fitted equation. If one value influences the mean, and hence the fit, excessively, it can be seen. In addition, a better overall measure of the residual mean square of the fit is obtained.

Example to measure the precision of calculations

Background of problem

In the September 1967 issue of the *Journal of the American Statistical Association* (Vol. 62, No. 319), James W. Longley of the Bureau of Labor Statistics presented the results of a study of the numerical accuracy of various least-squares programs run on a wide range of electronic computers. It had been observed that the results and conclusions drawn by various companies and government agencies using different computers and computer least-squares programs were not consistent. In similar stepwise regression programs different sets of variables had been chosen as important, and even the coefficients of some important variables had been reported as being opposite in sign.

Thus, the objective of this study was to determine the combined effect of computer precision and program technique on the coefficients of the resulting fitted equation. Other government agencies and some large companies with government contracts were asked to run the same specified problem. This problem, chosen by the Bureau of Labor Statistics, was economic in nature, involving price index, gross national product, unemployment, size of the armed forces, noninstitutional population, and time as the "independent" variables. Eight measures of employment, including total, agricultural, self-employed, family workers, domestic, industrial, and federal and state government workers, were used as the dependent variables.

Comparison of widely used computer programs

In order to check the precision of a computer program, it is necessary to know the correct answer to a sufficient number of digits. In this case a tremendous amount of effort was required to calculate the coefficients of the

TABLE 1

COMPARISON OF LINEAR LEAST-SQUARES PROGRAMS AND COMPUTERS
ACCURACY OF PRICE INDEX COEFFICIENT IN TOTAL EMPLOYMENT EQUATION

Program	Computer	Coefficient	No. of Digits Accuracy	% Error
Hand Calculation		15.06187227137		
Older Programs and Computers				
NIPD	IBM 7094	-41.	0	375
Step-Wise A	IBM 7094	-36.	0	340
Step-Wise B	IBM 7074	2.	0	82
UCLA, BMD	IBM 7094	27.	0	80
Johns Hopkins	IBM 1401	11.	1	23
Dartmouth	GE 235	13.	1	7
CORRE(SP)	IBM 360	15.00	3	4
CORRE(DP)	IBM 360	15.04	3	1
ORTHO	IBM 7074	15.065	4	0.03
LINWOOD(SP)	IBM 7044	15.066	4	0.03
LINWOOD(SP)	UNIVAC 1108	15.064	4	0.02
Current Programs and Computers				
BMDP1R	IBM 370/168	15.6	2	4
BMDP2R	IBM 370/168	14.7	1	2
SPSS	IBM 370/168	15.06184	6	10^{-4}
LINWOOD(DP)	HIS 635	15.061870	7	10^{-5}
LINWOOD(DP)	DEC 10, KA10	15.06187226	9	10^{-7}
LINWOOD(SP)	CDC 6400, 6600	15.061872272	10	10^{-8}
LINWOOD(DP)	AMDAHL 470/V6	15.0618722713	12	10^{-10}
LINWOOD(DP)	BUR 4700, 6700	15.0618722713	12	10^{-10}
LINWOOD(DP)	DEC 20, KL10	15.0618722713	12	10^{-10}
LINWOOD(DP)	IBM 370/168, 3033	15.0618722713	12	10^{-10}
LINWOOD(DP)	UNIVAC 1110	15.0618722713	12	10^{-10}

equations on desk calculators to 12 digits. As examples of the magnitude of the numbers involved, one determinant had 25 digits to the left of the decimal and another 18 digits to the right of the decimal.

As a basis for comparing the accuracy of the calculated coefficients we have chosen the coefficient of the first independent variable, price index, as being representative. This variable is used to measure the effect of inflation on total employment. The "correct" hand-calculated value of this coefficient is +15.061 872 271 373.

In the Bureau of Labor Statistics study, the percentage of error in the price index coefficient ranged from 0.03 to 375%. The first nine computer programs in Table 1 were from this study. Four programs did not agree even in the first digit, giving values of −36 and −41 instead of +15. Two were accurate to 1 digit, two to 3 digits, and one to 4 digits. The LINWOOD single precision program in comparable IBM 7044 and UNIVAC 1108 computers were accurate to 4 digits. Since accuracy is a prime concern in the use of these techniques, this problem provides a good basis for comparing computer precision and program technique.

In the lower portion of Table 1, comparisons are made using more recent computers. In an IBM 370/168 computer, the BMD Stepwise Regression (P2R) and Multiple Linear Regression (P1R) programs are accurate to 1 and 2 digits respectively. The SPSS Multiple Regression program is accurate to 6 digits and the LINWOOD program is accurate to 12 digits. In single precision, the LINWOOD program is accurate to 10 digits in the CDC 6400 and 6600 computers. In double precision, the accuracy ranges from 7 digits for the Honeywell 635 computer to 12 digits for the AMDAHL 480, Burroughs 4700, 6700, DEC 20, IBM 370/168, 3033, and UNIVAC 1110 computers.

To obtain a print out of this precision, alternate formats are provided in the subroutine FIT(CF3), statement numbers 950 and 3012. A printout of this problem is shown in Figures 17–23.

This example is included in the program's test problems to check the precision of each installation's computer and compiler. Whether the data are accurate enough to make the results meaningful is discussed in the February 1976 issue of the *Journal of the American Statistical Association* (Vol. 7, pp. 158–168) by A. E. Beaton, D. B. Rubin, and J. L. Barone.

LINEAR LEAST-SQUARES CURVE-FITTING PROGRAM

PRECISION - EMPLOY

DATA INPUT   6 INDEPENDENT VARIABLES   8 DEPENDENT VARIABLE(S)

OBSV.	SEQ.	1-11-21	2-12-22	3-13-23	4-14-24	5-15-25	6-16-26	7-17-27	8-18-28	9-19-29	10-20-30
1947	0	83.000	234289.000	2356.000	1590.000	107608.000	1947.000	60323.000	8256.000	6045.000	427.000
	1	1714.000	38407.000	1892.000	3582.000						
1948	0	88.500	259426.000	2325.000	1456.000	108632.000	1948.000	61122.000	7960.000	6139.000	401.000
	1	1731.000	39241.000	1863.000	3787.000						
1949	0	88.200	258054.000	3682.000	1616.000	109773.000	1949.000	60171.000	8017.000	6208.000	396.000
	1	1772.000	37922.000	1908.000	3948.000						
1950	0	89.500	284599.000	3351.000	1650.000	110929.000	1950.000	61187.000	7497.000	6069.000	404.000
	1	1995.000	39196.000	1928.000	4098.000						
1951	0	96.200	328975.000	2099.000	3099.000	112075.000	1951.000	63221.000	7048.000	5869.000	400.000
	1	2055.000	41460.000	2302.000	4087.000						
1952	0	98.100	346999.000	1932.000	3594.000	113270.000	1952.000	63639.000	6792.000	5670.000	431.000
	1	1922.000	42216.000	2420.000	4188.000						
1953	0	99.000	365385.000	1870.000	3547.000	115094.000	1953.000	64989.000	6555.000	5794.000	423.000
	1	1985.000	43587.000	2305.000	4340.000						
1954	0	100.000	363112.000	3578.000	3350.000	116219.000	1954.000	63761.000	6495.000	5880.000	445.000
	1	1919.000	42271.000	2188.000	4563.000						
1955	0	101.000	397469.000	2904.000	3048.000	117388.000	1955.000	66019.000	6718.000	5886.000	524.000
	1	2216.000	43761.000	2187.000	4727.000						
1956	0	104.600	419180.000	2822.000	2857.000	118734.000	1956.000	67857.000	6572.000	5936.000	581.000
	1	2359.000	45131.000	2209.000	5069.000						
1957	0	108.400	442769.000	2936.000	2798.000	120445.000	1957.000	68169.000	6222.000	6089.000	626.000
	1	2328.000	45278.000	2217.000	5409.000						
1958	0	110.800	444546.000	4681.000	2637.000	121950.000	1958.000	66513.000	5844.000	6185.000	605.000
	1	2456.000	43530.000	2191.000	5702.000						
1959	0	112.600	482704.000	3813.000	2552.000	123366.000	1959.000	68655.000	5836.000	6298.000	597.000
	1	2520.000	45214.000	2233.000	5957.000						
1960	0	114.200	502601.000	3931.000	2514.000	125368.000	1960.000	69564.000	5723.000	6367.000	615.000
	1	2489.000	45850.000	2270.000	6250.000						
1961	0	115.700	518173.000	4806.000	2572.000	127852.000	1961.000	65331.000	5463.000	6388.000	662.000
	1	2594.000	45397.000	2279.000	6548.000						
1962	0	116.900	554894.000	4007.000	2827.000	130081.000	1962.000	70551.000	5190.000	6271.000	623.000
	1	2626.000	46652.000	2340.000	6849.000						

Figure 17

399

DATA TRANSFORMATIONS

POSITION	CODE	OPERATION
1		NONE
2		NONE
3		NONE
4		NONE
5		NONE
6		NONE
7		NONE
8		NONE
9		NONE
10		NONE
11		NONE
12		NONE
13		NONE
14		NONE

CONSTANT	LOCATION	OMIT	VARIABLE	NAME
	1	0	1	PRICE
	2	0	2	GNP
	3	0	3	UNEMPY
	4	0	4	ARMY
	5	0	5	NONIST
	6	0	6	TIME
	7	0	7	TOTAL
	8	1		FARM
	9	1		SELF
	10	1		UNPAID
	11	1		HOUSE
	12	1		PRIVAT
	13	1		FED GO
	14	1		LOCALG

FIRST OBSERVATION AFTER TRANSFORMATIONS   THE FITTED EQUATION HAS   6 INDEPENDENT VARIABLES, 1 DEPENDENT VARIABLE(S)

	PRICE	GNP	UNEMPY	ARMY	NONIST	TIME	TOTAL
OBSV.	1-11-21	2-12-22	3-13-23	4-14-24	5-15-25	6-16-26	7-17-27
						8-18-28	9-19-29	10-20-30
1947	8.30000D+01	2.34289D+05	2.35600D+03	1.59000D+03	1.07608D+05	1.94700D+03	6.03230D+04

MEANS OF VARIABLES
1.01681D+02 3.87658D+05 3.19331D+03 2.60669D+03 1.17424D+05 1.95450D+03 6.53170D+04

ROOT MEAN SQUARES OF VARIABLES
1.07916D+01 9.93949D+04 9.34464D+02 6.95920D+02 6.95610D+03 4.76695D+00 3.51197D+03

SIMPLE CORRELATION COEFFICIENTS, R(I,I PRIME)

1	1.000						
2	0.992	1.000					
3	0.621	0.604	1.000				
4	0.455	0.446	-0.177	1.000			
5	0.979	0.991	0.687	0.364	1.000		
6	0.991	0.995	0.668	0.417	0.994	1.000	
7	0.971	0.984	0.502	0.457	0.960	0.571	1.000

Figure 18

400

LINEAR LEAST-SQUARES CURVE FITTING PROGRAM

PRECISION - EMPLOY   DEP VAR 1: TOTAL

MIN Y = 6.0170+04   MAX Y = 7.055D+04   RANGE Y = 1.038D+04

Y = B(0) + B(1)X1 + E(2)X2 + E(3)X3 + E(4)X4 + B(5)X5 + B(6)X6
Y = EMPLOYMENT
X1 = PRICE INDEX
X2 = GROSS NATIONAL PRODUCT
X3 = UNEMPLOYMENT
X4 = SIZE OF ARMED FORCES
X5 = NONINSTITUTICNAL POPULATION
X6 = TIME IN YEARS

IND.VAR(I)	NAME	COEF.B(I)	S.E. COEF.	T-VALUE	R(I)SQRD	MIN X(I)	MAX X(I)	RANGE X(I)	REL.INF.X(I)
0		-3.48226D+06							
1	PRICE	1.50619D+01	8.49C+01	0.2	0.9926	8.3CD+C1	1.169D+02	3.390D+01	0.05
2	GNP	-3.58152D-02	3.35D-02	1.1	0.9994	2.343D+05	5.549D+05	3.206D+05	1.11
3	UNEMPL	-2.02023C+00	4.38D-01	4.1	0.9703	1.870D+03	4.806D+03	2.936D+03	0.57
4	ARMY	-1.03323D+00	2.14D-01	4.8	0.7214	1.456D+03	3.594D+03	2.138D+03	0.21
5	NONINST	-5.11041D-02	2.2D-01	0.2	0.9975	1.076D+05	1.301D+05	2.247D+04	0.11
6	TIME	1.82915D+03	4.55D+02	4.0	0.9987	1.947D+03	1.962D+03	1.500D+01	2.64

NO. OF OBSERVATIONS   16
AC. OF IND. VARIABLES   6
RESIDUAL DEGREES OF FREEDOM   9
F-VALUE   330.3
RESIDUAL ROOT MEAN SQUARE   304.85407355
RESIDUAL MEAN SQUARE   92936.00610252
RESIDUAL SUM OF SQUARES   836424.05540265
TOTAL SUM OF SQUARES   185008825.99999999
MULT. CORREL. COEF. SQUARED   .5555

		ORDERED BY COMPUTER INPUT					ORDERED BY RESIDUALS				
IDENT.	OBSV.	WSS DISTANCE	OBS. Y	FITTED Y	RESIDUAL		OBSV.	OBS. Y	FITTED Y	ORDERED RESID.	STUD.RESID.	SEQ
JASA P	1947	2396.	60323.000	6C055.660	267.340		1956	67857.000	67401.606	455.394	1.8	1
JASA R	1948	1799.	61122.000	61216.014	-94.014		1961	69331.000	68989.068	341.932	1.4	2
JASA E	1949	1345.	60171.000	6C124.713	46.287		1951	63221.000	62911.285	309.715	1.6	3
JASA C	1950	889.	61187.000	61597.115	-410.115		1947	60323.000	6C055.660	267.340	1.2	4
JASA I	1951	545.	63221.000	62911.285	309.715		1949	60171.000	60124.713	46.287	0.2	5
JASA S	1952	325.	63639.000	6388E.311	-249.311		1955	66019.000	66004.695	14.305	0.1	6
JASA S	1953	175.	64989.000	65153.049	-164.049		1954	63761.000	63774.180	-13.180	-0.1	7
JASA O	1954	30.	63761.000	63774.180	-13.180		1957	68169.000	68186.269	-17.269	-0.1	8
JASA N	1955	16.	66019.000	66004.695	14.305		1958	66513.000	66552.055	-39.055	-0.2	9
JASA	1956	102.	67857.000	67401.606	455.394		1960	69564.000	69649.671	-85.671	-0.3	10
JASA T	1957	271.	68169.000	6818E.269	-17.269		1948	61122.000	61216.014	-94.014	-0.5	11
JASA E	1958	584.	66513.000	66552.055	-39.055		1959	68655.0C0	68810.550	-155.550	-0.6	12
JASA S	1959	872.	68655.000	68810.550	-155.550		1953	64989.000	65153.049	-164.049	-0.8	13
JASA T	1960	1297.	69564.000	6964S.671	-65.671		1962	70551.000	70757.758	-206.758	-1.2	14
JASA	1961	1874.	69331.000	68989.068	341.932		1952	63639.000	63888.311	-249.311	-1.0	15
JASA	1962	2446.	7C551.000	7C757.758	-206.758		1950	61187.000	61597.115	-410.115	-1.7	16

Figure 19

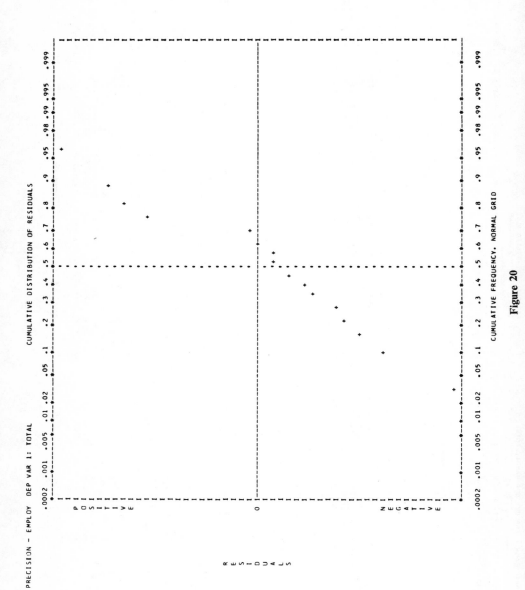

Figure 20

PRECISION - EMPLOY DEP VAR 1: TOTAL

RESIDUAL VS. FITTED Y

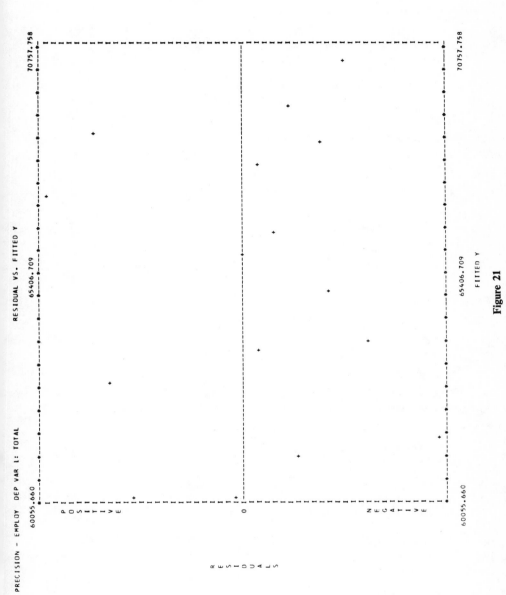

FITTED Y

Figure 21

PRECISION - EMPLOY   DEP VAR 1: TOTAL   SEARCH NUMBER 1

SELECTION OF VARIABLES FOR BASIC EQUATION

CP =*****,   VARIABLE ADDED   4   ARMY

CP =864.3,   VARIABLE ADDED   3   UNEMPY

CP = 6.2,   VARIABLE ADDED   6   TIME

CP = 3.2,   VARIABLE ADDED   2   GNP

CP = 5.0,   VARIABLE ADDED   5   NONIST

NUMBER OF OBSERVATIONS   16
NUMBER OF VARIABLES IN FULL EQUATION   6
NUMBER OF VARIABLES IN BASIC EQUATION   2
REMAINDER OF VARIABLES TO BE SEARCHED   4

EQUATION   P   CP   VARIABLES IN EQUATION

           7   7.0   FULL EQUATION

           3 864.3 BASIC SET OF VARIABLES   3   4
                                             UNEMPY   ARMY
                     RMS =* ***********

      1   5   4.6   BASIC SET PLUS   5   6
                                          NONIST TIME
                     RMS =89610.87709

      2   6   6.1   BASIC SET PLUS   1   5   6
                                          PRICE NONIST TIME
                     RMS =94273.03144

      3   7   7.0   BASIC SET PLUS   1   2   5   6
                                          PRICE   GNP   NONIST TIME
                     RMS =92936.00617

      4   6   5.0   BASIC SET PLUS   2   5   6
                                            GNP NONIST TIME
                     RMS =83934.80319

      5   5   3.2   BASIC SET PLUS   2   6
                                            GNP   TIME
                     RMS =78061.85508

      6   6   5.1   BASIC SET PLUS   1   2   6
                                          PRICE   GNP   TIME
                     RMS =84117.30036

      7   4   6.2   BASIC SET PLUS   6
                                           TIME
                     RMS =* ***********

SWEEP NUMBER   15, DELTA Z(K,K) = 6.0-17

ESTIMATE OF VARIANCE

RMS OF FULL EQUATION   92936.00616252
RMS OF SUBSET EQUATION   5 78061.85507543

RMS OF SUBSET EQUATION USED TO RECALCULATE AND PLOT CP VALUES.

                       P   CP
FULL EQUATION   7   8.7
SUBSET EQUATION   5   5   5.0

SEQ   EQUATION P   ORDERED CP
 1   5   5   5.0
 2   1   5   6.6
 3   4   6   6.8
 4   6   6   6.8
 5   2   6   8.1
 6   3   7   8.7
 7   7   4   9.0

Figure 22

404

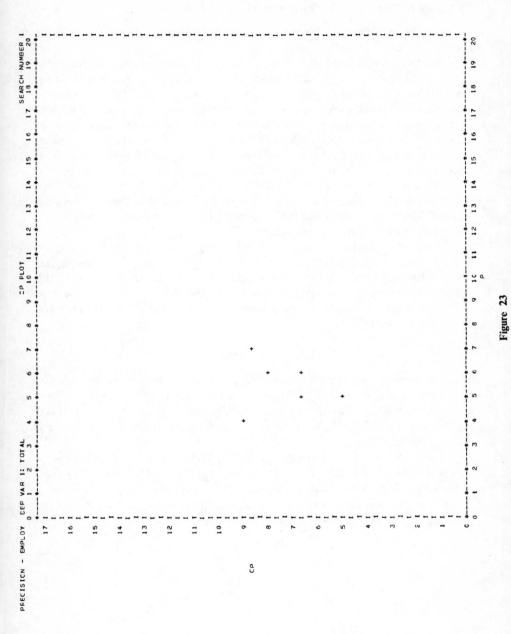

Figure 23

405

Example of using plots in choosing the form of equation

This example is taken with permission of *Technometrics* from F. S. Wood's paper, Volume 15, Number 4.

With a single independent variable, x, one can usually choose an acceptable form of the equation by making a plot of x versus y, the dependent variable. With two independent variables, x_1 and x_2, plots of x_1 versus y or x_2 versus y may be informative only if one independent variable is dominant or if there are several x_1 values at a fixed x_2 value or several x_2 values at a fixed x_1 value. If this is not the case (or if there are more than two influential independent variables operating), confusion is the principal product of such plots. With *any* equation with two or more influential independent variables, the plot of x_1 versus y can be made to take *any* shape one wishes by the appropriate selection of the values of the other independent variables. Being dependent on the allocation of the observations, such plots are of little aid in selecting the appropriate form of the equation.

Plots of the *residuals* versus each of the independent variables, however, are useful. They show the distribution of the residuals at each x_i, where i goes from 1 to the number of terms in the equation. The pattern produced by the residuals may suggest another functional form of that variable in the equation.

To provide an example, 40 observations, shown in Table 2, were generated from the equation

$$(1) \qquad y = \beta_0 + \beta_1 x_1 + \beta_2 x_2 + \beta_3 x_1{}^2 + \epsilon.$$

The values of x_1 and x_2 used to generate observations *1–20* were repeated for observations *21–40*. The first 20 values for the error term ϵ were taken from page 4, column 4, and the second 20 from page 12, column 2, of Rand Corporation (1970), *A Million Random Digits*, with mean zero and standard deviation one.

Figures 24 and 25 are plots of y versus x_1 and x_2 respectively. Other than general trends in the data, no distinct pattern is discernible from either plot. This supports our belief that such plots are of little value.

In order to determine whether the residual plots would show the need for the squared term, a least-squares fit was made with equation

$$(2) \qquad Y = b_0 + b_1 x_1 + b_2 x_2.$$

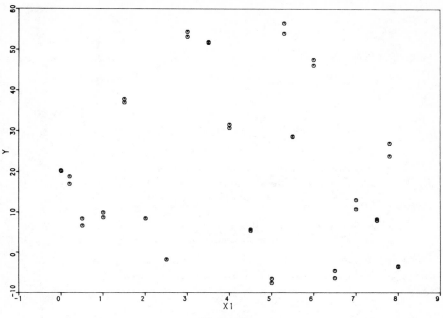

Figure 24

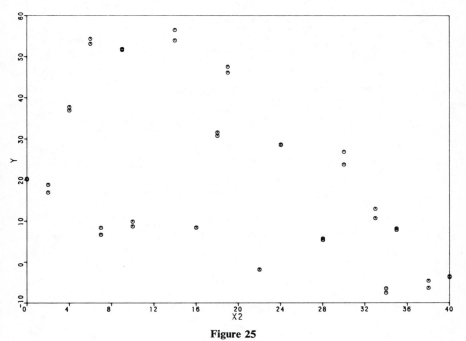

Figure 25

407

TABLE 2
DATA FOR TWO-VARIABLE EXAMPLE

OBSV.	X1	X2	Y	OBSV.	X1	X2	Y
1	0.0	0.	20.164	21	0.0	0.	20.283
2	0.2	2.	18.847	22	0.2	2.	16.970
3	0.5	7.	6.678	23	0.5	7.	8.379
4	1.0	10.	8.709	24	1.0	10.	9.876
5	1.5	4.	36.934	25	1.5	4.	37.715
6	2.0	16.	8.397	26	2.0	16.	8.445
7	2.5	22.	-1.778	27	2.5	22.	-1.841
8	3.0	6.	53.136	28	3.0	6.	54.339
9	3.5	9.	51.883	29	3.5	9.	51.646
10	4.0	18.	31.564	30	4.0	18.	30.738
11	4.5	28.	5.739	31	4.5	28.	5.373
12	5.0	34.	-6.466	32	5.0	34.	-7.537
13	5.3	14.	56.494	33	5.3	14.	53.574
14	5.5	24.	28.523	34	5.5	24.	28.636
15	6.0	19.	47.458	35	6.0	19.	46.065
16	6.5	38.	-4.575	36	6.5	38.	-6.327
17	7.0	33.	10.730	37	7.0	33.	13.001
18	7.5	35.	8.214	38	7.5	35.	7.855
19	7.8	30.	23.790	39	7.8	30.	26.865
20	8.0	40.	-3.518	40	8.0	40.	-3.659

As seen in Figure 26, a multiple correlation coefficient ($R_y{}^2$) of 0.91 was obtained with a residual root mean square of 6.2. However, the estimated standard deviation of measurement error obtained from "near neighbors" (shown in Figure 27) was slightly less than 1, indicating a considerable lack of fit, 6.2 versus 1.0. In this example, the simple correlation coefficient between x_1 and x_2 is 0.75, indicating that a moderately high correlation exists between the observations of x_1 and x_2. Unless the observations are taken in a designed experiment, the values of x_1 and x_2 are seldom uncorrelated (or orthogonal to one another). A plot of the residuals versus x_1, Figure 28, shows a horseshoe pattern suggesting that a squared term is needed for x_1. In Figure 29, with x_2, the pattern is not so pronounced. Since there is a moderately high correlation between x_1 and x_2, no change may be needed in the function of x_2.

Figure 30 is a plot of the component and the component-plus-residuals versus x_1. We can see that although the squared term is needed, its influence compared with the overall component effect of x_1 is small. A similar plot with x_2, Figure 31, shows that no change in the function of x_2 is needed.

In Pass 2 we fit equation

$$(3) \qquad Y = b_0 + b_1(x_1 - \bar{x}_1) + b_2 x_2 + b_3(x_1 - \bar{x}_1)^2.$$

The mean is subtracted from x_1 in order to reduce the correlation between x_1 and $x_1{}^2$. From the fit of this equation we obtain, in Figure 32, an $R_y{}^2$ of 0.9978 with a root mean square of 0.98 (close to our estimated standard deviation of measurement error). Figures 33 and 34 are plots of the residuals versus x_1 and x_2. Now the residuals are widely dispersed with no clear pattern. Figure 35 is a plot of each component plus its residual versus x_1, showing the excellent fit. Figure 36 is a similar plot for x_2.

LINEAR LEAST-SQUARES CURVE FITTING PROGRAM

2 VARIABLES PASS 1   DEP VAR 1: Y   MIN Y = -7.537D+00   MAX Y = 5.649D+01   RANGE Y = 6.403D+01

EQUATION: Y = 20 + 20X1 - 3X2 - 1X1SQRD + E2X2
FITTED WITH Y = B0 + B1X1 + E2X2
ERROR FROM "A MILLION RANDOM DIGITS", RAND CORP., FREE PRESS, 1955.
1-20 OBSV. FROM TABLE PAGE 4, COL 4; 21-40 OBSV. FROM PAGE 12, COL 2.
OBSERVATIONS 21-40 ARE REPLICATES OF OBSERVATIONS 1-20.
     STD. DEVIATION ESTIMATED FROM NEAR NEIGHBORS (1.C), IS MUCH LESS
THAN RMS OF FIT, SUGGESTING A BETTER FORM OF EQUATION IS NEEDED.
NEED FOR SQUARED TERM CAN BE SEEN FROM RESIDUAL VS. INDEPENDENT
VARIABLE X1 PLOT. SLIGHT EFFECT ON CURVATURE CAN BE SEEN FROM
COMPONENT AND COMPONENT-PLUS-RESIDUAL VS. INDEPENDENT VARIABLE X1
PLOT. FROM "THE USE OF INDIVIDUAL EFFECTS AND RESIDUALS IN FITTING
EQUATIONS TO DATA" BY F.S.WOOD, TECHNOMETRICS VOL.15, NO 4, NOV. 1973.

IND.VAR(I)	NAME	COEF.B(I)	S.E. COEF.	T-VALUE	R((I)SQRD	MIN X(I)	MAX X(I)	RANGE X(I)	REL.INF.X(I)
0		2.96276D+01							
1	X1	1.22333D+01	7.63D-01	16.0	0.7540	0.0	8.000D+00	8.000D+00	1.53
2	X2	-3.04944D+00	1.57D-01	19.4	0.7540	0.0	4.000D+01	4.000D+01	1.90

NO. OF OBSERVATIONS   40
NO. OF IND. VARIABLES   2
RESIDUAL DEGREES OF FREEDOM   37
F-VALUE   188.9
RESIDUAL ROOT MEAN SQUARE   6.16279160
RESIDUAL MEAN SQUARE   37.98000027
RESIDUAL SUM OF SQUARES   1405.26001001
TOTAL SUM OF SQUARES   15755.02817577
MULT. CORREL. COEF. SQUARED   .9108

REQUIRED X(I) PRECISION
(DIGIT RIGHT OF DECIMAL POSITIVE,
LEFT OF DECIMAL NEGATIVE)

IND.VAR(I)	DIGIT
1	-1
2	-2

Figure 26

LINEAR LEAST-SQUARES CURVE FITTING PROGRAM

2 VARIABLES PASS 1   DEP VAR 1: Y   RESIDUAL RCCT MEAN SQUARE OF FITTED EQUATION: 6.16

STANDARD DEVIATION ESTIMATED FROM RESIDUALS OF NEIGHBORING OBSERVATIONS (CBSERVATIONS 1 TO 4 APART IN FITTED Y ORDER).

NO.	CUMULATIVE STD DEV	WSSD	OBSV.	OBSV.	DEL RESIDUALS	WSSD	DEL RESIDUALS	FITTED Y	OBSV.	SEQ.
1	2.32	0.0	33	13	2.52	0.0	1.C7	-12.89	32	1
2	1.72	0.0	28	8	1.20	59.38	1.39	-12.89	12	2
3	1.58	0.0	35	15	1.43	C.0	0.06	-6.88	27	3
4	1.24	0.0	29	9	0.24	125.72	4.65	-6.88	7	4
5	1.14	0.0	25	5	C.75	0.0	1.75	-6.73	36	5
6	1.42	0.0	39	19	3.C6	40.25	3.92	-6.73	16	6
7	1.23	0.0	21	1	0.12	0.0	0.37	-0.71	31	7
8	1.30	0.0	22	2	1.88	59.38	3.30	-0.71	11	8
9	1.16	0.0	34	14	0.11	C.0	0.C5	5.30	26	9
10	1.12	0.0	30	10	0.83	282.88	12.27	5.30	6	10
11	1.05	0.0	38	18	0.36	C.0	0.14	5.52	40	11
12	1.14	0.0	37	17	2.27	413.43	7.54	5.52	20	12
13	1.17	0.0	23	3	1.7C	0.0	1.17	11.37	24	13
14	1.16	0.0	24	4	1.17	3.19	3.36	11.37	4	14
15	1.10	0.0	40	20	0.14	0.0	1.7C	14.40	23	15
16	1.03	C.0	26	6	0.05	331.99	6.09	14.40	3	16
17	0.99	0.0	31	11	0.37	C.0	2.27	14.63	37	17
18	1.02	C.0	36	16	1.75	1.96	2.89	14.63	17	18
19	0.97	0.0	27	7	0.C6	C.0	0.36	14.65	38	19
20	C.97	0.0	32	12	1.07	119.03	13.5C	14.65	18	20
21	0.95	1.14	22	1	0.46	0.0	0.83	23.67	30	21
22	1.00	1.14	2	1	2.34	17.68	2.98	23.67	10	22
23	0.97	1.14	22	21	0.34	0.0	0.11	23.72	34	23
24	1.02	1.14	2	21	2.22	229.19	13.80	23.72	14	24
25	1.15	1.96	37	18	4.80	C.0	1.88	25.98	22	25
26	1.20	1.56	17	18	2.53	1.14	2.22	25.98	2	26
27	1.33	1.96	37	38	5.16	C.0	0.12	29.63	21	27
28	1.38	1.96	17	38	2.89	460.09	2.77	29.63	1	28
29	1.39	3.19	9	6	1.78	3.C8	3.08	33.56	39	29
30	1.35	3.19	29	28	0.34	321.90	11.71	33.56	19	30
31	1.33	3.19	9	28	0.58	0.0	0.79	35.78	28	31
32	1.46	3.19	24	3	6.22	21.88	5.45	35.78	5	32
33	1.56	3.19	3	23	5.06	0.0	0.24	45.C0	29	33
34	1.64	3.19	24	23	4.53	49.11	5.91	45.00	9	34
35	1.68	3.19	4	23	3.36	C.0	1.43	45.09	35	35
36	1.69	6.C5	15	13	2.31	76.84	3.90	45.09	15	36
37	1.68	8.05	35	33	1.23	C.0	1.2C	48.03	28	37
38	1.64	8.05	15	33	0.21	36.51	2.90	48.C3	8	38
39	1.73	12.75	6	4	5.75	0.0	2.52	51.77	33	39
40	1.79	12.75	26	24	4.63	59.88	0.31	51.77	13	40

Figure 27

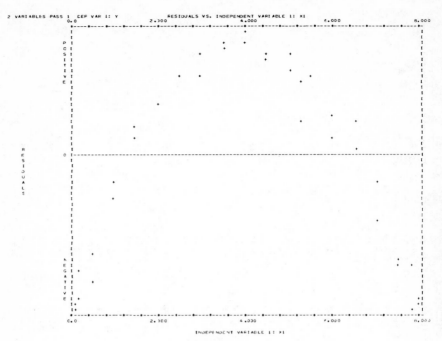

Figure 28

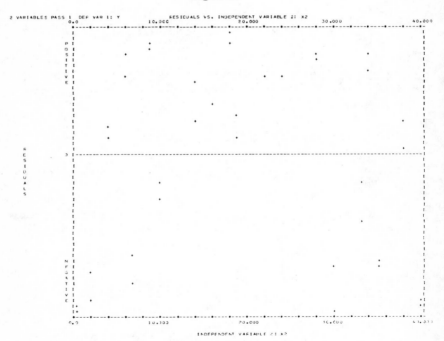

Figure 29

412

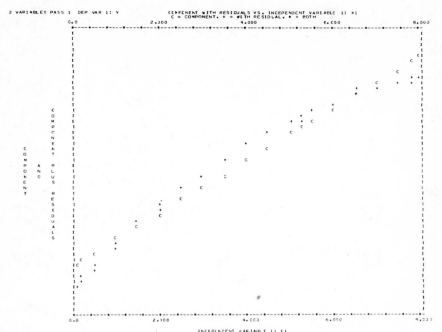

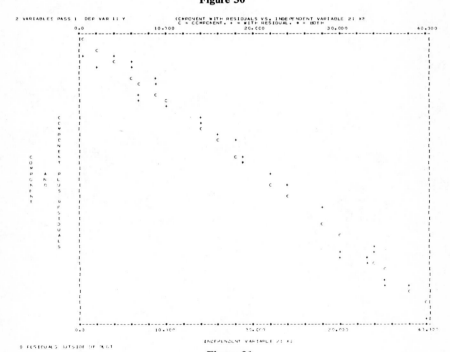

Figure 30

Figure 31

413

LINEAR LEAST-SQUARES CURVE FITTING PROGRAM

2 VARIABLES PASS 2 DEP VAR 1: Y MIN Y = -7.537D+00 MAX Y = 5.649D+01 RANGE Y = 6.403D+01

EQUATION: Y = 20 + 20X1 - 3X2 - 1X1SQRD + ERROR(0,1)
FITTED WITH Y = B0 + B1(X1-4.0) + B2X2 + B3(X1-4.0)SQRD
ERROR FROM "A MILLION RANDOM DIGITS", RAND CORP., FREE PRESS, 1955.
1-20 OBSV. FROM TABLE PAGE 4, COL 4; 21-40 OBSV. FROM PAGE 12, COL 2.
OBSERVATIONS 21-40 ARE REPLICATES OF OBSERVATIONS 1-20.
STD. DEVIATION ESTIMATED FROM NEAR NEIGHBORS (1.0), IS CLOSE TO
RMS OF FIT.
UNCODED FITTED EQUATION: Y = 20.006 + 20.310X1 - 3.007X2 - 1.038X1SQRD

IND.VAR(I)	NAME	COEF.B(I)	S.E. COEF.	T-VALUE	R(I)SQRD	MIN X(I)	MAX X(I)	RANGE X(I)	REL.INF.X(I)
0		8.46383D+01							
1	X1-4.0	1.20060D+01	1.22D-01	98.8	0.7546	-4.000D+00	4.000D+00	8.000D+00	1.50
2	X2	-3.00741D+00	2.51D-02	120.0	0.7545	0.0	4.000D+01	4.000D+01	1.88
3	X1SQRD	-1.03800D+00	2.75D-02	37.8	0.0025	0.0	1.600D+01	1.600D+01	0.26

NO. OF OBSERVATIONS	40
NO. OF IND. VARIABLES	3
RESIDUAL DEGREES OF FREEDOM	36
F-VALUE	5455.3
RESIDUAL ROOT MEAN SQUARE	0.98008387
RESIDUAL MEAN SQUARE	0.96056440
RESIDUAL SUM OF SQUARES	34.58031826
TOTAL SUM OF SQUARES	15755.02617977
MULT. CORREL. COEF. SQUARED	.9978

ORDERED BY COMPUTER INPUT						ORDERED BY RESIDUALS					
IDENT.	OBSV.	WSS DISTANCE	OBS. Y	FITTED Y	RESIDUAL	OBSV.	OBS. Y	FITTED Y	ORDERED RESID.	STUD.RESID.	SEQ
1	1	6140.	20.164	20.006	0.158	39	26.869	25.050	1.819	2.0	1
2	2	5177.	18.847	18.012	0.835	25	37.719	36.106	1.613	1.7	2
3	3	3402.	6.678	8.850	-2.172	16	-4.575	-6.116	1.541	1.7	3
4	4	2257.	8.709	9.204	-0.495	10	31.564	30.505	1.059	1.1	4
5	5	3235.	36.934	36.106	0.828	37	13.001	12.070	0.931	1.0	5
6	6	760.	8.397	8.356	0.041	2	18.847	18.012	0.835	0.9	6
7	7	450.	-1.778	-1.869	0.091	5	36.934	36.106	0.828	0.9	7
8	8	1909.	53.136	53.550	-0.414	28	54.339	53.550	0.789	0.9	8
9	9	1122.	51.883	51.309	0.574	20	-3.518	-4.242	0.724	0.8	9
10	10	70.	31.564	30.505	1.059	24	9.876	9.204	0.672	0.7	10
11	11	762.	5.739	6.174	-0.435	40	-3.659	-4.242	0.583	0.7	11
12	12	2160.	-6.466	-6.646	0.180	9	51.883	51.309	0.574	0.6	12
13	13	536.	56.494	56.388	0.106	34	28.636	28.134	0.502	0.5	13
14	14	525.	28.523	28.134	0.389	14	28.523	28.134	0.389	0.4	14
15	15	572.	47.498	47.357	0.141	29	51.646	51.309	0.337	0.4	15
16	16	4130.	-4.575	-6.116	1.541	21	20.283	20.006	0.277	0.3	16
17	17	3028.	10.730	12.070	-1.340	30	30.738	30.505	0.233	0.2	17
18	18	4083.	8.214	8.684	-0.470	12	-6.466	-6.646	0.180	0.2	18
19	19	3210.	23.790	25.050	-1.260	1	20.164	20.006	0.158	0.2	19
20	20	5398.	-3.518	-4.242	0.724	15	47.498	47.357	0.141	0.2	20
1	21	6140.	20.283	20.006	0.277	13	56.494	56.388	0.106	0.1	21
2	22	5177.	16.970	18.012	-1.042	7	-1.778	-1.869	0.091	0.1	22
3	23	3402.	8.379	8.850	-0.471	26	8.445	8.356	0.089	0.1	23
4	24	2257.	9.876	9.204	0.672	6	8.397	8.356	0.041	0.0	24
5	25	3235.	37.719	36.106	1.613	27	-1.841	-1.869	0.028	0.0	25
6	26	760.	8.445	8.356	0.089	36	-6.327	-6.116	-0.211	-0.2	26
7	27	450.	-1.841	-1.869	0.028	8	53.136	53.550	-0.414	-0.4	27
8	28	1909.	54.339	53.550	0.789	11	5.739	6.174	-0.435	-0.5	28
9	29	1122.	51.646	51.309	0.337	18	8.214	8.684	-0.470	-0.5	29
10	30	70.	30.738	30.505	0.233	23	8.379	8.850	-0.471	-0.5	30
11	31	762.	5.373	6.174	-0.801	4	8.709	9.204	-0.495	-0.5	31
12	32	2160.	-7.537	-6.646	-0.891	31	5.373	6.174	-0.801	-0.9	32
13	33	536.	53.974	56.388	-2.414	38	7.855	8.684	-0.829	-0.9	33
14	34	525.	28.636	28.134	0.502	32	-7.537	-6.646	-0.891	-1.0	34
15	35	572.	46.065	47.357	-1.292	22	16.970	18.012	-1.042	-1.1	35
16	36	4130.	-6.327	-6.116	-0.211	19	23.790	25.050	-1.260	-1.4	36
17	37	3028.	13.001	12.070	0.931	35	46.065	47.357	-1.292	-1.4	37
18	38	4083.	7.855	8.684	-0.829	17	10.730	12.070	-1.340	-1.4	38
19	39	3210.	26.869	25.050	1.819	3	6.678	8.850	-2.172	-2.3	39
20	40	5398.	-3.659	-4.242	0.583	33	53.974	56.388	-2.414	-2.6	40

Figure 32

Execution Time

A comparison of the computer processing unit (CPU) time to run ten test problems in various computers is given in Table 3 for reference. The test problems that are provided with the computer program are used to exercise the various options of the program and to check its output. The IBM 3033 operating under OS-MVS and the IBM 370/168 operating under OS-MVT took the shortest time, four seconds, while the DEC KA10 took the longest, 47 seconds. It is interesting to note that the Burroughs 4700 model is almost three times as fast as the Burroughs 6700 model. Thus it is important to know the relative speed of a given model before comparing a specific model with those of another manufacturer.

TABLE 3
COMPARISON OF CPU TIMES TO RUN TEST PROBLEMS

Computer	Processor	CPU Time (sec)
IBM	3033 MVS	4
AMDAHL	470/V6 MVS	5
IBM	370/168 MVS	6
IBM	370/168 MVT	4
CDC	6600	6
CDC	6400	10
Burroughs	4700	13
DEC 20	KL10	20
UNIVAC	1110	24
Honeywell	635	30
Burroughs	6700	35
DEC 10	KA10	47

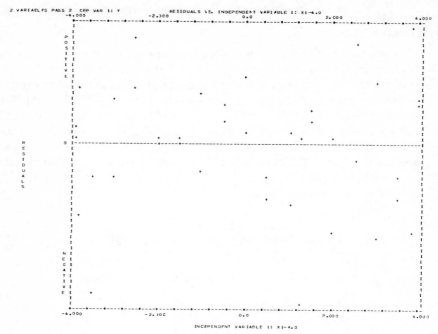

Figure 33

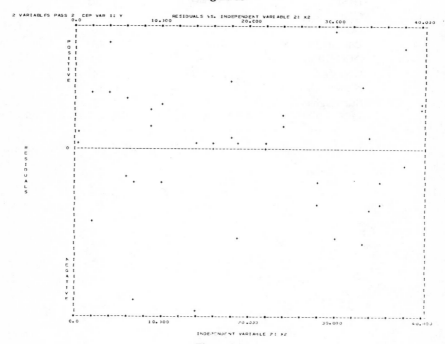

Figure 34

416

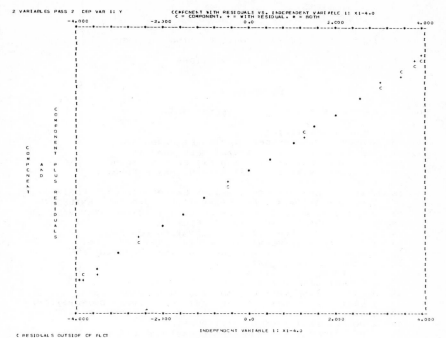

C RESIDUALS OUTSIDE OF PLOT

INDEPENDENT VARIABLE 1: X1-4.0

Figure 35

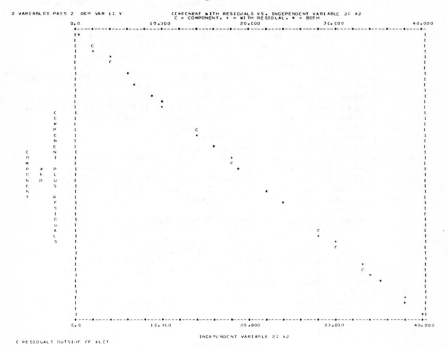

C RESIDUALS OUTSIDE OF PLOT

INDEPENDENT VARIABLE 2: X2

Figure 36

LINEAR LEAST-SQUARES CURVE-FITTING PROGRAM

1979 USER'S INSTRUCTIONS

CONTROL CARD

COLUMN   ----DESCRIPTION----   (NOTE: BLANKS = 0)

 1-18   PROBLEM IDENTIFICATION.
19-20   NUMBER OF INDEPENDENT VARIABLES READ IN.
21-22   NUMBER OF   DEPENDENT VARIABLES READ IN.
   24   E CAUSES RESIDUALS, Y AND FITTED Y TO TO BE
          LISTED WITH E FORMAT RATHER THAN F FORMAT.
   25   1 DO TRANSFORMATIONS.
   26   1 WEIGHT OBSERVATIONS. WEIGHT ENTERED IN LAST POSITION.
          2 WEIGHT BY 1/VAR. OF OBSERVATION. VARIANCE ENTERED.
   27   1 LET B(0) = ZERO, OTHERWISE B(0) IS CALCULATED VALUE.
   28   1 CHECK OBSERVATION NUMBER SEQUENCE.
   29   1 CHECK SEQUENCE WITHIN OBSERVATION NUMBER.
   30   1 DO BACK TRANSFORMATION OF DEPENDENT VARIABLE.
   31   1 PUNCH DATA FOR CONFIDENCE INTERVAL PROGRAM.
   32   0 LIST ALL INPUT DATA.
          1 LIST ONLY 1ST OBSERVATION.
          2 DO NOT LIST ANY INPUT DATA.
   33   0 LIST TRANSFORMATIONS AND ALL TRANSFORMED DATA.
          1 LIST TRANSFORMATIONS AND 1ST TRANSFORMED OBSERVATION.
          2 DO NOT LIST TRANSFORMATIONS OR TRANSFORMED DATA.
   34   1 DO NOT LIST SUMS OF VARIABLES.
   35   1 DO NOT LIST RAW SUMS AND CROSS PRODUCTS WHEN B(0)=0.
   36   1 DO NOT LIST RESIDUAL SUMS AND CROSS PRODUCTS.
   37   1 DO NOT LIST MEANS AND ROOT MEAN SQUARES OF VARIABLES.
   38   1 DO NOT LIST SIMPLE CORRELATION COEFFICIENTS.
   39   1 DO NOT LIST INVERSE MATRIX.
   40   1 DO NOT PRINT PLOTS OF RESIDUALS VS. FITTED Y.
          2 PLOT (A) RESIDUALS AND (B) COMPONENT AND COMPONENT-
            PLUS-RESIDUALS VS. EACH INDEPENDENT VARIABLE.
          3 PLOT (A) RESIDUALS ONLY.
          4 PLOT (B) COMPONENT AND COMPONENT-PLUS-RESIDUALS ONLY.
          5 PLOT (B) BUT EXPAND SCALE TO FILL EACH PLOT.
   41   0 READ DATA WITH STANDARD FORMAT (A6, I4, I2, 10F6.3).
          1 READ DATA WITH FORMAT TO BE READ, 1CARD(72COL)ASSUMED.
          2 READ DATA WITH READ DATA (REDATA) SUBROUTINE.
   42   0 DO NOT SAVE DATA FOR NEXT PROBLEM.
          1 SAVE DATA FOR NEXT PROBLEM.
          2 REUSE DATA FROM PREVIOUS PROBLEM.
43-44   NUMBER OF VARIABLES IN CP SEARCH IF GREATER THAN 12.
   45   NUMBER OF FORMAT CARDS IF GREATER THAN 1. (8 MAX.)
46-47   NUMBER OF INFORMATION CARDS TO BE READ FOR DISPLAY ON
          PRINTOUT IF DESIRED. 72 COL. EACH, 12 CARDS MAXIMUM.
   48   0 DO NOT READ NAMES OF VARIABLES FOR DISPLAY ON
            PRINTOUT.
          1 READ NAMES OF VARIABLES FROM CARDS. (NAMES IN SAME
            POSITION AS VARIABLES ON TRANSFORMATION CARDS -
            COLS. 1-6, 11-16, - - - , 61-66. (NAMES ARE NOT
            MOVED, ONLY LISTED OR DELETED.)
          2 REUSE NAMES OF VARIABLES FROM PREVIOUS PROBLEM.
   49   SEARCH FOR CANDIDATE EQUATIONS VIA CP. USE ESTIMATE
          OF VARIANCE (RMS) FROM
          1 FULL EQUATION.
          2 SUBSET EQUATION WITH MINIMUM RMS VALUE. OR
          3 ESTIMATE OF VARIANCE CARD, F16.8 FORMAT.

CONTROL CARD CONTINUED

COLUMN   ----DESCRIPTION----

50   NUMBER OF CP FACTORIAL SEARCH CARDS TO BE READ PER
      DEPENDENT VARIABLE. KEYPUNCH IN ASCENDING ORDER THE
      IDENTIFING NUMBER CF THE VARIABLES (AFTER TRANSFOR-
      MATIONS) TO BE USED IN THAT SEARCH, 18(2X,I2) FORMAT.
      ALL OTHER VARIABLES WILL BE PLACED IN BASIC EQUATION.

51   1 LIST COMPONENT-EFFECT TABLE, REQUIRED X(I) PRECISION,
        ESTIMATE OF STANDARD DEVIATION FROM NEAR NEIGHBORS,
        AND FUNCTIONS OF FITTED Y.
      2 LIST ONLY ESTIMATE OF STD. DEVIATION AND FUNCTIONS.
      3 LIST ONLY FUNCTICNS OF FITTED Y.

52-53 NUMBER OF FILE IF DATA ARE READ FROM SEPARATE FILE.
       WITH OR WITHOUT AN END CARD.

54-55 NUMBER OF INDEPENDENT VARIABLES GREATER THAN THE 99
       ALLOWED IN COLS 19-20.

56-58 TO REDUCE PRINTOUT WITH MANY OBSERVATIONS,
       NUMBER OF CENTRAL OBSERVATIONS AND RESIDUALS NOT
       TO BE PRINTED. THIS REQUIRES AN E IN COL 24.

59   CROSS VERIFICATION CF MODEL AND COEFFICIENTS WITH A
      SECOND SAMPLE OF DATA.
      1 WRITE B(0) AND B(I)S ON 4TH FILE IN ONE PROBLEM
        (TRANSFORMATIONS ARE NOT SAVED).
      2 READ B(0) AND B(I)S FROM 4TH FILE IN NEXT PROBLEM
        WITH SECOND SET OF OBSERVATIONS.

60   NUMBER OF DELETE CBSERVATION CARDS. KEYPUNCH
      OBSERVATION NUMBER OF DATA TO BE DELETED IN ORDER READ
      BY COMPUTER USING 7(I4,6X) FORMAT. 5 CARDS MAXIMUM.

61   NUMBER OF INDICATOR-VARIABLE-OBSERVATION CARDS.
      USING A 7(I4,2X,I3,1X) FORMAT, KEYPUNCH OBSERVATION
      NUMBER (IN ORDER READ) AND VARIABLE NUMBER (BEFORE
      TRANSFORMATICNS) INTO WHICH A 1 IS TO BE INSERTED.

ORDER OF CARDS FOR EACH PROBLEM

1   CONTROL CARD.
2   FORMAT CARD(S), IF ANY.
3   DELETE OBSERVATION CARD(S), IF ANY.
4   INDICATOR-VARIABLE-OBSERVATIONS CARD(S), IF ANY.
5   TRANSFORMATION CARD(S), IF ANY.
6   INFCRMATION CARD(S) FOR PRINTOUT, IF ANY.
7   NAMES OF VARIABLES CARD(S) FOR PRINTCUT, IF ANY.
8   CP FACTORIAL SEARCH CARC, IF ANY.
9   DATA CARDS (IF NOT READ FROM A FILE).
10 END CARD (END IN COL 1-3 OF IDENTIFICATION FIELD). THE NUMBER
    OF END CARDS "MUST" EQUAL THE NUMBER CF CARDS PER OBSERVATION.
    END CARDS ARE NOT NEEDED IF DATA ARE READ FROM A FILE.
11 ESTIMATE OF VARIANCE CARD, IF ANY.
12 ADDITIONAL CP FACTORIAL SEARCH AND ESTIMATE OF VARIANCE CARDS.

NOTE: FCRMAT, DATA, ENC AND NAME CARDS ARE NOT NEEDED IN
      SUBSEQUENT PROBLEMS IF SAME INPUT DATA ARE REUSED.

419

NONLINWOOD, A Computer Nonlinear Least-Squares Curve-Fitting Program

Abstract

This computer program allows the user to estimate the coefficients of a nonlinear equation such as $Y = A + B \log (x - C)$ and $Y = 1/(A + B^{-Cx})$ —equations that are nonlinear in the coefficients. An iterative technique is used; the estimates at each iteration are obtained by Marquardt's maximum neighborhood method which combines the Gauss (Taylor series) method and the method of steepest descent.

Since numerous forms of equations can be used, the user must specify the form by providing a subroutine to compute the values of the equation's coefficients. In addition, the user must provide a control card, a format card for reading data, and estimates of the starting values of the coefficients. Selected observations can be deleted if so desired. Information cards and coefficient name cards also can be read for display on the printout. Such displays are helpful to record the form of equation, the purpose of the run and any additional information that may help identify the printout in the future. Identification of the coefficients by name is particularly helpful when working with large or complex equations.

The output of the program is a printed report which includes a description of the problem, the starting values of the coefficients, the size of the incremental steps, a summary of each iteration and a summary of the final fit (in terms similar to those in the Linear Least-Squares Curve-Fitting Program). The statistics calculated include the number of observations, the number of

420

coefficients, the residual degrees of freedom, the maximum and minimum value of the dependent variable, as well as its range, the standard error and t-value for each coefficient, the residual sum of squares, the residual mean square, and the residual root mean square.

Listings are made of the observed and fitted values of the dependent variable—both in the sequence in which observations were given to the computer, and in the order of the magnitude of the differences between the observed and fitted values. Plots are made to indicate (1) whether these differences are normally distributed and (2) how they are distributed over all the fitted values of the dependent variable. Plots of these differences versus each of the independent variables can be used to choose the appropriate form of the equation and to visualize the distribution of the observations over the range of each independent variable.

Provisions are made to run multiple problems as well as different equations using the same data.

An example is given.

General Information

Background

This program was originally written by D. A. Meeter, University of Wisconsin, using D. W. Marquardt's Maximum Neighborhood Method (1963). The program was revised and programmed, using FORTRAN IV conventions, for IBM 370 System computers by F. S. Wood. The nomenclature, printout, and plots are consistent with the *Linear* Least-Squares Curve Fitting Program.

References

Marquardt, D. W., "An Algorithm for Least-Squares Estimation of Non-Linear Parameters," *J. Soc. Ind. App. Math.*, **11** 2 (June 1963), pp. 431–441.

Meeter, D. A., "Non-Linear Least Squares (GAUSHAUS)," University of Wisconsin Computing Center, 1964; program revised 1966.

Form of equation

Any that is nonlinear in the coefficients such as

$$Y = b_1 x_1{}^{b_2} + b_3 + b_4 x_2 + \cdots + b_K x_M$$

Computer Listing

where Y = the fitted value of the dependent variable,   FITTED Y
        $x_1, x_2, \ldots$ = independent variables, and   IND. VAR.
        $b_1, b_2, \ldots, b_i, \ldots, b_K$ = estimated coefficients,   COEF. B(I)

Note: the dependent and independent variable may be transformations of
the observed values.

Restrictions

Variables	A	B	C
Maximum (NVARX)*	80	20	15
Minimum			
Dependent	1		
Independent	1		
Coefficients			
Maximum (NCMAX)*	80	20	43
Observations			
Maximum (NOBMAX)*	170	170	170
MINIMUM	One greater than the number of coefficients being estimated		

* These restrictions may be altered by changing the values for NVARX, NCMAX, and
NOBMAX in the main program.

Input

Control card

Cols. 1–20. Problem Identification. These columns may contain any
grouping of letters, numbers, and blanks.

Col. 21. Source of Data.
0 (or blank) Read observations from data cards.
1 Reuse data from previous problem.

Cols. 23 and 24. Number of Coefficients to Be Estimated.

Cols. 25 and 26. Number of File if Data are Read from a Separate File.
END card(s) are not needed if only one set of data is on that file.

Cols. 27 and 28. Number of Independent Variables to Be Read.

Col. 32. Number of the Equation to Be Used. The program, as written,
allows the user to compile as many as five alternate equations at one time.
Each control card designates which of the five is to be used.

Cols. 33–36. Starting Value of Lambda. Lambda is used as a multiplier to scale the space or size of the steps taken. If not specified, the starting value for lambda will be set at 0.1. Occasionally a value of 1 will be required to control the initial iterations.

Cols. 37–40. Value of Nu. Nu is used as a divider or a multiplier to change the size of lambda depending on whether the sum of squares of that iteration is relatively close or far from the value of the previous iteration. It is a measure of whether that iteration is far from or near the minimum sum of squares.

Cols. 43 and 44. Maximum Number of Iterations. If not specified, the maximum number of Iterations is set at 20.

Cols. 45–48. Multiplier Used to Increment Value of Coefficients. This multiplier is used to increment the coefficients from one iteration to the next. If not specified, the value of the multiplier is set at 0.01.

Cols. 49–56. Magnitude of Sum of Squares Criterion. This is one of two criteria used to end converging iterations. If a value of 0.0001 is specified, the iterations will stop when there is a change of less than 0.0001 in the residual sum of squares from one iteration to the next. If the value is not specified, there will be *no* control by this criterion.

Cols. 57–64. Magnitude of Ratio of Coefficients Criterion. This is the second of two criteria used to end converging iterations. If a value of 0.001 is specified, the iterations will stop when there is a change of less than 0.001 in the ratio of *all* comparable coefficients from one iteration to the next. If a value is not specified, there will be *no* control by this criterion.

Cols. 65 and 66. Number of Information Card(s). If desired, the user can specify the number of 72 column information cards that are to be read for display on the computer printout. This information can include the form of the equation used, a statement of the purpose of the run or other pertinent information that will help identify the printout in the future. The information cards (12 max) follow the cards containing the starting values of the coefficients.

Col. 68. Names of Coefficients.

0 (or blank) Do not read names of coefficients for display on printout.

1 Read names of coefficients from cards for display on printout. The name of each coefficient is keypunched in the first 6 columns of a 10 column field, 7 per card (in the same positions as the corresponding numerical values keypunched on the starting coefficients cards). These coefficient name cards follow the information cards.

Col. 69. Plots of Residuals.

0 (or blank) Print both the cumulative distribution plot of the residuals versus their corresponding fitted Y values.

3 In addition, plot each independent variable versus the residuals.

Col. 70. Number of Delete-Observations Cards. Enter identifications of observations to be deleted in the order read by computer in columns 1–6, 11–16,..., 61–66. The delete-observations cards follow the format cards. If the data are reused, the same observations will be deleted automatically.

Format card

The format card indicates the fields in which information is to be found on the data cards. At least three fields are required; (1) identification of the observation, (2) the dependent variable, and (3) at least one independent variable. The format card is written using FORTRAN conventions; for example:

$$(A4, 6X, F10.2, 8F4.0/F4.2)$$

This indicates (*a*) the identification of the observation written in alphanumeric characters are in the first 4 columns, (*b*) skip six columns, (*c*) the dependent variable in columns 11–21 has a field of 10 with 2 numbers to the right of the decimal, (*d*) there are 8 independent variables in columns 22–53, each in a field of 4 columns with no number to the right of the decimal, and (*e*) the last independent variable is in the first 4 columns of the next card— so indicated by the slash. In this example there are two cards per observation so two END cards are required.

The format card is not needed in subsequent problems if the same data are reused. It should be noted that END card(s) are not needed if data are read for a separate file (see cols. 25–26).

Starting values of coefficients

The starting values (guesses) of the coefficients are keypunched on cards 10 columns per coefficient, 7 per card. These cards follow the format card.

Data cards

Any number of cards can be used per observation as long as it is so indicated on the format card. The first field of 4 columns contain the identification of the observation, the second field the dependent variable, and the third field the first independent variable. The identification of the observation field must be large enough to hold the word "END" in the first three positions of that field. The program counts the number of observations.

As with the format card, the data cards and the END card(s) are not needed on subsequent problems if the data are reused.

Order of cards

First Problem
1. Control card
2. Format card
3. Delete-observations card(s)
4. Starting-coefficients card(s)
5. Information card(s), if desired
6. Names-of-coefficients card(s), if desired
7. Data cards
8. END card(s)—the word END in the first three positions of the identification of observation field.

Second Problem (Different Data)
    Repeat 1 through 8 above

Third Problem (Same Data as in Second Problem)
1. Control card (1 in col. 21)
2. Starting-coefficients card(s)
3. Information card(s), if desired
4. Names-of-coefficients card(s), if desired

Control and data card entry forms

A Control Card Entry Form is given in Figure 1. The Data Card Entry Form is the same as used in the Linear Least-Squares Curve-Fitting Program.

Equation subroutine

The equations to be used are written in FORTRAN and placed in Subroutine MODEL1, MODEL2, MODEL3, MODEL4, or MODEL5 as follows:

Columns
4 5 6 7

```
     SUBROUTINE MODEL1 (NPROB, B, FY, NOB, NC, X, NVARX,
   1 NOBMAX, NCMAX)
     IMPLICIT REAL * 8(A-H, O-Z)*
```

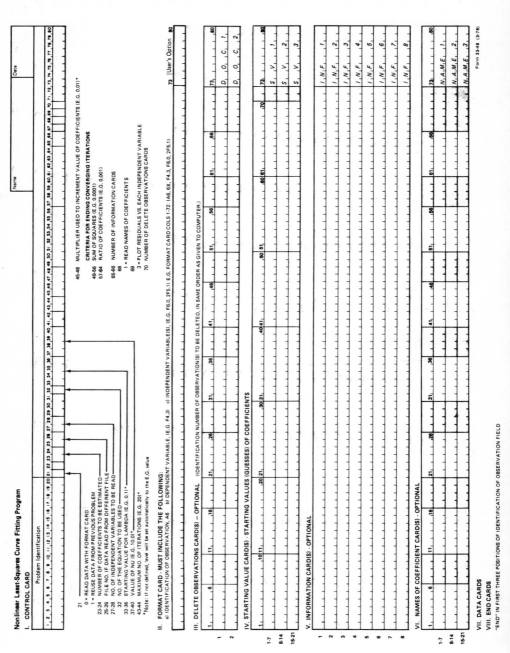

Figure 1

DIMENSION B(NCMAX), FY(NOBMAX), X(NVARX, NOBMAX)
DO 10 J = 1, NOB

Example of equation

$$FY(J) = B(1)*X(1,J)**B(2) + B(3)$$

---

1 0   CONTINUE
RETURN
END

Double precision programs only

Note: All equations are written so that the estimated coefficients are positive. A glossary of nomenclature follows:

B(I)	Array of coefficients indexed by I from 1 to NC
FY(J)	Fitted value of Y, array-indexed by J from 1 to NOB.
NC	Number of coefficients.
NCMAX	Maximum number of coefficients.
NOB	Number of observations.
NOBMAX	Maximum number of observations.
NPROB	Problem identification.
NVARX	Maximum number of variables.
X(L,J)	Double-indexed array of independent variables.
	L is index of the independent variable, and
	J is index of the observations from 1 to NOB

Example of Transformations

To perform transformations, add subroutine BLOCK DATA and appropriate cards in MODEL subroutines.

In this example the values of y are logged and the reciprocals of the x's are taken initially and then used for all subsequent iterations.

Columns

1 3 5 7

```
      BLOCK DATA
      IMPLICIT REAL*8(A–H,O–Z)
      COMMON /DATA05/ IOPEN
      DATA IOPEN /0/
      END
      SUBROUTINE MODEL 1 (NPROB, B, FY, NOB, NC, X, NVARX,
    1                      NOBMAX, NCMAX)
```

```
C
C    IN THIS EXAMPLE THE VALUES OF Y ARE LOGGED AND
C    THE RECIPROCAL OF X IS TAKEN INITIALLY AND THEN
C    USED FOR ALL SUBSEQUENT ITERATIONS
C
     IMPLICIT REAL*8(A-H,O-Z)
     COMMON /DATA01/ IDENT(170), Y(170)
     COMMON /DATA05/ IOPEN
     DIMENSION B(NCMAX), FY(NOBMAX), X(NVARX, NOBMAX)
     REAL*8 IDENT
C
     IF (IOPEN.NE.0) GO TO 7
     DO 5 J = 1, NOB
     Y(J) = DLOG(Y(J))
   5 X(1,J) = 1./X(1,J))
C
   7 DO 10 J = 1, NOB
     FY(J) = B(1)*X(1,J)**B(2) + B(3)
  10 CONTINUE
C
     IOPEN = 1
     RETURN
     END
```

Example 1

The problem for this example is from E. S. Keeping, *Introduction to Statistical Inference*, Van Nostrand Company, Princeton, 1962, p. 354.

The energy, y, radiated from a carbon filament lamp per cm^2 per second was measured at six filament temperatures, x. The observed data are given in the following table, where x is the absolute temperature of the filament in thousands of degrees Kelvin:

Observation	y	x
1	2.138	1.309
2	3.421	1.471
3	3.597	1.490
4	4.340	1.565
5	4.882	1.611
6	5.660	1.680

The experimenter wishes to fit the following equation:

$$Y = Ax^B$$

The information needed for keypunching is as follows:

Control Card

Column	Entry	Item
1–20	EXAMPLE 1, LAMP HEAT	Problem Identification
21	0	Read Data from Cards
24	2	No. Coefficients
28	1	No. Independent Variables
32	1	No. of Equation
33–36	0.10	Starting Value of Lambda
37–40	10.0	Value of Nu
43–44	20	Max No. of Iterations
45–48	0.01	Multiplier
49–56	0.0	Sum of Squares Criterion
57–64	0.000001	Ratio of Coefficients Criterion
65–66	3	No. of Information Cards
68	1	Read Names of Coefficients
69	3	Plot Residuals versus Independent Variables
70	1	Number of Delete-Observations Cards

Format Card

(A6, 6X, 2F6.0)

Delete-Observations Card(s)

To show how this option works, two observations with the identification of EXTRA and 642531 were added to the data set, to be deleted as follows:
Columns
1   *11*
EXTRA   642531

Starting values of coefficients card

From log–log plots of y versus x, A and B are approximately 0.7 and 4.0 respectively.
Columns
1   *11*
0.7   4.0

Information cards

Columns
1
Y = A(X)EXPONENT B, WHERE
Y = ENERGY RADIATED FROM CARBON FILAMENT / CMSQRD
    / SEC.
X = ABSOLUTE TEMPERATURE OF FILAMENT IN M DEGREES
    KELVIN.

Names of coefficients card

Columns
1   11
A   EXP. B

Data cards

		Columns	
Card	*1*	*14–18*	*20–24*
First	GAUS	2.138	1.309
Last	645321	5.660	1.680
	END		

Equation in subroutine MODEL 1

Columns
1   7
$$FY(J) = B(1)*X(1, J)**B(2)$$

Example on keypunch forms

The information needed for keypunching is shown in Figures 2 and 3. The printout of this problem is shown in Figures 4–9.

Examples 2 and 3

Pass 1 and Pass 10 of the cement problem, Chapter 9, are included as test problems with the computer programs (see Chapter 1 for listings of computer libraries).

Pass 1 has 15 independent variables (an indicator variable for each of the 14 cements and time) and estimates 43 coefficients (an exponential function of time, 14 coefficients of the initial acid heat of solution, 14 coefficients of the ultimate cumulative heat of hardening, and 14 coefficients of the equation constants). The control card and associated information are given in Figure 10 and the subroutine model in Figure 11.

Pass 10 has 6 variables (time and 5 compositions in weight percent transformed into mol percent) and 19 coefficients. The control card and associated information are given in Figure 12 and the subroutine model in Figure 13.

Background and results of the two fitted equations are given in Chapter 9.

Execution Time

In an IBM 360/168, the computer processing unit (CPU) time to run the 3 test problems is 26 seconds.

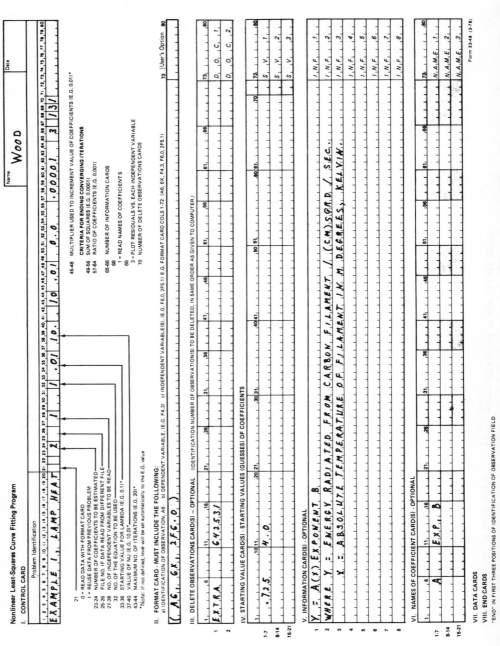

Figure 2

III STANDARD DATA-CARD ENTRY FORM

ENTER WEIGHTING FACTOR, IF ANY, AS LAST ENTRY.
"END" CARD MUST BE LAST CARD.

IDENT.	OBSV. NO.	SEQ NO.	x1–11–21	x2–12–22	x3–13–23	x4–14–24	x5–15–25	x6–16–26	x7–17–27	x8–18–28	x9–19–29	x10–20–30	USER'S OPTION
1 2 3 4 5 6	7 8 9 0	1 2	3 4 5 6 7 8	9 0 1 2 3 4	5 6 7 8 9 0	1 2 3 4 5 6	7 8 9 0 1 2	3 4 5 6 7 8	9 0 1 2 3 4	5 6 7 8 9 0	1 2 3 4 5 6	7 8 9 0 1 2	3 4 5 6 7 8 9 0
GALS			2.138	1.309									
HALS			3.421	1.471									
			3.597	1.490									
TEST			4.340	1.565									
			4.882	1.611									
PROB			5.660	1.680									
EXTRA			2.0	1.1									
642531			2.0	1.0									
END													

1 2 3 4 5 6 7 8 9 0 1 2 3 4 5 6 7 8 9 0 1 2 3 4 5 6 7 8 9 0 1 2 3 4 5 6 7 8 9 0 1 2 3 4 5 6 7 8 9 0 1 2 3 4 5 6 7 8 9 0 1 2 3 4 5 6 7 8 9 0 1 2 3 4 5 6 7 8 9 0

1 2 3 4 5 6 7 8

FORM B-56 REV. 3-69

Figure 3

NONLINEAR LEAST-SQUARES CURVE-FITTING PROGRAM

1979 VERSION OF THE NONLINWOOD 20 VARIABLE, 170 OBSERVATION PROGRAM

REFER TO FITTING EQUATIONS TO DATA BY DANIEL AND WOOD, SECOND EDITION, WILEY PUBLISHER.
FOR GLOSSARY OF TERMS, USER'S MANUAL, DETAILS OF CALCULATIONS AND INTERPRETATION OF RESULTS.

EXAMPLE 1, LAMP HEAT

ORDER OF CARDS

FIRST PROBLEM
1 CONTROL CARD.
2 FORMAT CARD (72 COLUMNS, ORDER: IDENT, Y, X(I),S).
   E.G. (A6, F6.0, (NOIND*F6.0)).
3 DELETE-OBSERVATIONS CARD(S), IF ANY.
4 STARTING VALUES (GUESSES) OF COEFFICIENTS
   (10 COLUMNS / COEFFICIENT, 7 COEFFICIENTS / CARD).
5 INFORMATION CARDS FOR PRINTOUT, IF ANY.
6 NAMES OF COEFFICIENTS CARD(S) FOR PRINTOUT, IF ANY.
7 DATA CARDS (IF NOT READ FROM A FILE).
8 END CARD (END IN COLUMNS 1 - 3 OF IDENTIFICATION
   FIELD). THE NUMBER OF END CARDS MUST EQUAL THE
   NUMBER OF CARDS PER OBSERVATION. (END CARDS
   ARE NOT NEEDED IF DATA ARE READ FROM A FILE).

SECOND PROBLEM
IF DATA ARE REUSED FROM FIRST PROBLEM,
   DELETE THE FORMAT, DATA AND END CARDS.
IF DIFFERENT DATA,
   REPEAT 1 - 8 ABOVE.

CONTROL CARD INFORMATION

COL.	INPUT	MAX	ITEM (NOTE: BLANK ON CARD = 0)
1-20			IDENTIFICATION OF PROBLEM.
21	0		0 OBSERVATIONS READ FROM CARDS OR FILE.
			1 REUSE DATA FROM PREVIOUS PROBLEM.
23-24	2	20	NUMBER OF COEFFICIENTS TO BE ESTIMATED.
25-26	0		FILE NUMBER IF DATA ARE TO BE READ FROM A SEPARATE FILE. NO END CARD(S) IF ONLY ONE SET OF DATA IS ON EACH FILE.
27-28	1	20	NUMBER OF INDEPENDENT VARIABLES TO BE READ IN.
31-32	1		NUMBER OF EQUATION TO BE USED.
33-36	0.010		STARTING VALUE FOR LAMBDA, F4.2, (E.G. 0.01), USED AS A MULTIPLIER TO SCALE THE SPACE OR SIZE OF STEPS TAKEN.
37-40	10.000		VALUE OF NU, F4.0, (E.G. 10.), DIVISOR AND MULTIPLIER TO CHANGE SIZE OF LAMBDA DEPENDING ON WHETHER SUM OF SQUARES OF ITERATION IS NEAR OR FAR FROM MINIMUM.
43-44	10		MAXIMUM NUMBER OF ITERATIONS, I2, (E.G. 20).
45-48	0.010		MULTIPLIER USED TO INCREMENT VALUE OF COEFFICIENTS, F4.3 (E.G. 0.01).

NOTE: IF VALUES IN COLUMNS 33-48 ARE NOT DEFINED
      ON THE CONTROL CARD, THEIR LEVEL WILL BE
      SET AUTOMATICALLY TO THE ABOVE E.G. VALUES.

CRITERIA FOR ENDING CONVERGING ITERATIONS.

COL.	INPUT	ITEM
49-56	0.0	SUM OF SQUARES CRITERION, F8.7. (E.G. 0.0001, A CHANGE OF LESS THAN 0.0001 IN THE RESIDUAL SUM OF SQUARES).
57-64	1.00-05	RATIO OF COEFFICIENTS CRITERION, F8.7, (E.G. 0.001, A CHANGE OF LESS THAN 0.001 IN THE RATIOS OF ALL COMPARABLE COEFFICIENTS).

NOTE: VALUES IN COLUMNS 49-64 CAN BE SET AT 0.0
      IF CONTROL OF EITHER OR BOTH IS NOT DESIRED.

```
65-66    3    12 NUMBER OF INFORMATION CARDS TO BE READ FOR DISPLAY
                  ON PRINTOUT IF DESIRED. 72 COLUMNS EACH.
  nd     1     1 READ NAMES OF CCEFFICIENTS FROM CARDS FOR
                  DISPLAY ON PRINTOUT. 1ST 6 OF 10 COLUMNS /
                  COEFFICIENT. 7 / CARD.
  69     3     3 PLOT RESIDUALS VS. EACH INDEPENDENT VARIABLE.
  70     1     2 NUMBER OF DELETE-OBSERVATIONS CARDS. OBSERVATION
                  IDENTIFICATION IN 1ST 6 OF 10 COLUMNS, 7 / CARD.

SUBROUTINE MODEL1 (NPROB, B, FY, NOB, NC, X, NVARX, NOBMAX, NCMAX)
DIMENSION E(NCMAX), FY(NOBMAX), X(NVARX,NOBMAX)

Y = A(X)EXPONENT B
WHERE Y = ENERGY RADIATED FROM CARBON FILAMENT / (CM)SQRD / SEC.
      X = ABSOLUTE TEMPERATURE OF FILAMENT IN M DEGREES, KELVIN.

DATA FORMAT ( A6, 6X, 2F6.0 )
```

OBSV. NO.	IDENT.	SEQ.	1-11-21	2-12-22	3-13-23	4-14-24	5-15-25	6-16-26	7-17-27	8-18-28	9-19-29	10-20-30
1	GAUS	1	2.138	1.309								
2	HAUS	1	3.421	1.471								
3		1	3.597	1.490								
4	TEST	1	4.340	1.565								
5		1	4.882	1.611								
6	PROB	1	5.660	1.680								

```
OBSERVATIONS DELETED
   EXTRA    642531
```

Figure 4

NONLINEAR ESTIMATION.   EXAMPLE 1. LAMP HEAT
NUMBER OF OBSERVATIONS   6
NUMBER OF COEFFICIENTS   2
STARTING LAMBDA   0.010
STARTING NU   10.000
MAX NO. OF ITERATIONS   10

INITIAL VALUES OF THE COEFFICIENTS
     1   2

7.25000D-01 4.00000D+00

PROPORTIONS USED IN CALCULATING DIFFERENCE QUOTIENTS
     1   2

1.00000D-02 1.00000D-02

INITIAL SUM OF SQUARES: 1.4721D-02

ITERATION NO. 1

EIGENVALUES OF MOMENT MATRIX - PRELIMINARY ANALYSIS

2.21970D+02 3.71620D-01

DETERMINANT: 2.0207D-02

ANGLE IN SCALED. COORD.=81.30 DEGREES

VALUES OF COEFFICIENTS
     1   2

7.62903D-01 3.87582D+00

LAMBDA: 1.0000-03   SUM OF SQUARES AFTER LEAST-SQUARES FIT: 4.4737D-03

ITERATION NO. 2

DETERMINANT: 1.8787D-02

ANGLE IN SCALED. COORD.=81.52 DEGREES

VALUES OF COEFFICIENTS
     1   2

7.68783D-01 3.86069D+00

LAMBDA: 1.0000-04   SUM OF SQUARES AFTER LEAST-SQUARES FIT: 4.3174D-03

436

ITERATION NO. 3

DETERMINANT: 1.8654D-02

ANGLE IN SCALED. COORD.=82.14 DEGREES

VALUES OF COEFFICIENTS
        1   2
7.68914D-01 3.86026D+00

LAMBDA: 1.000D-05   SUM OF SQUARES AFTER LEAST-SQUARES FIT: 4.3173D-03

ITERATION NO. 4

DETERMINANT: 1.8637D-02

ANGLE IN SCALED. COORD.=69.50 DEGREES

DETERMINANT: 1.8655D-02

ANGLE IN SCALED. COORD.=69.49 DEGREES

DETERMINANT: 1.8835D-02

ANGLE IN SCALED. COORD.=69.31 DEGREES

DETERMINANT: 2.0636D-02

ANGLE IN SCALED. COORD.=67.59 DEGREES

DETERMINANT: 3.8735D-02

ANGLE IN SCALED. COORD.=52.80 DEGREES

DETERMINANT: 2.2864D-01

ANGLE IN SCALED. COORD.=13.31 DEGREES

VALUES OF COEFFICIENTS
        1   2
7.68913D-01 3.86026D+00

LAMBDA: 1.000D-01   SUM OF SQUARES AFTER LEAST-SQUARES FIT: 4.3173D-03

ITERATION STOPS - RELATIVE CHANGE IN EACH COEFFICIENT LESS THAN: 1.0000D-05

CORRELATION MATRIX

DETERMINANT: 7.3241D+01
        1   2
1   1.0000
2 -0.9906   1.0000

Figure 5

437

NONLINEAR LEAST-SQUARES CURVE-FITTING PROGRAM

EXAMPLE 1. LAMP HEAT DEP.VAR.: MIN Y = 2.138D+00 MAX Y = 5.660D+00 RANGE Y = 3.522D+00

Y = A(X)EXPONENT B
WHERE Y = ENERGY RADIATED FROM CARBON FILAMENT / (CM)SQRD / SEC.
X = ABSOLUTE TEMPERATURE OF FILAMENT IN M DEGREES, KELVIN.
OBSERVATIONS DELETED: EXTRA 642531

IND.VAR(I)	NAME	COEF.B(I)	S.E. COEF	T-VALUE	95% CONFIDENCE LIMITS LOWER	UPPER
1	A	7.68913D-01	1.82D-02	42.4	7.19D-01	8.19D-01
2	EXP. B	3.86026D+00	5.09D-02	75.9	3.72D+00	4.00D+00

NO. OF OBSERVATIONS   6
NO. OF COEFFICIENTS   2
RESIDUAL DEGREES OF FREEDOM 4
RESIDUAL ROOT MEAN SQUARE   0.03285315
RESIDUAL MEAN SQUARE   0.00107933
RESIDUAL SUM OF SQUARES   0.00431732

------ORDERED BY COMPUTER INPUT------

OBS. NO.	OBS. Y	FITTED Y	RESIDUAL
GAUS	2.138	2.174	-0.036
HAUS	3.421	3.411	0.010
	3.597	3.584	0.013
TEST	4.340	4.333	0.007
GAUS	4.882	4.845	0.037
PROB	5.660	5.697	-0.037

------ORDERED BY RESIDUALS------

OBS. NO.	OBS. Y	FITTED Y	ORDERED RESID.	SEQ.
	4.882	4.845	0.037	1
	3.597	3.584	0.013	2
HAUS	3.421	3.411	0.010	3
TEST	4.340	4.333	0.007	4
GAUS	2.138	2.174	-0.036	5
PROB	5.660	5.697	-0.037	6

Figure 6

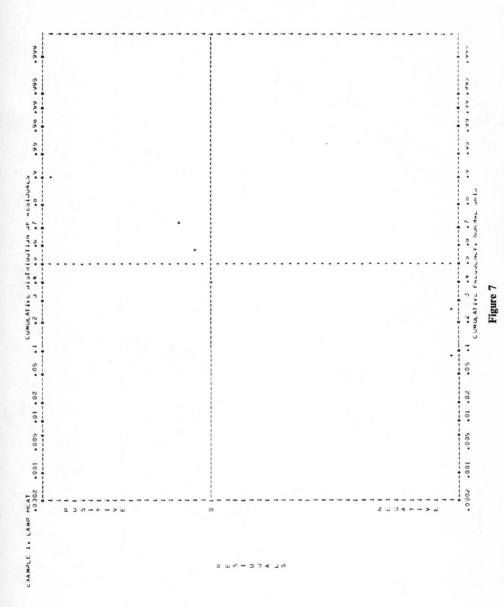

Figure 7

439

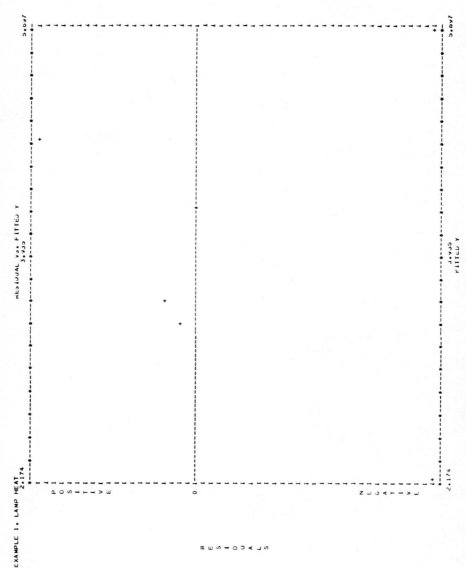

Figure 8

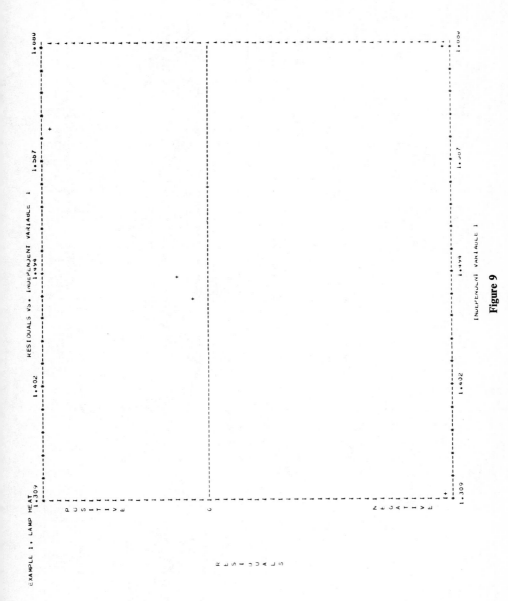

Figure 9

441

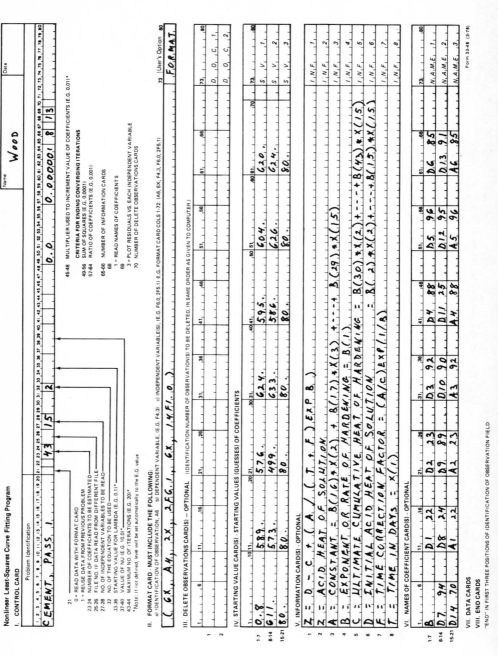

Figure 10

442

```
SUBROUTINE MODEL FOR CEMENT HEAT, PASS 1

SUBROUTINE MODEL2 (NPROB, B, FY, NOB, NC, X, NVARX, NOBMAX, NCMAX)

C --- 43 COEFFICIENTS, 14 INDICATOR VARIABLES FOR D, A AND C
C --- B IS COMMON.
C

IMPLICIT REAL*8(A-H,O-Z)
DIMENSION B(NCMAX), FY(NOBMAX), X(NVARX,NOBMAX)
C

RB = 1.0/B(1)
DO 10  J=1,NOB

D = B( 2)*X( 2,J) + B( 3)*X( 3,J) + B( 4)*X( 4,J) + B( 5)*X( 5,J)
1 + B( 6)*X( 6,J) + B( 7)*X( 7,J) + B( 8)*X( 8,J) + B( 9)*X( 9,J)
2 + B(10)*X(10,J) + B(11)*X(11,J) + B(12)*X(12,J) + B(13)*X(13,J)
3 + B(14)*X(14,J) + B(15)*X(15,J)
A = B(16)*X( 2,J) + B(17)*X( 3,J) + B(18)*X( 4,J) + B(19)*X( 5,J)
1 + B(20)*X( 6,J) + B(21)*X( 7,J) + B(22)*X( 8,J) + B(23)*X( 9,J)
2 + B(24)*X(10,J) + B(25)*X(11,J) + B(26)*X(12,J) + B(27)*X(13,J)
3 + B(28)*X(14,J) + B(29)*X(15,J)
C = B(30)*X( 2,J) + B(31)*X( 3,J) + B(32)*X( 4,J) + B(33)*X( 5,J)
1 + B(34)*X( 6,J) + B(35)*X( 7,J) + B(36)*X( 8,J) + B(37)*X( 9,J)
2 + B(38)*X(10,J) + B(39)*X(11,J) + B(40)*X(12,J) + B(41)*X(13,J)
3 + B(42)*X(14,J) + B(43)*X(15,J)
F = (A/C)**RB
FY(J) = D - C + ( A / ( (X(1,J) + F )**B(1) )   )
10 CONTINUE
RETURN
END
```

Figure 11

443

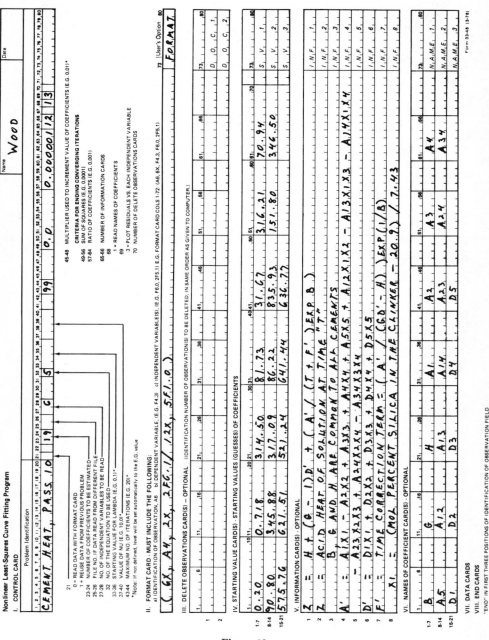

Figure 12

```
      SUBROUTINE MODEL FOR CEMENT HEAT, PASS 10

      SUBROUTINE MODEL5 (NPROB, B, FY, NOB, NC, X, NVARX, NOBMAX, NCMAX )
C           OXIDES MEASURED IN MOL PERCENTS.      D. = D1 TO D5
C --- 5 OXIDES( - MIN VALUES) VS D AND A.  B, G AND H ARE COMMON.
C
      IMPLICIT REAL*8(A-H,O-Z )
      DIMENSION  B(NCMAX), FY(NOBMAX), X(NVARX,NOBMAX )
      RB = 1.0/B(1)
      DO 10  J=1,NOB
      X1 = (X(2,J) - 20.9) / 7.43
      X2 = (X(3,J) - 2.03) / 7.43
      X3 = (X(4,J) - 0.44) / 7.43
      X4 = (X(5,J) - 66.1) / 7.43
      X5 = (X(6,J) - 3.10) / 7.43
      X12 = X1*X2
      X13 = X1*X3
      X14 = X1*X4
      X24 = X2*X4
      X23 = X2*X3
      X34 = X3*X4
      A =   B( 4)*X1   - B( 5)*X2   + B( 6)*X3   + B( 7)*X4   + B( 8)*X5
     1   + B( 9)*X12  - B(10)*X13  - B(11)*X14  - B(12)*X23  + B(13)*X24
     2   - B(14)*X34
      D = B(15)*X1   + B(16)*X2   + B(17)*X3   + B(18)*X4   + B(19)*X5
      C = B( 2)*D   - B( 3)
      F = (A/C)**RB
      FY(J) = D - C + ( A / ( (X(1,J) + F )**B(1) )  )
  10 CONTINUE
      RETURN
      END
```

Figure 13

445

NONLINEAR LEAST-SQUARES CURVE-FITTING PROGRAM

1979 USER'S INSTRUCTIONS

CONTROL CARD

COLUMN -----DESCRIPTION----   (NOTE: BLANKS = 0)

1-20 IDENTIFICATION OF PROBLEM.
21 0 OBSERVATIONS READ FROM CARDS OR FILE.
    1 REUSE DATA FROM PREVIOUS PROBLEM.
23-24 NUMBER OF COEFFICIENTS TO BE ESTIMATED.
25-26 FILE NUMBER IF DATA ARE TO BE READ FROM
      SEPARATE FILE. NO END CARD(S) IF ONLY ONE
      SET OF DATA IS ON EACH FILE.
27-28 NUMBER OF INDEPENDENT VARIABLES TO BE READ.
31-32 NUMBER OF THE EQUATION TO BE USED.

33-36 STARTING VALUE FOR LAMBDA, F4.2, (E.G. 0.1),
      USED AS A MULTIPLIER TO SCALE THE SPACE OR
      SIZE STEPS TAKEN.
37-40 VALUE OF NU, F4.0, (E.G. 10.),
      DIVISOR AND MULTIPLIER TO CHANGE SIZE OF
      LAMBDA DEPENDING ON WHETHER SUM OF SQUARES OF
      ITERATION IS NEAR OR FAR FROM MINIMUM.
43-44 MAXIMUM NUMBER OF ITERATIONS, I2, (E.G. 20).
45-48 MULTIPLIER USED TO INCREMENT VALUE OF
      COEFFICIENTS, F4.3 (E.G. 0.01).

NOTE: IF VALUES IN COLUMNS 33-48 ARE NOT DEFINED
      ON THE CONTROL CARD, THEIR LEVEL WILL BE SET
      AUTOMATICALLY TO THE ABOVE E.G. VALUES.

      CRITERIA FOR ENDING CONVERGING ITERATIONS.
49-56 SUM OF SQUARES CRITERION, F8.7,
      (E.G. 0.0001, A CHANGE OF LESS THAN 0.0001
      IN THE RESIDUAL SUM OF SQUARES).
57-64 RATIO OF COEFFICIENTS CRITERION , F8.7,
      (E.G. 0.001, A CHANGE OF LESS THAN 0.001 IN
      THE RATIOS OF ALL COMPARABLE COEFFICIENTS).

NOTE: VALUES IN COLS. 49-64 CAN BE SET AT 0.0 IF
      CONTROL OF EITHER OR BOTH IS NOT DESIRED.

65-66 NUMBER OF INFORMATION CARDS TO BE READ FOR
      DISPLAY ON PRINTOUT IF DESIRED, 72 COL. EACH.
  68 1 READ NAMES OF COEFFICIENTS FROM CARDS FOR
      DISPLAY ON PRINTOUT, 1ST 6 OF 10 COLS. /
      COEFFICIENT, 7 / CARD.
  69 3 PLOT RESIDUALS VS. EACH INDEPENDENT VARIABLE.
  70 NUMBER OF DELETE-OBSERVATIONS CARDS. OBSERVATION
      IDENTIFICATION IN 1ST 6 OF 10 COLS., 7 / CARD.

FORMAT CARD TO READ DATA

```
COL. 1-72 E.G. (A6, F6.0, (NOIND*F6.0)
          IDENT IDENTIFICATION OF OBSERVATION, A6
          Y(J) DEPENDENT VARIABLE, J-TH OBSERVATION, F6.0
          X(I,J) INDEPENDENT VARIABLE, I-TH VARIABLE, J-TH
                 OBSERVATION, NOIND*(F6.0)
```

EXAMPLE OF SUBROUTINE MODEL

EQUATIONS TO BE USED WILL BE WRITTEN IN FORTRAN AND PLACED
IN SUBROUTINE MODEL1, 2, 3, 4, OR 5 AS FOLLOWS:

```
SUBROUTINE MODEL1 (NPROB, B, FY, NOB, NC, X, NVARX, NOBMAX, NCMAX)
IMPLICIT REAL*8(A-H,O-Z)
DIMENSION  B(NCMAX), FY(NOBMAX), X(NVARX,NOBMAX)

DO 10  J = 1, NOB
               EXAMPLE OF EQUATION
FY(J) = B(1)*X(1,J)**B(2) + B(3)

10 CONTINUE
RETURN
END
```

       NOTE: ALL EQUATIONS ARE TO BE WRITTEN SO THAT THE
             COEFFICIENTS ESTIMATED AND CALCULATED WILL BE
             POSITIVE. IN SINGLE PRECISION PROGRAMS DELETE
             THE ABOVE IMPLICIT REAL*(A-H,O-Z) STATEMENT.

ORDER OF CARDS FOR EACH PROBLEM

1 CONTROL CARD.
2 FORMAT CARD (72COL). E.G. (A6,F6.0,(NOIND*F6.0)) TO READ DATA.
3 DELETE-OBSERVATIONS CARD(S), IF ANY
  (1ST 6 OF 10 COLS. / OBSERVATION DELETED, 7 / CARD).
4 STARTING VALUES (GUESSES) OF COEFFICIENTS (10COL/COEF, 7/CARD).
5 INFORMATION CARDS FOR PRINTOUT,IF ANY (72COL/CARD, 12 CARDS MAX).
6 NAMES OF COEFFICIENTS CARD(S) FOR PRINTOUT, IF ANY
  (1ST 6 OF 10 COLS. / COEFFICIENT, 7 / CARD).
7 DATA CARDS (IF NOT READ FROM A FILE).
8 END CARD (END IN FIRST 3 POSITIONS OF IDENTIFICATION).

     NOTE: FORMAT, DELETE, DATA AND END CARDS ARE NOT NEEDED IN
           SUBSEQUENT PROBLEMS IF SAME INPUT DATA ARE REUSED.

Bibliography

Acton, F. S., *The Analysis of Straight-Line Data*, Wiley, 1959.

Anderson, R. L., and T. A. Bancroft, *Statistical Theory in Research*, McGraw-Hill, 1952.

Anscombe, F. J., and J. W. Tukey, "The Examination and Analysis of Residuals," *Technometrics*, **5**, No. 2 (May 1963), pp. 141–160.

Bartlett, M. S., "Fitting a Straight Line When Both Variables Are Subject to Error," *Biometrics*, September 1949, p. 207.

Beaton, A. E., D. B. Rubin, and J. L. Barone, "The Acceptability of Regression Solutions: Another Look at Computational Accuracy," *J. Amer. Statist. Assoc.*, **71** (1976), pp. 158–168.

Behnken, D. W., "Estimation of Copolymer Reactivity Ratios: An Example of Non-Linear Estimation," *J. Polymer Sci.* Part A, **2** (1964), pp. 645–668.

Behnken, D. W., and N. R. Draper, "Residuals and Their Variance Patterns," *Technometrics*, **14**, No. 1 (1972), pp. 101–111.

Bennett, C. A., and N. L. Franklin, *Statistical Analysis in Chemistry and the Chemical Industry*, Wiley, 1954.

Berkson, J., "Are There Two Regressions?" *J. Amer. Statist. Assoc.*, **45**, (1950), pp. 164–180.

Bliss, C. I., *Statistics in Biology*, Vol. 2, McGraw-Hill, 1970, pp. 309–311.

Blom, G., *Statistical Estimates and Transformed Beta Variables*, Wiley, 1958.

Bogue, R. H., "Calculation of the Compounds in Portland Cement," *Ind. Eng. Chem. (Anal. Ed.)*, **1**, 4 (October 15, 1929).

Booth, G. W., and T. I. Peterson, "Non-Linear Estimation," IBM Share Program No. 687, 1958.

Box, G. E. P., "Fitting Empirical Data," *Ann. N.Y. Acad. Sci.*, **86**, (May 1960), Article 3, pp. 792–816.

Box, G. E. P., and J. S. Hunter, "The 2^{k-p} Fractional Factorial Designs I and II," *Technometrics*, **3**, No. 3 (1961), pp. 311–351, and No. 4 (1961), pp. 449–458.

Brownlee, K. A., *Statistical Theory and Methodology in Science and Engineering*, Wiley, Second Edition, 1965.

Chow, G. C., "Tests of Equality Between Sets of Coefficients in Two Linear Regressions," *Econometrics*, **28**, No. 3 (1960), pp. 591–605.

Cook, R. D., "Detection of Influential Observation in Linear Regression," *Technometrics*, **19**, No. 1 (1977), pp. 15–18.

Daniel, C., "Use of Half-Normal Plots in Interpreting Factorial Two-Level Experiments," *Technometrics*, **1**, No. 4 (1959), pp. 311–341.

Daniel, C., *Applications of Statistics to Industrial Experimentation*, Wiley, 1976.

Davies, O. L., Editor, *Statistical Methods in Research and Production*, Hafner, 1958.

Davies, O. L., Editor, *Design and Analysis of Industrial Experiments*, Hafner, 1960.

Denby, L., and C. L. Mallows, "Two Diagnostic Displays for Robust Regression Analysis," *Technometrics*, **19**, No. 1 (February 1977), pp. 1–13.

Dixon, W. J., and F. J. Massey, *Introduction to Statistical Analysis*, McGraw-Hill, 1951.

Draper, N. R., and H. Smith, *Applied Regression Analysis*, Wiley, 1966.

Ezekiel, M., "A method of handling curvilinear correlation for any number of variables," *J. Amer. Statist. Assoc.*, **19** (1924), pp. 431–453.

Ezekiel, M., and K. A. Fox, *Methods of Correlation and Regression Analysis*, Wiley, Third Edition, 1959.

Fisher, R. A., and F. Yates, *Statistical Tables for Biological, Agricultural and Medical Research*, Hafner, 1953, pp. 86–87.

Furnival, G. M., and R. W. Wilson, "Regression by Leaps and Bounds," *Technometrics*, **16**, No. 4 (November 1974), pp. 499–511.

Gorman, J. W., and J. E. Hinman, "Simplex Lattice Designs for Multicomponent Systems," *Technometrics*, **4**, No. 4 (November 1962), pp. 463–487.

Gorman, J. W., and R. J. Toman, "Selection of Variables for Fitting Equations to Data," *Technometrics*, **8**, No. 1 (February 1966), pp. 27–51.

Gorman, J. W., "Fitting Equations to Mixture Data with Restraints on Composition," *J. of Quality Technology*, **2**, No. 4 (October 1970), pp. 186–194.

Hader, R. J., and A. H. E. Grandage, "Simple and Multiple Regression Analyses," *Experimental Designs in Industry* (edited by Chew), Wiley, 1956, pp. 109–137.

Hald, A., *Statistical Theory and Engineering Applications*, Wiley, 1952.

Hamaker, H. C., "On Multiple Regression Analysis," *Statist. Neerlandica*, **16**, (1962), pp. 31–56, (in English).

Hansen, W. C., "Potential Compound Composition of Portland Cement Clinker," *J. Materials*, **3**, No. 1 (1968).

Hartley, H. O., "The Modified Gauss-Newton Method for the Fitting of Non-Linear Regression Functions by Least-Squares," *Technometrics*, **3**, No. 2 (1961), pp. 269–280.

Hocking, R. R., and R. N. Leslie, "Selection of the Best Subset in Regression Analysis," *Technometrics*, **9**, No. 4 (November 1967), pp. 531–540.

Hoerl, A. E., "Fitting Curves to Data," *Chemical Business Handbook* (edited by J. H. Perry), McGraw-Hill, 1954, Section 20, pp. 55–77.

Hoerl, A. E., and R. W. Kennard, "Ridge Regression: Applications to Nonorthogonal Problems," *Technometrics*, **12**, No. 1 (February 1970), pp. 69–82.

Keeping, E. S., *Introduction to Statistical Inference*, D. Van Nostrand, 1962.

La Motte, L. R., and R. R. Hocking, "Computational Efficiency in the Selection of Regression Variables," *Technometrics*, **12**, No. 1 (February 1970), pp. 83–93.

Larsen, W. A., and S. J. McCleary, "The Use of Partial Residual Plots in Regression Analysis," *Technometrics*, **14** (1972), pp. 781–790.

Lindley, D. V., and A. F. M. Smith, "Bayes Estimates for the Linear Model," *J. Roy. Statist. Soc. Ser. B*, **34**, No. 1 (1972), pp. 1–18.

Longley, J. W., "An Appraisal of Least-Squares Programs for the Electronic Computer from the Point of View of the User," *J. Amer. Statist. Assoc.* **62**, No. 319, September 1967.

Lund, R. E., "Tables for An Approximate Test for Outliers in Linear Models," *Technometrics*, **17**, No. 4 (1975), pp. 473–476.

Madansky, A., "The Fitting of Straight Lines When Both Variables Are Subject to Error," *J. Amer. Statist. Assoc.*, **54**, (1959), pp. 173–205.

Mallows, C. L., "Choosing Variables in a Linear Regression: A Graphical Aid," presented at the Central Regional Meeting of the Institute of Mathematical Statistics, Manhattan, Kansas, May 7–9, 1964.

Mallows, C. L., "Some Comments on C_p," *Technometrics*, **15**, No. 4 (November 1973), pp. 661–675.

Mandel, J., *The Statistical Analysis of Experimental Data*, Wiley, 1964.

Marquardt, D. W., "Solution of Non-Linear Chemical Engineering Models," *Chem. Eng. Progr.*, **55**, No. 6 (June 1959), pp. 65–70.

Marquardt, D. W., and T. Baumeister, "Least-Squares Estimation of Non-Linear Parameters," IBM Share Program No. 1428, 1962.

Marquardt, D. W., "An Algorithm for Least-Squares Estimation of Non-Linear Parameters," *J. Soc. Ind. Appl. Math.*, **11**, No. 2 (June 1963), pp. 431–441.

Mayer, L. S., and T. A. Wilke, "On Biased Estimation in Linear Models," *Technometrics*, **15**, No. 3 (August 1973), pp. 497–508.

McDonald, G. C., and D. J. Galarneu, "A Monte Carlo Evaluation of Some Ridge-Type Estimates," *J. Amer. Statist. Assoc.*, **70**, No. 350 (June 1975), pp. 407–416.

Meeter, D. A., "Non-Linear Least Squares (GAUSHAUS)," University of Wisconsin Computing Center, 1964; program revised 1966.

Moriceau, J., "Une Method Statistique d'Evaluation des Temps Operationels," *Rev. Statistique Appliquee*, **2**, No. 3 (1954), pp. 57–74.

National Bureau of Standards, *Fractional Factorial Experimental Designs for Factors at Two Levels*, Applied Mathematics Series 48, 1962.

Oosterhoff, J., "On the Selection of Independent Variables in a Regression Equation," Stiching Mathematisch Centrum, 2 Boerhaavestraat 49, Amsterdam, Report s319 (VP 23), December 1963 (in English).

Owens, D. B., *Statistical Tables*, Wesley, 1962, pp. 373–382.

Ralston, A., and H. S. Wilf, Editors, *Mathematical Methods for Digital Computers*, Wiley, 1960, pp. 191–203.

Rand Corporation, *A Million Random Digits with 100,000 Normal Deviates*, The Free Press, 1955.

Scheffé, H., "Experiments with Mixtures," *J. Roy. Statist. Soc.*, Series B, **20**, (1958), pp. 344–360.

Scheffé, H., *The Analysis of Variance*, Wiley, 1959.

Wood, F. S., "The Use of Individual Effects and Residuals in Fitting Equations to Data," *Technometrics*, **15**, No. 4 (1973), pp. 677–695.

Woods, H., H. H. Steinour, and H. R. Starke, "The Heat Evolved by Cement During Hardening," *Eng. News-Record*, October 6, 1932, pp. 404–407, and October 13, 1932, pp. 435–437.

Woods, H., H. H. Steinour, and H. R. Starke, "Effect of Composition of Portland Cement on Heat Evolved During Hardening," *Ind. Eng. Chem.*, **24**, No. 11 (November 1932), pp. 1207–1214.

Woods, H., H. H. Steinour, and H. R. Starke, "Heat Evolved by Cement in Relation to Strength," *Eng. News-Record*, April 6, 1933, pp. 431–433.

Yates, F., *The Design and Analysis of Factorial Experiments*, Technical Communication No. 35, Harpenden, England: Imperial Bureau of Soil Science, 1935.

Index